Piezoelectric and Perovskite Materials: Synthesis, Properties and Applications

Piezoelectric and Perovskite Materials: Synthesis, Properties and Applications

Edited by
Ian Ferrer

Willford Press
www.willfordpress.com

Published by Willford Press,
118-35 Queens Blvd., Suite 400,
Forest Hills, NY 11375, USA

ISBN: 978-1-64728-458-9

Cataloging-in-publication Data

Piezoelectric and perovskite materials : synthesis, properties and applications / edited by Ian Ferrer.
p. cm.
Includes bibliographical references and index.
ISBN 978-1-64728-458-9
1. Piezoelectric materials. 2. Perovskite materials. 3. Materials. I. Ferrer, Ian.
TK7872.P54 P54 2023
621.381 5--dc23

For information on all Willford Press publications
visit our website at www.willfordpress.com

Willford Press

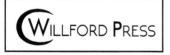

Contents

Preface

Piezoelectric materials or piezoelectrics are the materials that can produce electric energy when mechanical stress is applied on them. There are two broad categories of piezoelectric materials. First category includes piezoelectric materials that exist in nature and the second category comprises man-made or synthetic materials. Quartz, Rochelle salt, topaz, and tourmaline-group minerals are a few examples of naturally-existing piezoelectrics which are anisotropic dielectrics crystals having a non-centrosymmetric crystal lattice. Piezoelectric materials are widely used in electronic devices such as actuators, sensors, accelerators, ultrasonic motors, transducers, filters, resonators, and micro-electromechanical systems. A perovskite is any material that has a crystal structure that follows the basic chemical formula ABX3. Some of the prominent physical properties of these materials are magnetoresistance, superconductivity and ionic conductivity. They find a variety of applications as photovoltaics, lasers, scintillators and light-emitting diodes. This book outlines the synthesis, properties, and applications of piezoelectric and perovskite materials in detail. It aims to serve as a resource guide for students and researchers alike and contribute to the growth of study on these materials.

Significant researches are present in this book. Intensive efforts have been employed by authors to make this book an outstanding discourse. This book contains the enlightening chapters which have been written on the basis of significant researches done by the experts.

Finally, I would also like to thank all the members involved in this book for being a team and meeting all the deadlines for the submission of their respective works. I would also like to thank my friends and family for being supportive in my efforts.

Editor

1

Lead-Free Perovskite Nanocomposites: An Aspect for Environmental Application

Manojit De

Abstract

Perovskites possess an interesting crystal structure and its structural properties allow us to achieve various applications. Beside its ferroelectric, piezoelectric, magnetic, multiferroic, etc., properties, these branches of materials are also useful to develop materials for various environmental applications. As the population is increasing nowadays, different type of environmental pollution is one of the growing worries for society. The effort of researchers and scientists focuses on developing new materials to get rid of these individual issues. With modern advances in synthesis methods, including the preparation of perovskite nanocomposites, there is a growing interest in perovskite-type materials for environmental application. Basically, this chapter concludes with a few of the major issues in the recent environment: green energy (solar cell), fuel cell, sensors (gas and for biomedical), and remediation of heavy metals from industrial wastewater.

Keywords: perovskite, nanocomposite, fuel cell, sensors, solar cell, heavy metals, wastewater treatment

1. Introduction

Perovskites possess a very interesting crystal structure; are basically a combination of three basic crystal structures (simple cubic structure, body center cubic structure, and face-centered cubic structure). The extraordinary range of structure and properties interplay of perovskites makes them an exceptional research field for different branches like materials science, physics and solid-state chemistry. A wide range of unique functional materials and device ideas can be predicted through a basic understanding of the correlation between structural and chemical compatibility.

The perovskite structure is shown to be the single most adaptable ceramic host. Inorganic perovskite-type oxides are attractive compounds for varied applications due to its large number of compounds, they exhibit both physical and biochemical characteristics and their nano-formulation have been utilized as catalysts in many reactions due to their sensitivity, unique long-term stability, and anti-interference ability. Some perovskite materials are very hopeful applicants for the improvement of effective anodic catalysts performance. Depending on perovskite-phase metal oxides' distinct variety of properties they became useful for various applications they are newly used in electrochemical sensing of alcohols, glucose, hydrogen-peroxide, gases, and neurotransmitters. Perovskite organometallic halide showed efficient essential properties for photovoltaic solar cells.

1.1 The basic structure of perovskites

Figure 1 depicts the ideal perovskite structure. In the ideal crystal structure of perovskite with general formula ABX_3; where “A” and “B” are generally metal cations and “X” is an oxide or halide like Cl, Br, I, etc., “A” can be Ca, K, Na, Pb, Sr, and other rare-earth metals which occupy the 12-fold coordinated sites between the octahedra. “X” forms the BX_6 octahedra where B located at the center of octahedra. Perovskite can be described as consisting of corner-sharing [BX_6] octahedra with the A-cation occupying the 12-fold coordination site formed in the middle of the cube of eight such octahedra. In an ideal cubic unit cell of perovskite, Wyckoff positions for A- ion is at cube-corner positions (0, 0, 0); ion B sites at body center position (1/2, 1/2, 1/2) and ion X sits at face-centered positions (1/2, 1/2, 0). **Figure 2** shows the elements which can be in A-site, B-site from the periodic table.

In ideal perovskite such as $SrTiO_3$ [3], $CsSnBr_3$ [4], etc., there is no such distortion in the unit cell. There are many different types of lattice distortions that can happen due to the flexibility of bond angles within the ideal perovskite structure [5].

i. Distortion in BX_6 octahedra, by the Jahn-Teller effect.

ii. Off-center displacement of B-cation in BX_6 octahedra, this one of the causes for ferroelectricity in these type of materials.

iii. So-called tilting in octahedra framework, usually occurring as a result of too small A-cations at cuboctahedral site.

iv. Ordering of more than one type of cations A or B, or of vacancies.

v. Ordering of more than one kind of anions X, or of vacancies.

The different physical properties (mainly electronic, magnetic, dielectric, and piezoelectric properties) of perovskite materials are crucially dependent on these distortions. The distortion as a consequence of cationic substitution can be used to fine-tune physical properties exhibited by perovskite.

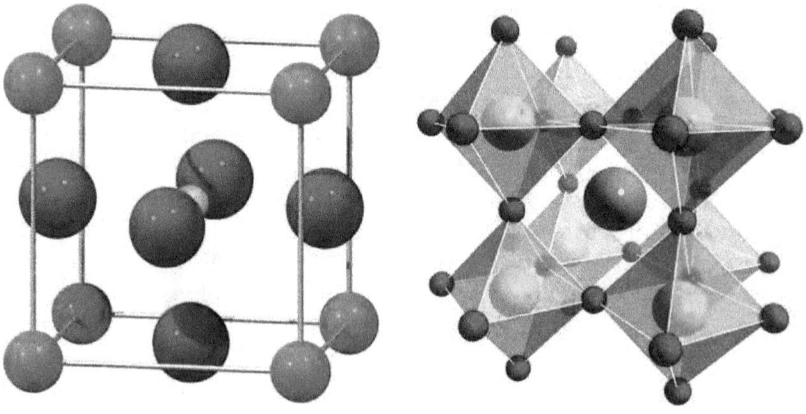

Figure 1.
The ideal structure for perovskite; blue balls represent the A-site, yellow ball shows the B-site, and magenta balls showing the position of X anion (face center position) [1].

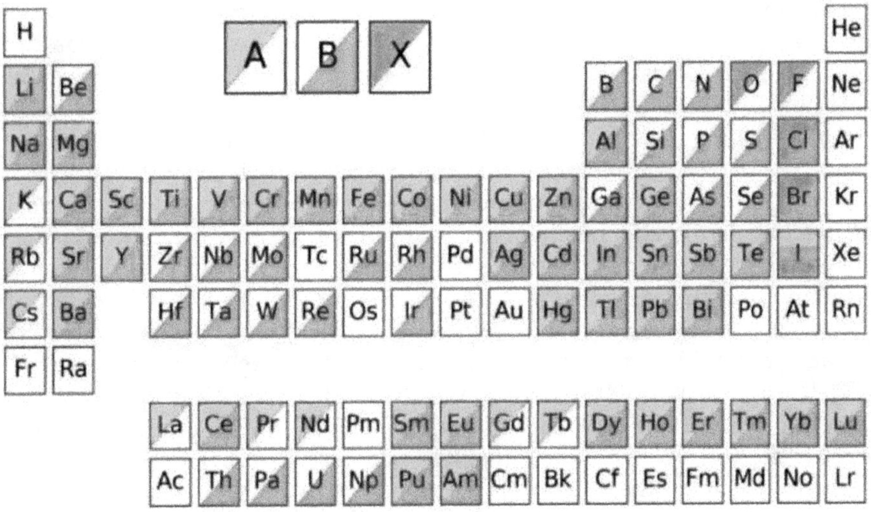

Figure 2.
A map of the elements in the periodic table which can occupy the A, B, and/or X sites [2].

In the case of perovskite structure (closed packed), A-cation must fit among four BX_6 octahedra. Each A-cation is surrounded by 12 nearest X-anions (12-fold coordination). Therefore A-cations have limited space to accommodate itself in the interstitial position. In the case of ideal perovskite structure, the cell axis (a) is geometrically related to the ionic radii (r_A, r_B and r_X) as described in the following equation.

$$a = \sqrt{2}\left(r_A + r_X\right) = 2\left(r_B + r_X\right) \tag{1}$$

The ratio of the two expressions for the cell length is called Goldschmidt's tolerance factor (t) and it allows us for evaluating the degree of distortion in the unit cell. The expression for Goldschmidt's tolerance factor [6] is as follows.

$$t = \frac{\left(r_A + r_X\right)}{\sqrt{2}\left(r_B + r_X\right)} \tag{2}$$

where r_A is the radius of A-cation is, r_B is the radius for B-cation and r_X is the radius for X-anion.

1.2 Why lead-free?

Lead (and its oxide form) is highly hazardous and its harmfulness is further improved due to its volatilization at high temperature mainly during calcination and sintering causing environmental pollution during different sample preparation techniques [7]. According to the European Union (EU), hazardous substances like lead and other heavy metals is planning to strictly prohibit [8, 9].

1.2.1 Toxicity effects of lead

The main indications of lead poisoning are tiredness, muscles and joints pain, abdominal uneasiness, etc. Sometimes the deposition of lead sulfide can be found out in the dental margin of the gums of the patients having poor dental hygiene. Lead harming has been considered as a health hazard, for its bad effects on neurological and cerebral development [10–12]. The main route of absorption in adults is the respiratory region where 30–70% of inhaled lead (typically the inorganic form like oxides and salts) goes into the cardiovascular system. The maximum tolerance of lead in blood ranges from 1.45 to 2.4 mol L^{-1} (30–50 g 100 mL^{-1}) with a provision of 6 monthly observations [13]. Basically, lead has few significant biochemical properties that give toxic effects on the human biological system. (i) As lead is electropositive in nature, it shows a very high affinity for the enzymes, which are necessary for the synthesis of hemoglobin. (ii) The divalent lead behaves similarly to calcium preventing mitochondrial oxidative phosphorylation as a result intelligence quotient (IQ) got reducing. (iii) The transcription of DNA can also disturb by lead by interacting with binding protein and nucleic acids [14, 15]. **Figure 3** illustrated the adverse effect of lead on the human body.

Bearing in mind the hazardous effect of Pb in Pb-based compounds, the research communities focused on designing the materials which are basically Pb-free. Hence, this chapter concludes with some Pb-free perovskite-type materials for the environmental application point of view.

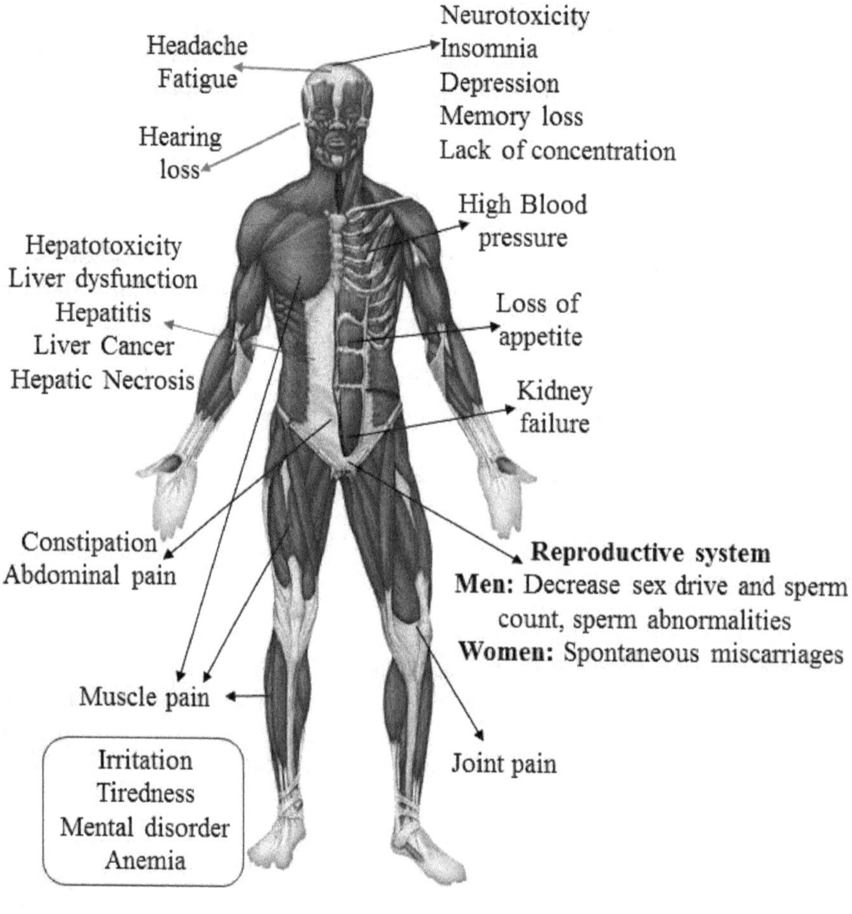

Figure 3.
Schematic diagram for toxic effect of lead on human body.

2. Application of Pb-free perovskites from a different aspect of environmental

2.1 Perovskites as solid oxide fuel cells

Recently, the inorganic perovskite-type of oxide nanomaterials have been widely applied in the processing of chemically modified electrodes [16, 17]. They have acknowledged considerable attention in the last few decades because of their catalytic activity in diverse processes like purification of waste gas and catalytic combustion.

In the fuel cell, there is a direct conversion of chemical energy into electrical energy similar to a battery. These are attractive because of their great efficiency, low emission, almost zero pollution (basically noise pollution). The solid oxide fuel cells (SOFCs) have come into the picture as effective substitutions to the combustion engines due to their prospective to minimize the environmental impact of the use of conventional fossil fuels. Perovskite oxides exhibited attractive properties like a high electrical and ionic conductivity similar to that of metals and the perfect mix of these two types [18]. This mixed conduction properties of perovskite oxides are advantageous for electrochemical reaction. The working principle of a SOFC is depicted by **Figure 4** [19]. The perovskite $Ba_{0.5}Sr_{0.5}Co_{0.8}Fe_{0.2}O_{3-\delta}$ used as an effective cathode for intermediary SOFC reported by Shao and Haile. This cathode unveiled the maximum power density of 402 and 1010 mW cm^{-2} at 500 and 600°C, respectively [20]. The combination of single and double perovskite oxide $Ba_{0.5}Sr_{0.5}(Co_{0.7}Fe_{0.3})_{0.6875}W_{0.3125}O_{3-\delta}$ (B-SCFW) was investigated by Shin et al. [21] for self-assembled perovskite composites for SOFC. In contrast, Goodenough reported that the double perovskite $Sr_2MgMnMoO_{6-\delta}$ can act as an anode material for SOFC with dry methane as the fuel and it shows maximum power density of 438 mW cm^{-2} at 800°C. This anode material exhibited long-term stability and having oxygen insufficiency, as well as some good environmental effects like tolerance to sulfur, stability in reducing atmosphere [22]. **Table 1** enlisted with some perovskites used as anode and cathode for SOFCs.

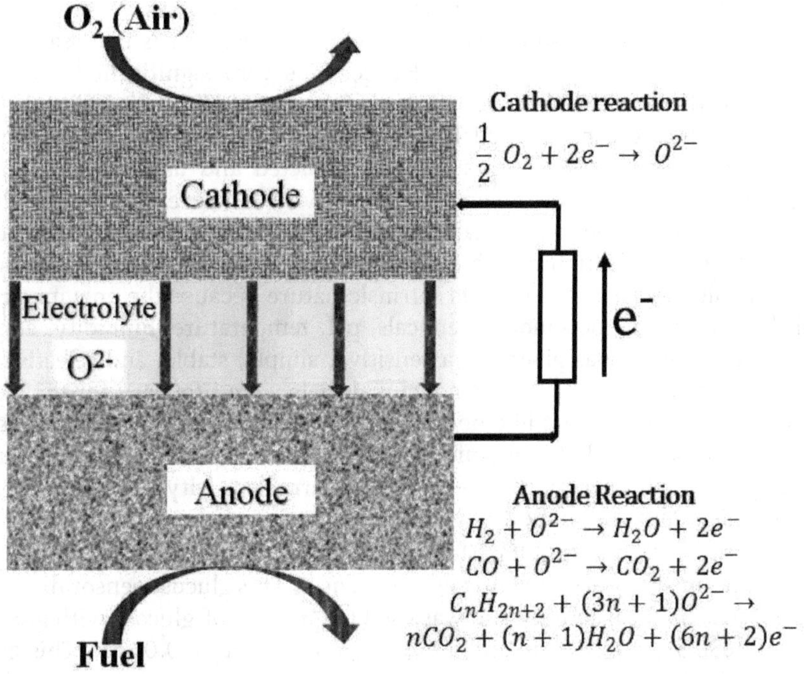

Figure 4.
Diagram illustrates the working principle of SOFC [19].

Perovskite compositions	Anode/ cathode In cell	Fuel used	Operating temperature (°C)	Maximum power density ($mW\ cm^{-2}$)	Reference
$Ba_{0.5}Sr_{0.5}Co_{0.8}Fe_{0.2}O_{3-\delta}$	Cathode	Humidified H_2 (~3% H_2O)	500	402	[20]
			600	1010	
$Ba_{0.5}Sr_{0.5}Co_{0.2}Fe_{0.8}O_{3-\delta}$	Cathode	Humidified H_2	800	266	[23]
$NdFeO_3$	Anode	Sulfur vapor or SO_2	620	0.154	[24]
			650	0.265	
$La_{0.6}Sr_{0.4}Fe_{0.8}Co_{0.2}O_3$	Cathode	Glycerol	800	Not reported	[25]
$Sm_{0.5}Sr_{0.5}CoO_{3-\delta}$	Cathode	Not reported	700	936	[26]
$La_{0.8}Sr_{0.2}Cr_{0.97}V_{0.03}O_3$	Anode	Dry methane	800	Not reported	[27]
$La_{0.75}Sr_{0.25}Cr_{0.5}Mn_{0.5}O_3$	Anode	Methane	Not reported	Not reported	[28]

Table 1.
Summary report of few Pb-free perovskites used in SOFCs.

2.2 Perovskites as sensor

2.2.1 Perovskites as glucose sensor

It is very essential to determine hydrogen peroxide (H_2O_2) and glucose analytically in any aspect of our daily life. In environmental waste management, chemical and food industries, and medical diagnostics H_2O_2 widely used as one of the most important oxidizing agents [29]. On the other hand, glucose represents a fundamental component in human blood that delivers energy through the metabolic process. If in human blood, the glucose concentration fluctuates than the normal range of 80–120 mg dL^{-1} (4.4–6.6 mM) is related to the metabolic disorder from insulin insufficiency and hyperglycemia, the so-called diabetes mellitus [30]. To perform the diagnosis and supervision of such health issue it is necessary a tight observation of glucose level of blood. Hence, it is very significant to make the biosensors for the sensitive determination of glucose and H_2O_2. Basically, there are two types of glucose sensors available: enzymatic and non-enzymatic. Different types of enzymatic glucose sensors were constructed and used in the literature exhibiting the advantages of simplicity and sensitivity. However, enzymatic glucose sensors suffered from the lack of stability and the difficult procedures required for the effective immobilization of the enzyme on the electrode surface. The lack of enzyme stability was attributed to its intrinsic nature because the enzyme activity was highly affected by poisonous chemicals, pH, temperature, humidity, etc. As a result, most attention was given for a sensitive, simple, stable, and selective non-enzymatic glucose sensor. Different novel materials were proposed for the electrocatalytic oxidation of glucose like noble nanometals, nanoalloys, metal oxides, and inorganic perovskite oxides. Inorganic perovskite oxides as nanomaterials exhibited fascinating properties for glucose sensing like ferroelectricity, superconductivity, charge ordering, high thermopower, good biocompatibility, catalytic.

Wang et al. utilized a carbon paste electrode (CPE) modified with $LaNi_{0.5}Ti_{0.5}O_3$ (LNT) as a promising nonenzymatic glucose sensor. This glucose sensor displayed a perfect electrochemical activity and was used to quantify of glucose with great sensitivity of 1630.57μA mM^{-1} $cm^{-2}$and a low detection limit of 0.07μM. This glucose sensor also demonstrated an excellent reproducibility, long-term immovability, as well as outstanding selectivity with no interference from the common interfering substances such as dopamine, ascorbic acid, and uric acid [31]. The perovskite-spinel type composite oxide $LaNi_{0.5}Ti_{0.5}O_3$-$NiFe_2O_4$with different compositions was demonstrated as the glucose sensor by Wang et al. This material also exhibits admirable reproducibility, stability and selectivity in glucose sensitivity with a linear signal-to-glucose concentration range of 0.5–10 mM and a detection limit (S/N = 3) of 0.04 mM [32]. Furthermore, $La_xSr_{1-x}Co_yFe_{1-y}O_{3-\delta}$ (x = 0.6; y = 0.0 and 0.2) perovskites were studied as electro-catalytic materials for H_2O_2 and glucose electrochemical sensors by Liotta et al. [33]. The group of Atta et al. has reformed $SrPdO_3$ perovskite with gold nanoparticles to be employed as a non-enzymatic voltammetric glucose sensor. This nanocomposite disclosed an excellent performance to glucose sensing in terms of highly reproducible response, high sensitivity, low detection limit, appreciable selectivity, long-standing stability [34]. He et al. depicted that the perovskite oxide $La_{0.6}Sr_{0.4}CoO_{3-\delta}$can provide superior electro-oxidation activities (to $H_2O_2$and glucose) over $La_{0.6}Sr_{0.4}Co_{0.2}Fe_{0.8}O_{3-\delta}$ and $LaNi_{0.6}Co_{0.4}O_3$ that translates to good H_2O_2or glucose detection performance. They have modified the sensor by making composite with reduced graphene oxide (RGO) and $La_{0.6}Sr_{0.4}CoO_{3-\delta}$ for exhibiting higher detection properties and improved selectivity [35].

2.2.2 Perovskites as a gas sensor

The clean air is undoubtedly most necessary than water for human health, but unfortunately, human activities accompanying socioeconomic developments are the vital pollution sources. So, it is very important to closely observe the quality of the air, including the indoor air quality (IAQ) as we spent most of our time (~90%) of our time in the indoor climate, to prevent different unusual symptoms [35–37]. Thus, researchers and scientists across the whole globe have been developing new and advanced material based innovative methods for consistent and careful detection of gases and volatile organic compounds (VOCs) hazardous to human and environmental health [38, 39]. The environmental worries about health hazards due to the existence of poisonous gases, for example, CO, CO_2, NO_2, O_3, etc., and subsequent safety regulations have demanded the enhanced use of sensors in various sceneries from the industrial sites to automobiles, the different workplaces and even homes. Among the several toxic gases, CO and NO_2 are the most harmful air pollutants and are dangerous for animals, plants and as well as human beings. The Occupational Safety and Health Administration (OSHA) have also announced the limit lowest tolerance for these type of gases in a particular time period, for example, the limits for CO and NO_2gases are ~20 ppm and ~5 ppm over the period of 8 h respectively. Over-acquaintance to these gases could be the reason for diseases and in dangerous cases even loss of human life [40]. There are a number of features that the materials can have to be utilized as gas sensors, explicitly, an excellent similitude with the target gases, easy to synthesize, thermal stability, appropriate electronic structure, and adaptation with present technologies. The perovskite oxides are interesting materials as gas sensors because of their ideal bandgap, excellent thermal stability and the size difference between the A- and B-sites cations, tolerating different dopants addition for monitoring the catalytic properties and their semiconducting properties. Lots of perovskites were synthesized to utilize as gas sensors for detecting different hazardous gases. **Table 2** enlisted with different perovskite oxides for various gas sensing applications.

2.3 Perovskites as solar cell materials

The consumption of energy has been continuously increasing globally and limitations of sources of fossil fuels leading to perform the research on sustainable, environment-friendly, and renewable energy sources. Due to the abundance of sun rays on

Perovskites	Sensing for	Response ratio %	Reference
$LaCoO_3$	CO	Under 5000 ppm CO at 500°C, the thick film sensor achieved a high sensing response of ~279.86	[41]
$La_{0.9}Ce_{0.1}CoO_3$	CO	240% with respect to 100 ppm CO in air	[42]
Ca modified $LaFeO_3$	SO_2	Maximum resistive response of 3 ppm SO_2 was detected at ~275°C by the $LaCaFeO_3$ samples, and it shows 15% higher efficiency than $LaFeO_3$	[43]
$NdFe_{1-x}Co_xO_3$	CO	1215% at 170°C for 0.03% CO gas	[44]
$LaFeO_3$	Ethanol	Not reported	[45]
Ag-$LaFeO_3$	Methanol	The maximum response to the other test gases is 8	[46]
Ag-$LaFeO_3$	Formaldehyde	Best response to 0.5 ppm formaldehyde (24.5) at 40°C	[47]
$SrFeO_3$	Ethanol	Not reported	[48]
$LaFeO_3$	NO_2	Not reported	[49]
$GaFeO_3$	Ethanol	Ethanol sensing down to 1 ppm at 350°C	[50]
$SrTi_{1-x}Fe_xO_{3-\delta}$	Hydrocarbons	Not reported	[51]
α-Fe_2O_3/$LaFeO_3$	Acetone	Response 48.3% at 100 ppm concentration at 350°C	[52]
$YCo_{1-x}Pd_xO_3$	CO, NO_2	Different for the different composition	[53]
$ZnSnO_3$	n-Propanol gas	The detection limit of $ZnSnO_3$ nanospheres to 500 ppb n-propanol gas could reach 1.7	[54]
$LaFeO_3$ and rGO-$LaFeO_3$	NO_2 and CO	Response 183.4% for 3 ppm concentration of NO_2 at a 250°C	[55]
$BaTiO_3$/$LaFeO_3$ nanocomposite	Ethanol	Response 102.7% to 100 ppm ethanol at 128°C	[56]
Ba-$BiFeO_3$	Ethanol	Temperature dependent sensing performance toward 100 ppm ethanol gas; maximum sensitivity at 400°C	[57]
La doped $BiFeO_3$	Acetone	The morphotropic phase boundary (MPB) phase $Bi_{0.9}La_{0.1}FeO_3$ shows ultra-low concentration detection of 50 ppb acetone	[58]
Pr doped $BiFeO_3$	Formaldehyde	50 ppm, 190°C, R_{gas}/R_{air} = 17.6	[59]
$BaTiO_3$ thick films	H_2S	$BaTiO_3$ sensor operated at 350°C	[60]
Sr doped $BaTiO_3$	NH_3 and NO_2	0.2 mol% doping of Sr showed enhanced performance for sensing of both NO_2 and NH_3 gases at room temperature	[61]
$BaSnO_3$	SO_2	10 ppm of SO_2	[62]
$BaSnO_3$	LPG	Addition of noble metal Pt, the operating temperature decreases and the sensitivity improved but also imparted partial selectivity for the detection of LPG	[63]
$LaBO_3$ (B = Fe, Co)	Acetone	At a low operating temperature of 120°C showed that the $LaFeO_3$ NFs based sensor displayed high stable and selective response toward 40 ppm acetone with fast response and recovery time of 14 and 49 s	[64]

Table 2.
Tabulated with different perovskite oxides for various gas sensing applications.

our globe, the transformation of sunlight into electricity is one of the most favorable studies for increasing energy demands without having any adverse effect on the global climate. Solar cell technology offers an eco-friendly and renewable energy path to convert photon energy into electricity openly [65]. Nowadays a large effort has been put in the research to develop high efficiency, low-cost photovoltaic devices but regrettably did not succeed yet. During the last decade, research into perovskite solar cells (PSCs) has increased and it also been nominated as a runner-up for the top 10 breakthroughs research of 2013 by the editors of Science [66]. The organic-inorganic perovskite having the general formula $ABX_3$where A is cesium (Cs), methylammonium (MA), or formamidinium (FA); B is Pb or Sn; and X is Cl, Br, or I, have recently appeared as an exciting class of semiconductors which can act as solar cell materials [67]. These organic-inorganic halide perovskite solar cells have shown substantial improvement of power conversion efficiency (PCE) from the preliminary efficiency of 3.8% [68] to about 22.1% [69]. The maximum theoretical power conversion efficiency accomplished by perovskite ($CH_3NH_3PbI_3$) is about 31.4% [70]. The organic-halide perovskites owing extraordinary performance because of some unique properties like (i) high absorbing coefficient, (ii) high charge carrier mobility, (iii) long diffusion length, (iv) direct bandgap which can be engineered easily and (v) moreover easy to fabricate [71]. The normally used Pb-based perovskites have numerous advantages such as (i) large diffusion length, (ii) absorption range, (iii) low exciton binding energy, and (iv) high carrier mobility. Conversely, the Pb-based perovskite solar cell has a serious toxic issue on both humans and the environment [72]. Generally, PV panels are positioned on the roof of houses or in the open field, their exposure to rainfall is inescapable. In the manifestation of rain and moisture the degradation of PbI_2 may cause mild to acute health issues, like effects on cardiovascular, neurological, reproductive system [73] mainly it is carcinogenic [72, 74] Additionally, lead pollution has severe effects on water and soil resources and emission of greenhouse gasses [75, 76]. Hence, in PSCs, it is essential to replace Pb for economical green energy conversion devices which may use mankind in future endeavors [77].

To develop Pb-free PSCs,$^{2+}$Sn metal cation was the first another candidate to replace Pb^{2+}as of its comparable electronic configuration and effective ionic radius (Sn^{2+}: 115 pm) to lead (Pb^{2+}: 119 pm) [78]. The hybrid organic-inorganic halide perovskites having the chemical formula of $AM^{IV}X_3^{VII}$,where A represents a small monovalent organic molecule, M^{IV} is a divalent group-IVA cation and $X^{VII}$is a halogen anion, have recently attracted remarkable attention in the photovoltaic community $MASnI_3$and MASn-$(I_{3-x}Br_x)$have been shown to the efficiencies of 6.4% [79] and 5.73% [80] respectively. $CH_3NH_3SnI_3$, $HC(NH_2)_2SnI_3$, $NH_2NH_3SnI_3$, and $NH_2(CH_2)_3SnI_3$is the promising candidate to be a light sensitizer with suitable inorganic hole-transport material to achieve cost-effective and efficient lead-free perovskite solar cell [81, 82]. Gagandeep et al. uses the graphene as the layer for charge transport and is demonstrate the structure like n-Graphene/$CH_3NH_3SnI_3$/p-Graphene which shows the efficiency of 10.67–13.28% [83]. Giorgi et al. substituted Pb by TlBi ($MATl_{0.5}Bi_{0.5}I_3$) and InBi ($MAIn_{0.5}Bi_{0.5}I_3$) and depicted that these systems are quietly equivalent to $MAPbI_3$and can be good replacements for PSCs [84]. Germanium is also assumed as a possible candidate for Pb substitution in halide perovskites. The theoretical structure and electronic properties of $AGeI_3$(A = MA, FA, Cs) were investigated by Krishnamoorthy et al. [85]. There are several elements like: Bi [86–88], Sb [89–91], Ti [92], Cu [93, 94] that use to substitute Pb to decrease the lead pollution. The group of Song et al. has reported the photovoltaic application of Sn-based halide perovskite materials having the general formula $ASnI_3$ (A = Cs, methylammonium and formamidinium tin iodide as the representative light absorbers). Among all the perovskites, $CsSnI_3$ devices accomplished a maximum power conversion efficiency of 4.81% [95]. Nishimura et al. synthesized

GeI_2 doped $FA_{0.98}EDA_{0.01}SnI_3$ and GeI_2 doped $EA_{0.98}EDA_{0.01}SnI_3$ PSCs and shows the power conversion efficiency of 13.24% for lead-free perovskite solar cell has been demonstrated with mixed cation and surface passivation [96].

2.4 Removal of heavy metals from wastewater

To survive on this planet the clean air, water, and foods are essential to all forms of life. The surface and the groundwaters are only the sources of clean water which help to all living systems as well as human activities such as consuming, irrigation of crops, industrial application, etc. [97]. Water pollution is one of the most world-wide common issues as the population outbursts and industrial evolutions are there. Day by day, the heavy metals (maybe in the form of ions) are released into water bodies by various industries [98] and are exceedingly water-soluble, non-decom-posable, oncogenic agents and cause adverse health complications on the animals as well as human beings. Wastewaters coming out from various industries contain many heavy metal ions, for example, Cu^{2+}, As^{5+}, Ni^{2+}, Sb^{5+}, Zn^{2+}, Cd^{2+}, and Pb^{2+} [99]. In addition to heavy metal ions, the different organic and inorganic dyes are alternative pollutant releases from different industries for example papers, textiles, and plastics where the dyes are used for coloring their product and also generate significant volumes of wastewater. Many of these dyes containing heavy metal ions have a tendency to store in the living entities causing a different type of diseases and disorders [100–102]. Hence, it is essential to purify the metal-contaminated water before its discharge to the environment. Among all compare to current methods to remove heavy metal from the contaminated water [100, 101] adsorption method is the most likely one because it low cost-effective,high efficiency and simple to run.

2.4.1 *Heavy metals from contaminated water*

The group of Zhang et al. synthesized the porous nano-calcium titanate micro-spheres via a citric acid assisted modified sol-gel method and used for absorption of heavy metals like lead, cadmium, and zinc [103]. Haron et al. reported that the nano-crystalline $LaGdO_3$perovskite was synthesized by the co-precipitation method could adsorb heavy metal ions ($Cd^{2+}$and Pb^{2+}) which should be the attention in an application such as wastewater treatment [104]. Zhang et al. synthesized porous nano-barium-strontium titanate via sol-gel method using sorghum straw as a template and investigate about adsorption mechanism of Pb, Zn, and Cd from contaminated water [105]. $LaFeO_3$ nanoparticles were synthesized by Rao et al. by the sol-gel method in presence of different chelating agents and these nanoparticles utilized for an adsorbent of the removal of heavy metal ions in particular cadmium ion. The $LaFeO_3$sample prepared with succinic acid (SA) as a chelating agent shows a higher removal efficiency of Cd^{2+}ions from aqueous systems [106]. Zhang et al. investigated Sr modified $LaFeO_3$ and its structural and catalytic activity. $La_{0.8}Sr_{0.2}FeO_3$contributed significantly enhanced activity in methane combustion and CO oxidation because the oxygen vacancies accelerated the dissociation of gaseous oxygen on the surface in CO oxidation and facilitated the diffusion of lattice oxygen from the bulk to the surface during CH_4combustion [107]. The perovskite $LaAlO_3$was manufactured using the co-precipitation method by Haron et al. The structural and efficiency of removal of heavy metal ($Cd^{2+}$and Pb^{2+}) were extremely investigated by them. The adsorption performance was studied which fit with the Langmuir isotherm. The results disclosed that $LaAlO_3$perovskite showed high efficiency as heavy metal ions remover from the contaminated water. This adsorbent could be recycled with an EDTA solution and reprocessed with only slightly less efficient than that of the fresh sample [108].
The group of Chen et al. synthesized ternary photocatalyst $ZnTiO_3/Zn_2Ti_3O_8/ZnO$

heterojunction which displays excellent performance for the degradation of organic pollutants as well as reduction of heavy metal Cr(VI) ions from wastewater [109]. **Figure 5** schematically represent the heavy metal ion (Cd^{2+}) adsorb with $LaFeO_3$ perovskite prepared with the different chelating agent.

2.4.2 Heavy metals in wastewater from the textile industry

The growing population in our globe, demands to clothe and increase with the taming sense of fashion and lifestyle thus textiles are contrived to meet the growing demands. In several countries such as India and Sri Lanka; the production of textile becomes their source of income that subsidizes their gross domestic product (GDP). However, this has brought both significances to such countries either in a positive way which is an enhancement of the economy or in a negative way indorsed to environmental pollution. The textile industries have been adapted as the worst reprobates of pollution contributors [110]. Especially, in India, according to the Central Pollution Control Board [111], a total of 2324 textile industries are set up. The textile industries employ different types of dyes for the manufacturing of various fabric materials. In reality, about 1 million different dyes are found in the market [112] and roughly 700,000 tons of artificial dyes are produced per year [113]. The disposal of dyes in waters exemplifies a severe environmental issue due to the coinciding presence of various types of pollutants [114–116]. All traditional methods used for the treatment of dyes and/or heavy metals have limitations because of cost, efficiency and operational complications. Among all of them, adsorption was exposed as one of the most effective methods due to its simplicity in operation, adaptability, high-treatment efficiency and low cost, and hence it is extensively applied for wastewaters treatment [117–121].

The perovskite oxide $La_{0.9}Sr_{0.1}FeO_3$, capped with cetyl trimethyl ammonium bromide (CTAB) cationic surfactant, and used as a sorbent for the removal of the anionic Congo red (CR) dye from aqueous solutions was reported by Ali et al. [122]. The group of Chu et al. demonstrated the efficiency of Ag-$La_{0.8}Ca_{0.2}Fe_{0.94}O_{3-\delta}$ for the removal of organic and bacterial pollutants by catalytic peroxymonosulfate (PMS) activation. The oxygen vacancies in the B-site of perovskite enhances PMS activation and The SO_4• and •OH radicals enhance the biocidal activity [123]. Nanocrystalline $LaAlO_3$:Sm^{3+}:Bi^{3+} composites are used to adsorb Direct Blue-53 (DB-53) dye was reported by Pratibha et al. [124]. This adsorbent is good and promising in the adsorption capacity and is advantageous in the elimination of toxic and non-biodegradable pollutants from water. The group of Dong et al. hydrothermally synthesized perovskite $BaZrO_3$in the form of hollow micro- and nano-sphere. This size-tunable $BaZrO_3$hollow nanospheres exhibited an excellent adsorption performance for reactive dyes in acidic conditions and can be used as excellent circular adsorbents for removing reactive dyes. They show the adsorption capacities are over

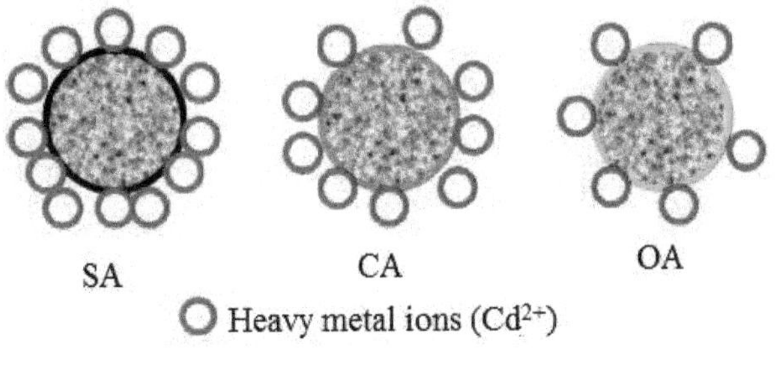

Figure 5.
Representation of the adsorption process over $LaFeO_3$ nanoparticles surfaces prepared using succinic acid (SA), citric acid (CA), and oxalic acid (OA) [106].

160 mg g^{-1} for different investigated dyes at a pH value of 2. The adsorbents were easily recovered by using a basic solution with the adsorption performance persistent and the desorption rate is more than 97 wt% [125]. Siddharth et al. synthesized the perovskite structure of Ti-doped $BaMnO_3$ ($BaMn_{0.85}Ti_{0.15}O_{2.93}$) and its enhanced photocatalytic degradation (~99%) as compared to $BaMnO_3$ toward toxic water impurities like RhB and MB dyes within 270 and 150 min under sunlight [126].

3. Conclusion

Nowadays, because of the increasing population and demand for natural resources is a global concern, hence, it is one of the origins of environmental pollution. Among all the pollutants, lead is one of the most hazardous materials. For the reason that the toxic effect of lead in lead-based materials, the researcher communities and scientists concentrated on synthesized lead-free materials possessing the same properties. Since the last two decades, the investigation on such materials enhances unexpectedly. Lead-free perovskite materials with excellent dielectric and piezoelectric properties belonging to the ferroelectric family and these reduce the adverse effect on the human body as well as the environment. Here this chapter concludes with different applications like SOFCs, sensors, solar cells, wastewater treatment of lead-free perovskite materials.

Author details

Manojit De[1,2]

1 Department of Pure and Applied Physics, Guru Ghasidas Vishwavidyalaya, Bilaspur, India

2 Department of Applied Physics, Chouksey Engineering College, Bilaspur, India

*Address all correspondence to: manojit.manojit.de1@gmail.com

References

[1] Levy MR. Crystal Structure and Defect Property Predictions in Ceramic Materials. London: Department of Materials, Imperial College of Science, Technology and Medicine. January 2005. Available from: https://spiral.imperial.ac.uk/bitstream/10044/1/11804/2/Levy-MR-2005-PhD-Thesis.pdf

[2] New Tolerance Factor to Predict the Stability of Perovskite Oxides and Halides. Available from: https://arxiv.org/ftp/arxiv/papers/1801/1801.07700.pdf

[3] Glazer AM. The classification of tilted octahedra in perovskites. Acta Crystallographica. Section B: Structural Crystallography & Crystal Chemistry. 1972;**28**:3384-3392

[4] Scaife DE, Weller PF, Fisher WG. Crystal preparation and properties of cesium tin(II) trihalides. Journal of Solid State Chemistry. 1974;**9**:308-314

[5] Aleksandrov KS. The sequences of structural phase transitions in perovskites. Ferroelectrics. 1976;**14**:801-805

[6] Goldschmidt VM. Die Gesetze der Krystallochemie. Die Naturwissenschaften. 1926;**21**:477-485

[7] Yugong W, Zhang H, Zhang Y, Jinyi M, Daohua X. Lead-free piezoelectric ceramics with composition of (0.97−x) $Na_{1/2}Bi_{1/2}TiO_3$-$0.03NaNbO_3$-$xBaTiO_3$. Journal of Materials Science. 2003;**38**:987

[8] United Nations Environment Programme. Nairobi; 2019. Available from: https://www.unenvironment.org/explore-topics/transport/what-we-do/partnership-clean-fuels-and-vehicles/lead-campaign

[9] Takenaka T, Nagata H. Current status and prospects of lead-free piezoelectric ceramics. Journal of the European Ceramic Society. 2005;**25**:2693-2700

[10] Gordon JN, Taylor A, Bennett PN. Lead poisoning: Case studies. British Journal of Clinical Pharmacology. 2002;**53**:451

[11] Barltrop D, Smith AM. Kinetics of lead interaction with human erythrocytes. Postgraduate Medical Journal. 1985;**51**:770

[12] Rabinowitz MB, Wetherill GW, Kopple JD. Kinetic analysis of lead metabolism in healthy humans. The Journal of Clinical Investigation. 1976;**58**:260

[13] Courtney D, Meekin SR. Changes in blood lead levels of solderers following the introduction of the control of lead at work regulations. Occupational Medicine. 1985;**35**:128

[14] Goering PL. Lead-protein interactions as a basis for lead toxicity. Neurotoxicology. 1993;**14**:45

[15] Chisolm JJ Jr. Treatment of lead poisoning. Modern Treatment. 1971;**8**:593-611

[16] Galal A, Atta NF, Ali SM. Optimization of the synthesis conditions for $LaNiO_3$ catalyst by microwave assisted citrate method for hydrogen production. Applied Catalysis, A: General. 2011;**409-410**:202-208

[17] Giang HT, Duy HT, Ngan PQ, Thai GH, Thu DTA, Thu DT, et al. Hydrocarbon gas sensing of nano-crystalline perovskite oxides $LnFeO_3$ (Ln = La, Nd and Sm). Sensors and Actuators B: Chemical. 2011;**158**:246

[18] Ishihara T. In: Ishihara T, editor. Perovskite Oxide for Solid Oxide Fuel Cells, Fuel Cells and Hydrogen Energy. Springer Science Business Media, LLC; 2009. pp. 1-16 Chapter 1

[19] Atta NF, Galal A, El-Ads EH. Perovskite nanomaterials–Synthesis, characterization, and applications. In: Pan L, Zhu G, editors. Perovskite Materials–Synthesis, Characterisation, Properties, and Applications. Rijeka: IntechOpen, DOI: 10.5772/61280. Available from: https://www.intechopen.com/books/perovskite-materials-synthesis-characterisation-properties-and-applications/perovskite-nanomaterials-synthesis-characterization-and-applications

[20] Shao Z, Haile SM. A high-performance cathode for the next generation of solid-oxide fuel cells. Nature. 2004;**431**:170-173

[21] Shin J, Xu W, Zanella M, Dawson K, Savvin SN, Claridge JB, et al. Self-assembled dynamic perovskite composite cathodes for intermediate temperature solid oxide fuel cells. Nature Energy. 2017;**2**:16214

[22] Huang YH, Dass RI, Xing ZL, Goodenough JB. Double perovskites as anode materials for solid-oxide fuel cells. Science. 2006;**312**:254-257

[23] Song K, Lee K. Characterization of $Ba_{0.5}Sr_{0.5}M_{1-x}Fe_xO_{3-\delta}$ (M = Co and Cu) perovskite oxide cathode materials for intermediate temperature solid oxide fuel cells. Ceramics International. 2012;**38**:5123-5131

[24] Tongyun C, Liming S, Feng L, Weichang Z, Qianfeng Z, Xiangfeng C. $NdFeO_3$ as anode material for S/O_2 solid oxide fuel cells. Journal of Rare Earths. 2012;**30**(11):1138-1141

[25] Conceiçăo LD, Silva AM, Ribeiro NFP, Souza MMVM. Combustion synthesis of $La_{0.7}Sr_{0.3}Co_{0.5}Fe_{0.5}O_3$ (LSCF) porous materials for application as cathode in IT-SOFC. Materials Research Bulletin. 2011;**46**:308-314

[26] Wang F, Chen D, Shao Z. $Sm_{0.5}Sr_{0.5}CoO_{3-\delta}$ -infiltrated cathodes for solid oxide fuel cells with improved oxygen reduction activity and stability. Journal of Power Sources. 2012;**216**:208-215

[27] Vernoux P, Guillodo M, Fouletier J, Hammou A. Alternative anode material for gradual methane reforming in solid oxide fuel cells. Solid State Ionics. 2000;**135**:425-431

[28] Tao SW, Irvine JTS. A redox-stable efficient anode for solid-oxide fuel cells. Nature Materials. 2003;**2**:320-323

[29] Deganello F, Liotta LF, Leonardi SG, Neri G. Electrochemical properties of Ce-doped $SrFeO_3$ perovskites-modified electrodes towards hydrogen peroxide oxidation. Electrochimica Acta. 2016;**190**:939-947

[30] Wang J. Electrochemical glucose biosensors. Chemical Reviews. 2008;**108**:814-825

[31] Wang Y, Xu Y, Luo L, Dinga Y, Liu X, Huang A. A novel sensitive nonenzymatic glucose sensor based on perovskite $LaNi_{0.5}Ti_{0.5}O_3$-modified carbon paste electrode. Sensors and Actuators B. 2010;**151**:65-70

[32] Wang Y, Xu Y, Luo L, Ding Y, Liu X. Preparation of perovskite-type composite oxide $LaNi_{0.5}Ti_{0.5}O_3$–$NiFe_2O_4$ and its application in glucose biosensor. Journal of Electroanalytical Chemistry. 2010;**642**:35-40

[33] Liotta LF, Puleo F, La Parola V, Leonardi SG, Donato N, Aloisio D, et al. $La_{0.6}Sr_{0.4}FeO_{3-\delta}$ and $La_{0.6}Sr_{0.4}Co_{0.2}Fe_{0.8}O_{3-\delta}$ perovskite materials for H_2O_2 and glucose electrochemical sensors. Electroanalysis. 2014;**27**:1-10

[34] El-Ads EH, Galal A, Atta NF. Electrochemistry of glucose at gold nanoparticles modified graphite/$SrPdO_3$ electrode—Towards a novel non-enzymatic glucose sensor. Journal

of Electroanalytical Chemistry. 2015;**749**:42-52

[35] Wolkoff P. Volatile organic compounds sources, measurements, emissions, and the impact on indoor air quality. Indoor Air. 1995;**5**:5

[36] Boor BE, Spilak MP, Laverge J, Novoselac A, Xu Y. Human exposure to indoor air pollutants in sleep microenvironments: A literature review. Building and Environment. 2017;**125**:528

[37] Joshi N, Hayasaka T, Liu Y, Liu H, Oliveira ON, Lin L. A review on chemiresistive room temperature gas sensors based on metal oxide nanostructures, graphene and 2D transition metal dichalcogenides. Microchimica Acta. 2018;**185**:213

[38] Zheng H, Ou JZ, Strano MS, Kaner RB, Mitchell A, Kalantarzadeh K. Nanostructured tungsten oxide—Properties, synthesis, and applications. Advanced Functional Materials. 2011;**21**:2175

[39] Bai J, Zhou B. Titanium dioxide nanomaterials for sensor applications. Chemical Reviews. 2014;**114**:10131

[40] Thirumalairajan S, Girija K, Mastelaro VR, Ponpandian N. Surface morphology-dependent room-temperature $LaFeO_3$ nanostructure thin films as selective NO_2 gas sensor prepared by radio frequency magnetron sputtering. ACS Applied Materials & Interfaces. 2014;**6**:13917-13927

[41] Jun-Chao D, Hua-Yao L, Ze-Xing C, Xiao-Dong Z, Guo X. $LaCoO_3$–based sensors with high sensitivity to carbon monoxide. RSC Advances. 2015;**5**:65668-65673

[42] Ghasdi M, Alamdari H, Royer S, Adnot A. Electrical and CO gas sensing properties of nanostructured $La_{1-x}Ce_xCoO_3$ perovskite prepared by activated reactive synthesis. Sensors and Actuators B. 2011;**156**:147-155

[43] Sowmya P, Kaushik SD, Siruguri V, Diptikanta S, Viegas AE, Chandrabhas N, et al. Investigation of Ca substitution on the gas sensing potential of $LaFeO_3$ nanoparticles towards low concentration SO_2 gas. Dalton Transactions. 2016;**45**(34):13547-13555

[44] Zhang R, Jifan H, Zhouxiang H, Ma Z, Zhanlei W, Yongjia Z, et al. Electrical and CO-sensing properties of $NdFe_{1-x}Co_xO_3$ perovskite system. Journal of Rare Earths. 2010;**28**(4):591-595

[45] Benali A, Azizi S, Bejar M, Dhahri E, Graça MFP. Structural, electrical and ethanol sensing properties of double-doping $LaFeO_3$ perovskite oxides. Ceramics International. 2014;**40**(9):14367-14373

[46] Rong O, Zhang Y, Hu J, Li K, Wang H, Chen M, et al. Design of ultrasensitive Ag-$LaFeO_3$ methanol gas sensor based on quasi molecular imprinting technology. Scientific Reports. 2018;**8**:14220

[47] Yumin Z, Qingju L, Jin Z, Qin Z, Zhu Z. High sensitive and selective formaldehyde gas sensor using molecular imprinting technique based on Ag-$LaFeO_3$. Journal of Materials Chemistry C. 2014;**2**:10067

[48] Wang Y, Chen J, Wu X. Preparation and gas-sensing properties of perovskite-type $SrFeO_3$ oxide. Materials Letters. 2001;**49**:361-364

[49] Carotta MC, Butturi MA, Martinelli G, Sadaoka Y, Nunziante P, Traversa E. Microstructural evolution of nanosized $LaFeO_3$ powders from the thermal decomposition of a cyano-complex for thick film gas sensors. Sensors and Actuators B. 1997;**44**:590-594

[50] Sen S, Chakraborty N, Rana P, Narjinary M, Mursalin SD, Tripathy S, et al. Nanocrystalline gallium ferrite: A novel material for sensing very low concentration of alcohol vapour. Ceramics International. 2015;**41**:10110

[51] Sahner K, Moos R, Matam M, Tunney JJ, Post M. Hydrocarbon sensing with thick and thin film p-type conducting perovskite materials. Sensors and Actuators B. 2005;**108**:102-112

[52] Zhang D et al. Microwave-assisted synthesis of porous and hollow α-Fe_2O_3/$LaFeO_3$ nanostructures for acetone gas sensing as well as photocatalytic degradation of methylene blue. Nanotechnology. 2020;**31**:215601

[53] Francesco B, Ada F, Valerio V, Marco M, Rossella B. Optimization of perovskite gas sensor performance: Characterization, measurement and experimental design. Sensors. 2017;**17**:1352

[54] Yaoyu Y, Yanbai S, Pengfei Z, Rui L, Ang L, Sikai Z, et al. Fabrication, characterization and n-propanol sensing properties of perovskite-type $ZnSnO_3$ nanospheres based gas sensor. Applied Surface Science. 2020;**509**:145335

[55] Neeru S, Himmat KS, Sharma SK, Sachdev K. Fabrication of $LaFeO_3$ and rGO-$LaFeO_3$ microspheres based gas sensors for detection of NO_2 and CO. RSC Advances. 2020;**10**:1297-1308

[56] Wang H, Guo Z, Wentao H, Li S, Yongjia Z, Ensi C. Ethanol sensing characteristics of $BaTiO_3$/$LaFeO_3$ nanocomposite. Materials Letters. 2019;**234**:40-44

[57] Dong G, Huiqing F, Hailin T, Jiawen F, Qiang L. Gas-sensing and electrical properties of perovskite structure p-type barium-substituted bismuth ferrite. RSC Advances. 2015;**5**:29618-29623

[58] Silu P, Min M, Weijie Y, Ziqian W, Zihan W, Jian B, et al. Acetone sensing with parts-per-billion limit of detection using a $BiFeO_3$-based solid solution sensor at the morphotropic phase boundary. Sensors and Actuators B: Chemical. 2020;**313**:128060

[59] Tie Y, Ma SY, Pei ST, Zhang QX, Zhu KM, Zhang R, et al. Pr doped $BiFeO_3$ hollow nanofibers via electrospinning method as a formaldehyde sensor. Sensors and Actuators B: Chemical. 2020;**308**:127689

[60] Jain GH, Patil LA, Wagh MS, Patil DR, Patil SA, Amalnerkar DP. Surface modified $BaTiO_3$ thick film resistors as H_2S gas sensors. Sensors and Actuators B. 2006;**117**:159-165

[61] Patil RP, Gaikwad SS, Karanjekar AN, Khanna PK, Jain GH, Gaikwad VB, et al. Optimization of strontium- doping concentration in $BaTiO_3$ nanostructures for room temperature NH_3 and NO_2 gas sensing. Materials Today Chemistry. 2020;**16**:100240

[62] Artem M, Marina R, Alexander B, Alexander G. Nanocrystalline $BaSnO_3$ as an alternative gas sensor material: Surface reactivity and high sensitivity to SO_2. Materials. 2015;**8**:6437-6454

[63] Gopal Reddy CV, Manorama SV, Rao VJ. Preparation and characterization of barium stannate: Application as a liquefied petroleum gas sensor. Journal of Materials Science: Materials in Electronics. 2001;**12**:137-142

[64] Katekani S, Swart HC, Mhlongo GH. $LaBO_3$ (B = Fe, Co) nanofibers and their structural, luminescence and gas sensing characteristics. Physica B: Physics of Condensed Matter. 2020;**578**:411883

[65] Shaikh JS, Shaikh NS, Sheikh AD, Mali SS, Kale AJ, Kanjanaboos P, et al.

Perovskite solar cells: In pursuit of efficiency and stability. Materials and Design. 2017;**136**:54

[66] Newcomer juices up the race to harness sunlight. Science. 2013;**342**:1438-1439. DOI: 10.1126/science.342.6165.1438-b

[67] Gholipour S, Saliba M. From exceptional properties to stability challenges of perovskite solar cells. Small. 2018;**14**:1802385

[68] Kojima A, Teshima K, Shirai Y, Miyasaka T. Organometal halide perovskite as visible-light sensitizer for photovoltaic cells. Journal of the American Chemical Society. 2009;**131**:6050-6051

[69] Yang WS, Noh JH, Jeon NJ, Kim YC, Ryu S, Seo J, et al. High-performance photovoltaic perovskite layers fabricated through intramolecular exchange. Science. 2015;**348**:1234-1237

[70] Yin WJ, Yang JH, Kang J, Yan Y, Wie SH. Halide perovskite materials for solar cells: A theoretical review. Journal of Materials Chemistry. 2015;**3**:8926-8942

[71] Green MA, Baillie AH, Snaith HJ. The emergence of perovskite solar cells. Nature Photonics. 2014;**8**:506-514

[72] Babayigit A, Ethirajan A, Muller M, Conings B. Toxicity of organometal halide perovskite solar cells. Nature Materials. 2016;**15**:247-251

[73] Flora G, Gupta D, Tiwari A. Toxicity of lead: A review with recent updates. Interdisciplinary Toxicology. 2012;**5**:47-58

[74] Benmessaoud IR, Mahul-Mellier A-L, Horv'ath E, Maco B, Spina M, Lashuel HA, et al. Health hazards of methylammonium lead iodide based perovskites: Cytotoxicity studies. Toxicological Research. 2016;**5**:407-419

[75] Gong J, Darling SB, You F. Perovskite photovoltaics: Lifecycle assessment of energy and environ mental impacts. Energy & Environmental Science. 2015;**8**:1953-1968

[76] Fthenakis VM, Kim HC, Alsema E. Emissions from photovoltaic life cycles. Environmental Science & Technology. 2008;**42**:2168-2174

[77] Antonio A. Perovskite solar cells go lead free. Joule. 2017;**1**:659-664

[78] Hoefler SF, Trimmel G, Rath T. Progress on lead-free metal halide perovskite for photovoltaic applications: A review. Monatshefte fuer Chemie. 2017;**148**:795-826

[79] Noel NK, Stranks SD, Abate A, Wehrenfennig C, Guarnera S, Haghighirad AB, et al. Lead-free organic-inorganic tin halide perovskites for photovoltaic applications. Energy & Environmental Science. 2014;**7**:3061-3068

[80] Hao F, Stoumpos CC, Cao DH, Chang RP, Kanatzidis H. Lead-free solid-state organic-inorganic halide perovskite solar cells. Nature Photonics. 2014;**8**(8):489-494

[81] Mandapu U, Vedanayakam SV, Thyagarajan K, Reddy MR, Babu BJ. Design and simulation of high efficiency tin halide perovskite solar cells. International Journal of Renewable Energy Research. 2017;**7**:1603-1612

[82] Ngoc TV, Doan HT. Lead-free hybrid organic-inorganic perovskites or solar cell applications. Journal of Chemical Physics. 2020;**152**:014104

[83] Gagandeep, Mukhtiyar S, Ramesh K, Vinamrita S. Graphene as charge transport layers in lead free perovskite solar cell. Materials Research Express. 2019;**6**:115611

[84] Giorgi G, Yamashita K. Alternative, lead-free, hybrid organic-inorganic perovskites for solar applications: A DFT analysis. Chemistry Letters. 2015;**44**:826-828

[85] Krishnamoorthy T, Ding H, Yan C, Leong WL, Baikie T, Zhang Z, et al. Lead-free germanium iodide perovskite materials for photovoltaic applications. Journal of Materials Chemistry A. 2015;**3**:23829-23832

[86] Yang B, Chen J, Hong F, Mao X, Zheng K, Yang S, et al. Lead-free, air-stable all-inorganic cesium bismuth halide perovskite nanocrystals. Angewandte Chemie International Edition. 2017;**56**:12471-12475

[87] Lou Y, Fang M, Chen J, Zhao Y. Formation of highly luminescent cesium bismuth halide perovskite quantum dots tuned by anion exchange. Chemical Communications. 2018;**54**:3779-3782

[88] Nelson RD, Santra K, Wang Y, Hadi A, Petrich JW, Panthani MG. Synthesis and optical properties of ordered-vacancy perovskite cesium bismuth halide nanocrystals. Chemical Communications. 2018;**54**: 3640-3643

[89] Zuo C, Ding L. Lead-free perovskite materials $(NH_4)_3Sb_2I_xBr_{9-x}$. Angewandte Chemie International Edition. 2017;**56**:6528-6532

[90] Saparov B, Hong F, Sun J-P, Duan H-S, Meng W, Cameron S, et al. Thin-film preparation and characterization of $Cs_3Sb_2I_9$: A lead-free layered perovskite semiconductor. Chemistry of Materials. 2015;**27**:5622-5632

[91] Hebig J-C, Kühn I, Flohre J, Kirchartz T. Optoelectronic properties of $(CH_3NH_3)_3Sb_2I_9$ thin films for photovoltaic applications. ACS Energy Letters. 2016;**1**:309-314

[92] Chen M, Ju M-G, Carl AD, Zong Y, Grimm RL, Gu J, et al. Cesium titanium(IV) bromide thin films based stable lead-free perovskite solar cells. Joule. 2018;**2**:558-570

[93] Li X, Zhong X, Hu Y, Li B, Sheng Y, Zhang Y, et al. Organic–inorganic copper(II)-based material: A low-toxic, highly stable light absorber for photovoltaic application. Journal of Physical Chemistry Letters. 2017;**8**:1804-1809

[94] Yang P, Liu G, Liu B, Liu X, Lou Y, Chen J, et al. All-inorganic Cs_2CuX_4 (X = Cl, Br, and Br/I) perovskite quantum dots with blue-green luminescence.Chemical Communicatio ns. 2018;**54**:11638-11641

[95] Tze-Bin S, Takamichi Y, Shinji A, Kanatzidis Mercouri G. Performance enhancement of lead-free tin-based perovskite solar cells with reducing atmosphere-assisted dispersible additive. ACS Energy Letters. 2017;**2**(4):897-903

[96] Nishimura K, Kamarudin MA, Hirotani D, Hamada K, Shen Q, Iikubo S, et al. Lead-free tin-halide perovskite solar cells with 13% efficiency. Nano Energy. 2020;**74**:104858

[97] World Health Organization. Guidelines for Drinking Water Quality. 2nd ed. Geneva, Switzerland: Author; 1997

[98] Fu F, Wang Q. Removal of heavy metal ions from wastewaters: A review. Journal of Environmental Management. 2011;**92**:407-418

[99] Biswas AK, Tortajada C. Water quality management: An introductory frame work. Water Resources Management. 2011;**27**:5-11

[100] Taseidifar M, Makavipour F, Pashley RM, Rahman AM. Removal of heavy metal ions from water using ion flotation. Environmental Technology and Innovation. 2017;**8**:182-190

[101] Mohammed AA, Ebrahim SE, Alwared AI. Flotation and sorptive-flotation methods for removal of lead ions from wastewater using SDS as surfactant and barley husk as biosorbent. Journal of Chemistry. 2013:1-6

[102] Rauf MA, Qadri SM, Ashraf S, Al-Mansoori KM. Sorption and desorption of Pb^{2+} ions by dead *Sargassum* sp. biomass. Chemical Engineering Journal. 2009;**150**:90-95

[103] Zhanga D, Chun-li Z, Zhou P. Preparation of porous nano-calcium titanate microspheres and its adsorption behavior for heavy metal ion in water. Journal of Hazardous Materials. 2011;**186**:971-977

[104] Haron W, Sirimahachai U, Wisitsoraat A, Wongnawa S. Removal of Cd^{2+} and Pb^{2+} from water by $LaGdO_3$ perovskite. SNRU Journal of Science and Technology. 2017;**9**(3):544-551

[105] Zhang D, Wang M, Tan Y. Preparation of porous nano-barium-strontium titanate by sorghum straw template method and its adsorption capability for heavy metal ions. Acta Chimica Sinica: Chinese Edition. 2010;**68**(16):1641-1648

[106] Rao MP, Musthafa S, Wu JJ, Anandan S. Facile synthesis of perovskite $LaFeO_3$ ferroelectric nanostructures for heavy metal ion removal applications. Materials Chemistry and Physics. 2019;**232**:200-204

[107] Zhang X, Li H, Li Y, Shen W. Structural properties and catalytic activity of Sr-substituted $LaFeO_3$ perovskite. Chinese Journal of Catalysis. 2012;**33**:1109-1114

[108] Haron W, Wisitsoraat A, Sirimahachai U, Wongnawa S. Removal of toxic heavy metal ions from water with $LaAlO_3$ perovskite. Songklanakarin Journal of Science and Technology. 2018;**40**(5):993-1001

[109] Chen F, Yu C, Wei L, Fan Q, Ma F, Zeng J, et al. Fabrication and characterization of $ZnTiO_3/Zn_2Ti_3O_8/ZnO$ ternary photocatalyst for synergetic removal of aqueous organic pollutants and Cr(VI) ions. Science of The Total Environment. 2020;**706**:136026

[110] Normala H, Soo YRG. Removal of heavy metals from textile wastewater using zeolite. Environment Asia. 2010;**3**:124-130

[111] CPCB (Central Pollution Control Board). Advance Methods for Treatment of Textile Industry Effluents. Publication No. RERES/7/2007. New Delhi: Government of India; 2007. Available from: https://www.researchgate.net/publication/280572135_advance_methods_for_treatment_of_textile_industry_effluents

[112] Asad S, Amoozegar MA, Pourbabaee AA, Sarbolouki MN, Dastgheib SMM. Decolorization of textile azo dyes by newly isolated halophilic and halotolerant bacteria. Bioresource Technology. 2007;**98**:2082-2088

[113] Ananthanarayanan Y, Natchimuthu K, Sudipta T, Soundarapandian K, Ramasundaram T. Environment-friendly management of textile mill wastewater sludge using epigeic earthworms: Bioaccumulation of heavy metals and metallothionein production. Journal of Environmental Management. 2020;**254**:109813

[114] Toledano Garcia D, Ozer LY, Parrino F, Ahmed M, Brudecki GP, Hasan SW, et al. Photocatalytic ozonation under visible light for the remediation of water effluents and its integration with an electro-membrane bioreactor. Chemosphere. 2018;**209**:534-541

[115] Matei E, Covaliu CI, Predescu A, Berbecaru A, Tarcea C, Predescu C. Remediation of wastewater with ultraviolet irradiation using a novel titanium (IV) oxide photocatalyst. Analytical Letters. 2019;**52**:2180-2187

[116] Neagu S, Cojoc R, Enache M, Mocioiu OC, Precupas A, Popa VT, et al. Biotransformation of waste glycerol from biodiesel industry in carotenoids compounds by halophilic microorganisms. Waste and Biomass Valorization. 2019;**10**:45-52

[117] Kamińska G, Dudziak M, Kudlek E, Bohdziewicz J. Preparation, characterization and adsorption potential of grainy halloysite-CNT composites for anthracene removal from aqueous solution. Nanomaterials. 2019;**9**:890

[118] Wang J, Zhuang S. Removal of various pollutants from water and wastewater by modified chitosan adsorbents. Critical Reviews in Environmental Science and Technology. 2017;**47**:2331-2386

[119] Xu S, Zhang S, Chen K, Han J, Liu H, Wu K. Biosorption of La^{3+} and Ce^{3+} by *Agrobacterium* sp. HN1. Journal of Rare Earths. 2011;**29**:265-270

[120] Yagub MT, Sen TK, Afroze S, Ang HM. Dye and its removal from aqueous solution by adsorption: A review. Advances in Colloid and Interface Science. 2014;**209**:172-184

[121] Almasri DA, Saleh NB, Atieh MA, McKay G, Ahzi S. Adsorption of phosphate on iron oxide doped halloysite nanotubes. Scientific Reports. 2019;**9**:3232

[122] Ali SM, Eskandrani AA. The sorption performance of cetyl trimethyl ammonium bromide-capped $La_{0.9}Sr_{0.1}FeO_3$ perovskite for organic pollutants from industrial processes. Molecules. 2020;**25**(7):1640

[123] Chu Y, Tan X, Shen Z, Liu P, Han N, Kang J, et al. Efficient removal of organic and bacterial pollutants by Ag-$La_{0.8}Ca_{0.2}Fe_{0.94}O_{3-\delta}$ perovskite via catalytic peroxymonosulfate activation. Journal of Hazardous Materials. 2018;**356**:53-60

[124] Pratibha S, Dhananjaya N, Manjunatha CR, Narayana A. Fast adsorptive removal of direct blue-53 dye on rare-earth doped Lanthanum aluminate nanoparticles: Equilibrium and kinetic studies. Materials Research Express. 2019;**6**:1250i5

[125] Dong Z, Ye T, Zhao Y, Yu J, Wang F, et al. Perovskite $BaZrO_3$ hollow micro- and nanospheres: Controllable fabrication, photoluminescence and adsorption of reactive dyes. Journal of Materials Chemistry. 2011;**21**:5978

[126] Siddharth S, Soumitra M, Sonia R, Hari R, Bisht RS, Panigrahi SK, et al. Ti doped $BaMnO_3$ perovskite structure as photocatalytic agent for the degradation of noxious air and water pollutants. SN Applied Sciences. 2020;**2**:310

2

Thermodynamics of Perovskite: Solid, Liquid, and Gas Phases

Sergey Shornikov

Abstract

The present work is devoted to the review of experimental data on thermodynamic properties of perovskite in the condensed state, as well as the gas phase components over perovskite and its melts at high temperatures.

Keywords: perovskite, thermodynamics

1. Introduction

Calcium titanate ($CaTiO_3$) or perovskite was found by Rose [1] in the Ural Mountains in 1839 and named after the Russian Statesman Count Lev Perovski. Perovskite is a relatively rare mineral, which is a promising material for use as matrices for safe long-term storage of actinides and their rare earth analogs that are present in radioactive waste [2]. It is of particular interest for petrology and cosmochemical research as a mineral which is a part of refractory Ca-Al inclusions often found in carbonaceous chondrites, which are the earliest objects of the solar system with unusual isotopic characteristics [3–5].

In addition to perovskite, two more calcium titanates, $Ca_3Ti_2O_7$ [6] and $Ca_4Ti_3O_{10}$ [7], melting incongruently, were found in the CaO-TiO_2 system. The other calcium titanates, Ca_4TiO_6 [8], Ca_3TiO_5 [9], $Ca_8Ti_3O_{14}$ [8, 10], Ca_2TiO_4 [9], $Ca_5Ti_4O_{13}$ [11], $Ca_2Ti_3O_8$ [6], $CaTi_2O_5$ [12, 13], $CaTi_3O_7$ [14], $Ca_2Ti_5O_{12}$ [15], and $CaTi_4O_9$ [16], are mentioned in the literature. They also seem to be unstable, which does not exclude their possible existence [17]. Compiled in this paper on the basis of data [7, 18], as well as the results of recent studies by Gong et al. [19], the phase diagram of the CaO-TiO_2 system in the high-temperature region is shown in **Figure 1**.

2. The thermodynamic properties of perovskite solid phase

Thermochemical data on perovskite [20–24] are based on calorimetric measurements of entropy of perovskite formation $\Delta S_{298}(CaTiO_3)$, obtained by Shomate [25], and high-temperature heat capacity of perovskite $C_p(CaTiO_3)$ in the temperature ranges of 15–398 K [26], 293–773 K [27], 376–1184 K [28], 383–1794 K [29], and 413–1825 K [30]. The differences between the data do not exceed 5 J/(mol K) up to a temperature of 1200 K, but they are quite contradictory at higher temperatures (**Figure 2**).

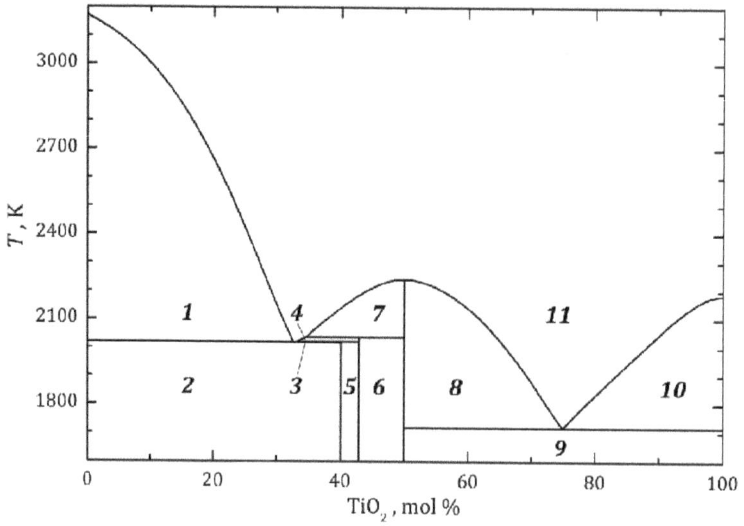

Figure 1.
The phase diagram of the CaO-TiO_2 system [7, 18, 19]: (1) CaO + liquid; (2) CaO + $Ca_3Ti_2O_7$; (3) $Ca_3Ti_2O_7$ + liquid; (4) $Ca_4Ti_3O_{10}$ + liquid; (5) $Ca_3Ti_2O_7$ + $Ca_4Ti_3O_{10}$; (6) $Ca_4Ti_3O_{10}$ + $CaTiO_3$; (7, 8) $CaTiO_3$ + liquid; (9) $CaTiO_3$ + TiO_2; (10) TiO_2 + liquid; (11) liquid.

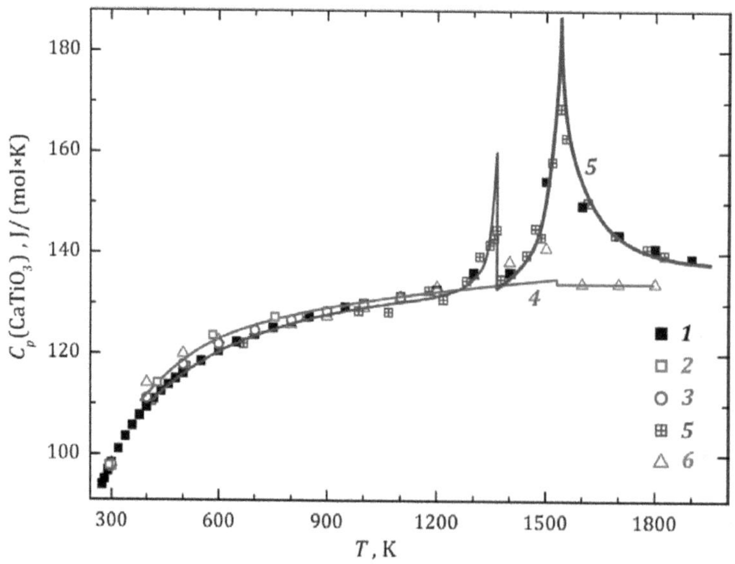

Figure 2.
Heat capacity of perovskite: (1–5) determined via high-temperature calorimetry [26–30], respectively, and (6) taken from [22].

Naylor and Cook [29] determined the enthalpy of perovskite phase transition: 2.30 ± 0.07 kJ/mol at 1530 ± 1 K. The overlapping phase transitions in perovskite observed by Guyot et al. [30] (**Figure 2**) were explained as consequences of a structural change, i.e., a transition from orthorhombic (*Pbnm*) to orthorhombic (*Cmcm*) structure at 1384 ± 10 K and the overlapping of transitions from orthorhombic to tetragonal (*I4/mcm*) and tetragonal to cubic phase ($Pm\bar{3}m$) at 1520 ± 10 K with heat effects of 1.0 ± 0.5 and 5.5 ± 0.5 kJ/mol, respectively. The considerable anomaly of perovskite heat capacity above 1520 K could be caused by strong disordering of the cubic phase up to the temperature of perovskite melting.

However, based on diffraction data about perovskite structure obtained in the

Experimental approach	T, K	ΔH_T, kJ/mol	ΔS_T, J/(mol K)	Refs.
HF/HCl solution calorimetry	298		1.05 ± 0.21	[25]
HF/HCl solution calorimetry	298	-40.48 ± 0.42	1.86 ± 0.71	[33]
Bomb calorimetry	298	-41.84 ± 1.88		[32]
Adiabatic calorimetry	298	-40.48 ± 1.65	2.40 ± 0.35	[26]
Adiabatic calorimetry	298	-47.11 ± 1.41	2.72 ± 0.23	[28]
$Na_6Mo_4O_{15}$ solution calorimetry	298	-42.98 ± 1.96		[34]
EMF	888–972	-37.05 ± 3.28	4.62 ± 3.54	[35]
EMF	900–1250	-40.07 ± 0.05	3.15 ± 0.05	[36]
$Pb_2B_2O_5$ solution calorimetry	973	-38.73 ± 1.34		[37]
$Na_6Mo_4O_{15}$ solution calorimetry	975	-42.25 ± 1.05		[38]
$Na_6Mo_4O_{15}$ solution calorimetry	975	-41.88 ± 1.36		[39]
$Na_6Mo_4O_{15}$ solution calorimetry	976	-42.86 ± 1.71		[40]
$Pb_2B_2O_5$ solution calorimetry	1046	-42.38 ± 1.82		[37]
(Li,Na)BO_2 solution calorimetry	1068 ± 2	-40.45 ± 1.15		[41]
$Pb_2B_2O_5$ solution calorimetry	1073	-40.43 ± 1.87	4.37 ± 1.23	[42]
$Pb_2B_2O_5$ solution calorimetry	1074	-43.78 ± 1.73		[43]
$Pb_2B_2O_5$ solution calorimetry	1078	-42.83 ± 3.12		[44]
EMF	1180–1290	-26.88 ± 8.05	11.07 ± 6.44	[45]
Raman spectroscopy	1300		2.82 ± 1.29	[46]
Thermochemical calculations	1600–1800	-37.19 ± 0.14	5.85 ± 0.09	[22]
Thermochemical calculations	1600–2100	-38.23 ± 0.04	5.00 ± 0.02	[23]
Thermochemical calculations	1600–2100	-37.47 ± 0.03	5.60 ± 0.02	[24]
DTA	1740 ± 20	-37.55 ± 3.76		[47]
Knudsen mass spectrometry	1791–2241	-39.98 ± 0.54	3.15 ± 0.28	[48]
Knudsen mass spectrometry	2241–2398	7.73 ± 1.76	24.39 ± 0.76	[48]

Table 1.
Enthalpy and entropy of perovskite formation from simple oxides (calculated per 1 mol of compound).

temperature range of 296 – 1720 K, Yashima and Ali [31] concluded that the *Cmcm* phase does not exist and claimed that the first transition is the (*Pbnm*) → (*I4/mcm*) at 1512 ± 13 K, followed by the (*I4/mcm*) → ($Pm\overline{3}m$) transition at 1635 ± 2 K.

Panfilov and Fedos'ev [3 2] determined the enthalpy of the reaction with a calorimetric bomb by burning stoichiometric mixtures of rutile TiO_2 and calcium carbonate $CaCO_3$ (here and below, the square brackets denote the condensed phase; the parentheses denote the gas phase):

$$[CaCO_3] + [TiO_2] = [CaTiO_3] + (CO_2). \quad (1)$$

They then calculated the enthalpy of perovskite formation $\Delta H_{298}(CaTiO_3)$: -41.84 ± 1.88 kJ/mol (**Table 1**). Although the obtained value was determined with poor accuracy, due to the difficulty of determining the amounts of substances in the reaction products (1), it corresponded satisfactorily to the more accurate results of Kelley et al. [33], who determined this value according to the reaction:

$$[CaO] + [TiO_2] = [CaTiO_3] \quad (2)$$

by solution calorimetry in a mixture of hydrofluoric and hydrochloric acids. The reactions of dissolution were more complete than combustion reaction (1).

Navrotsky et al. [19, 26, 34, 37–43] performed a number of studies by various calorimetric methods, using adiabatic calorimetry [26] and solution calorimetry in the (Li, Na)BO_2 [41], $Pb_2B_2O_5$ [37, 42, 43], and $Na_6Mo_4O_{15}$ [19, 34, 38–40] salts in the temperature range 973 – 1074 K. They determined the value of $\Delta H_T(CaTiO_3)$, which lies in the range of −44 to −39 kJ/mol; the accuracy of measurements was 2 kJ/mol (**Table 1**). The differences could be due to the properties of the solvents that were used. Perovskite and its oxides are poorly soluble in oxide solvent $Pb_2B_2O_5$; $Na_6Mo_4O_{15}$ liquid alloy is quite volatile and cannot be used at tempera-tures above 1000 K; (Li,Na)BO_2 solution is hygroscopic, which creates difficulties in synthesizing [49]. The $\Delta H_{1078}(CaTiO_3)$ value determined by Koito et al. [44] by similar method (solution calorimetry in $Pb_2B_2O_5$ salt) is less accurate but close to the results obtained by Navrotsky et al. [19, 34, 37–43].

The data obtained by Sato et al. [28] using adiabatic calorimetry deviate negligibly (by as much as 7 kJ/mol) in the enthalpy values of $(H_T – H_{298})$ in the temperature range above 1000 K. Approximately the same systematic deviation is observed in determination of enthalpy of perovskite formation from oxides (per 1 mole of the compound) due to its use in calculations of the rough semiempirical approximation proposed in [50]. At the same time, the entropy of perovskite formation determined by Sato et al. [28] is in satisfactory agreement with the results obtained by Kelley and Mah [20], Woodfield et al. [26], and Prasanna and Navrotsky [42] and calculated by Gillet et al. [46] based on information obtained on Raman spectra (**Table 1**).

Golubenko and Rezukhina [45] studied the heterogeneous reaction using a solid electrolyte galvanic cell (EMF method) in the temperature range of 1180–1290 K:

$$[CaO] + ½[Ti_2O] + ¾(O_2) = [CaTiO_3]. \qquad (3)$$

A mixture of FeO and Fe (or NbO and Nb) was used as the reference electrode, and a mixture of La_2O_3-ThO_2 crystals was used as the solid electrolyte. The Gibbs energy o f perovskite $\Delta G_T(CaTiO_3)$ was calculated based on a compilation of fairly approximate literature data and their own estimates of the thermodynamic properties of the [Ti_2O] compound. This produced a considerable error in determining this value (**Figure 3**). Rezukhina et al. [35] later made more precise measurements of $\Delta G_T(CaTiO_3)$ in the temperature range of 888–972 K, inside a galvanic cell with CaF_2 as the electrolyte (**Figure 3**). However, the non-systematic errors in determining $\Delta H_T(CaTiO_3)$ and $\Delta S_T(CaTiO_3)$ values were considerable (**Table 1**).

Taylor and Schmalzried [51] (at 873 K) and Jacob and Abraham [36] (at 900–1250 K) also determined the perovskite Gibbs energy via EMF using the same solid electrolyte. The obtained ΔG_T ($CaTiO_3$) values were close to the results of Rezukhina et al. [35].

Klimm et al. [47] used differential thermal analysis (DTA) to determine the enthalpy of reaction (2) at 1740 ± 20 K. The obtained value is consistent with the results of thermochemical calculations, although it has a significant error.

Suito et al. [54, 55] has studied the equilibrium at 1873 K:

$$[Ti] + 2(O) = [TiO_2] \qquad (4)$$

in CaO-TiO_x -(or CaO_xTiO_2 Al O_3) - slags with liquid nickel, relative to oxygen and nitrogen, depending on the content of Ti (or Al) in the metal. Crucibles made of CaO or Al_2O_3 were used. The activity (a_i) of titanium oxide was estimated indirectly, depending on the content of Al, Ti, and O in the slag, and using data on the Gibbs energies of oxide formation (**Figure 4**).

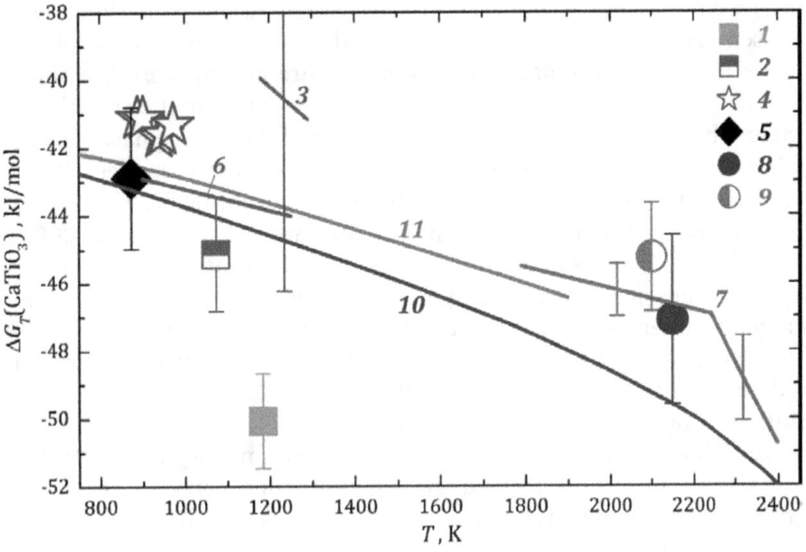

Figure 3.
The Gibbs energy of the formation of perovskite, determined via (1, 2) calorimetry [28, 42], respectively; (3–6) EMF [35, 36, 45, 51], respectively; (7–9) Knudsen effusion mass spectrometric method in [48, 52, 53], respectively; (10, 11) calculated from thermochemical data in [24, 26], respectively.

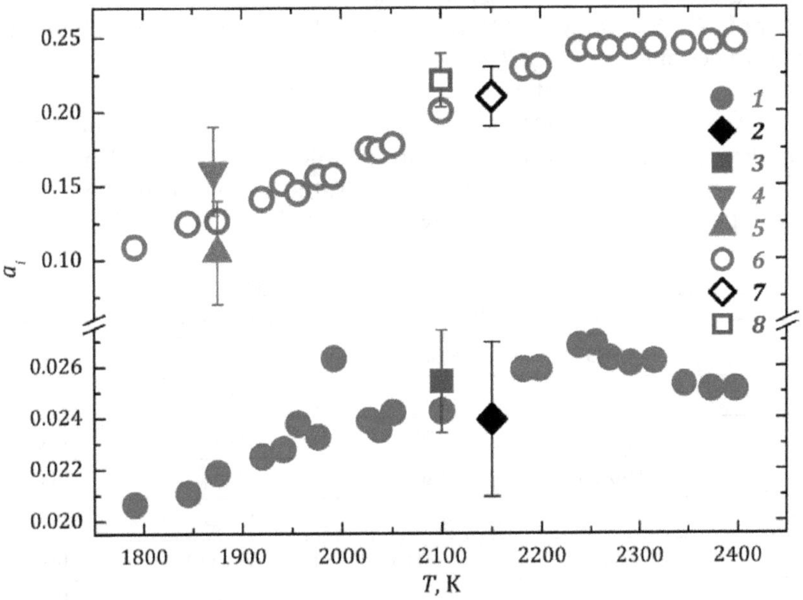

Figure 4.
Activities of CaO (1–3) and TiO_2 (4–8) in perovskite, determined (1–3, 6–8) via Knudsen effusion mass spectrometry [48, 52, 53] and (4, 5) in studying the heterogeneous equilibria of multicomponent melts [54, 55], respectively.

Banon et al. [52] (at 2150 K) and Shornikov et al. [48, 53, 56] (at 1791–2398 K) determined the values of CaO and TiO_2 activities (**Figure 4**) and $\Delta G_T(CaTiO_3)$ via Knudsen effusion mass spectrometric method. The resulting values correlated with one another within the experimental error (up to 2.5 kJ/mol); however, the $\Delta G_T(CaTiO_3)$ was obtained at a temperature ~1000 K higher than in the earlier results (**Figure 3**).

As it is seen in **Figure 4**, the values of oxide activities in perovskite determined via Knudsen effusion mass spectrometric method agree with one another in the investigated temperature range. As the temperature grows, there is a slight trend toward the higher activities of calcium and titanium oxides in the crystalline perovskite phase. This trend is less noticeable in the area of the liquid phase. The activities of titanium oxide calculated based on studying of equilibria in slags [54, 55] are fairly approximate but not inconsistent with the results in [48, 52, 53].

The values of Gibbs energy of perovskite formation determined at 800–1200 K via EMF [35, 36, 51] and solution calorimetry [42] correlate satisfactorily with the obtained via Knudsen effusion mass spectrometric method [48, 52, 53, 56]. **Figure 3** shows a good agreement between these data and the results from thermochemical calculations performed by Woodfield et al. [26]. The difference between our findings and the thermochemical data calculated by Bale et al. [24] is large in the perovskite melting area, but is still less than 3 kJ/mol.

The $\Delta H_T(CaTiO_3)$ and ΔS_Ta(C TiO_3) values determined in [48] correlate with the results from studies performed via solution calorimetry [37, 41, 42, 44] and EMF [35, 36] at lower temperatures and by Raman spectroscopy [46] and DTA [47] at similar temperatures (**Table 1**).

3. Melting of perovskite

The data characterizing the melting of simple oxides [23, 24, 57–60] are quite rough (**Table 2**). According to different thermochemical data, perovskite's melting temperature lies in the range of 2188 to 2243 K.

Klimm et al. [47] estimated the perovskite melting enthalpy as 56.65 ± 11.33 kJ/mol at 2220 ± 20 K (**Table 2**) which i s close to earlier thermochemical estimates [24, 59].

Shornikov [48] based on his own data (**Table 1**) has obtained more accurate values, characterizing the perovskite melting (**Table 2**). They coincide satisfactorily with the experimental data obtained by Klimm et al. [47] and the thermochemical

Compound	T, K	ΔH_{melt}, kJ/mol	ΔS_{melt}, J/(mol K)	References
CaO	2843	52.00	18.29	[59]
	2845	79.50	27.94	[24]
	3200 ± 50	79.50	24.84	[58]
	3210 ± 10	55.20	17.20	[60]
$CaTiO_3$	2188	41.84	19.12	[61]
	2220 ± 20	56.65 ± 11.33	25.52 ± 5.10	[47]
	2233	53.32	23.88	[24]
	2241 ± 10	47.61 ± 1.84	21.24 ± 0.81	[48]
	2243	63.65	28.38	[59]
TiO_2	2103	66.90	31.81	[59]
	2130 ± 20	66.94 ± 16.70	31.43 ± 7.84	[23, 58]
	2130	46.02	21.61	[24]
	2185 ± 10	68.00 ± 8.00	31.12 ± 3.66	[57]

Table 2.
Temperatures, enthalpies, and entropies of the melting of compounds in the CaO-TiO_2 system (calculated for 1 mol of compound)

estimates made by Bale et al. [24]. The enthalpy of perovskite melting estimated using Walden's empirical rule [62] is also close to the result obtained by Shornikov [48]: ΔH_{melt} = 8.8 [J/(g-at K)] T_{melt} = 49.39 kJ/mol.

4. The thermodynamic properties of perovskite melts

Thermodynamic information about the CaO-TiO_2 melts is quite scarce and limited by the results of only a few experimental studies. Consider the available experimental data, obtained by the Knudsen effusion mass spectrometry.

Banon et al. [52] investigated the evaporation of 24 compositions of the $CaTiO_3$-$Ti_2O_3TiO_2$- system from molybdenum containers at 1900–2200 K. The synthesized compositions contained up to 90.2 mol% $Ti_2O_3$and up to 42 mol% TiO_2 as well as $CaTiO_3$ compound. Based on the partial vapor pressures (Ca), (TiO), and (TiO_2) over melts at 2150 K, the authors calculated the Ti, TiO, Ti_2O_3, TiO_2, and $CaTiO_3$ activities, as well as mixing energies in the melts. In the case of the $CaTiO_3$-TiO_2 melts, the TiO_2 and $CaTiO_3$ activities were calculated by extrapolation from the data relating to the $CaTiO_3$-Ti_2O_3-TiO_2 system and thus had, according to the authors themselves, low accuracy, which apparently was caused by inconsistency with different versions of the CaO-TiO_2 phase diagram [7, 9, 11, 18]. Nevertheless Banon et al. [52], interpreting the obtained high values of TiO_2 activities in the region close to titanium dioxide (**Figure 5**), assumed the presence of immiscibility of the CaO-TiO_2 melts in this region.

Stolyarova et al. [63] investigated the properties of the gas phase over 14 compositions of the CaO-TiO_2-SiO_2 system and also determined the values of oxide activity and melt mixing energy by high-temperature mass spectrometry during the evaporation of melts from tungsten effusion containers at 1800–2200 K. The synthesized compositions contained up to 70 mol% CaO, up to 69 mol% SiO_2, and up to 40 mol% TiO_2. As it is shown in **Figure 5**, one of the two studied compositions of the CaO-TiO_2 system at 2057 K was in the "CaO + liquid" region, and thus its value

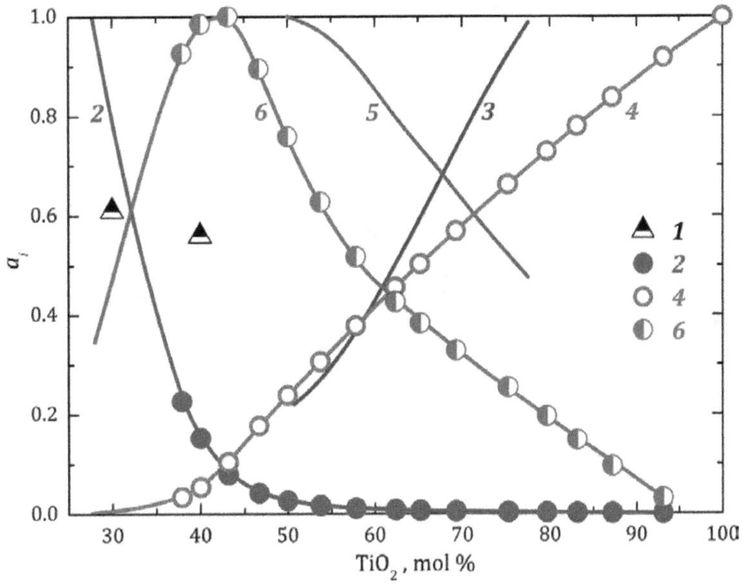

Figure 5.
The activities of CaO (1, 2), TiO_2 (3, 4), and $CaTiO_3$ (5, 6) in the CaO-TiO_2 melts, determined at 2057 K (1) in [63], at 2150 K (3, 5) in [52], and at 2250 K (2, 4, 6) in [64].

should be close to 1 . The second composition was in the region of "$Ca_4Ti_3O_{10}$ + liquid," according to the information presented in [7, 18], or in the region of "$Ca_3Ti_2O_7$ + liquid," as follows from the data presented by Tulgar [11]. However, the calculated values are quite close (**Figure 5**), which contradicts the CaO-TiO_2 phase diagram (**Figure 1**). A possible reason for the discrepancies seems to be a significant error in the measurements of CaO activities in the melt, which may be, in our opinion, more than 50%.

Shornikov [64] investigated the evaporation from molybdenum containers of more than 200 compositions of the CaO-TiO_2 system containing from 34 to 98 mol % TiO_2 at 2241–2441 K. The studied compositions were the CaO-TiO_2-SiO_2 residual melts containing up to 1 mol% SiO_2 that was lost during high-temperature evaporation. The determined composition of the gas phase over the CaO-TiO_2 melts allowed to conclude that evaporation reactions are typical for individual oxides predominate.

The oxide activities in the CaO-TiO_2 melts were calculated according to Lewis equations [65]:

$$a_i = p_i/p_i^\circ, \tag{5}$$

where p_i° and p_i are the partial pressures of vapor species over individual oxide and melt, respectively. However, it is preferable to calculate the values of oxide activities using the Belton-Fruehan approach [66] via the following equation:

$$\ln a_i = -\int x_j d\ln\left(a_j/a_i\right), \tag{6}$$

in which the ratio of the oxide activities in the melt could be easily converted to the ratio of the partial pressures, proportional to the ion currents (I_i):

$$\begin{aligned}\ln a_{\mathrm{TiO_2}} &= -\int x_{\mathrm{CaO}} d\ln\left(p_{\mathrm{CaO}}/p_{\mathrm{TiO_2}}\right) = -\int x_{\mathrm{CaO}} d\ln\left(p_{\mathrm{Ca}}p_{\mathrm{O}}/p_{\mathrm{TiO}}p_{\mathrm{O}}\right)\\ &= -\int x_{\mathrm{CaO}} d\ln\left(I_{\mathrm{Ca}}/I_{\mathrm{TiO}}\right),\end{aligned} \tag{7}$$

and thus to evade the needs in additional thermochemical data, used in Eq. (5).

The consistency of the values of TiO_2 activities calculated by relation (7) was verified using the Gibbs-Duhem equation [67]:

$$\ln a_{\mathrm{CaO}} = -\int \frac{x_{\mathrm{TiO_2}}}{x_{\mathrm{CaO}}} d\ln a_{\mathrm{TiO_2}}, \tag{8}$$

Values of chemical potentials ($\Delta\mu_i$), partial enthalpy (ΔH_i), and entropy (ΔS_i) of oxides in the CaO-TiO_2 melts were calculated by known equations [67]:

$$\Delta\mu_i = RT\ln a_i \tag{9}$$

$$\Delta\mu_i = \Delta H_i - T\Delta S_i \tag{10}$$

$$\Delta H_i = \frac{d(\Delta\mu_i/T)}{d(1/T)} = R\frac{d\ln a_i}{d(1/T)} \tag{11}$$

$$\Delta S_i = -d\Delta\mu_i/dT, \tag{12}$$

which are related to the corresponding integral thermodynamic mixing functions:

$$\Delta G_T^m = \sum_i x_i \Delta \mu_i \tag{13}$$

$$\Delta H_T = \sum_i x_i \Delta H_i \tag{14}$$

$$\Delta S_T = \sum_i x_i \Delta S_i \tag{15}$$

$$\Delta G_T^m = \Delta H_T - T\Delta S_T \tag{16}$$

and are represented in **Figure 6**.

The results presented by Banon et al. [52] correlate with the data found in [64]. Some difference in values, as mentioned above, is probably due to the procedures for extrapolating information obtained by Banon et al. [52] for compositions of the $CaTiO_3$-Ti_2O_3-TiO_2 triple system, which could reduce their accuracy. The observed behavior of TiO_2 activity in melts in the concentration region close to rutile may indicate some immiscibility of the melt, which follows from the observed inflection of the concentration dependence (**Figure** 5, line 3). However, in our opinion, the behavior of TiO_2 and $CaTiO_3$ activities (**Figure** 5, lines 4 and 6) are close to the ideal. The maximum value corresponds to the area of compositions close to

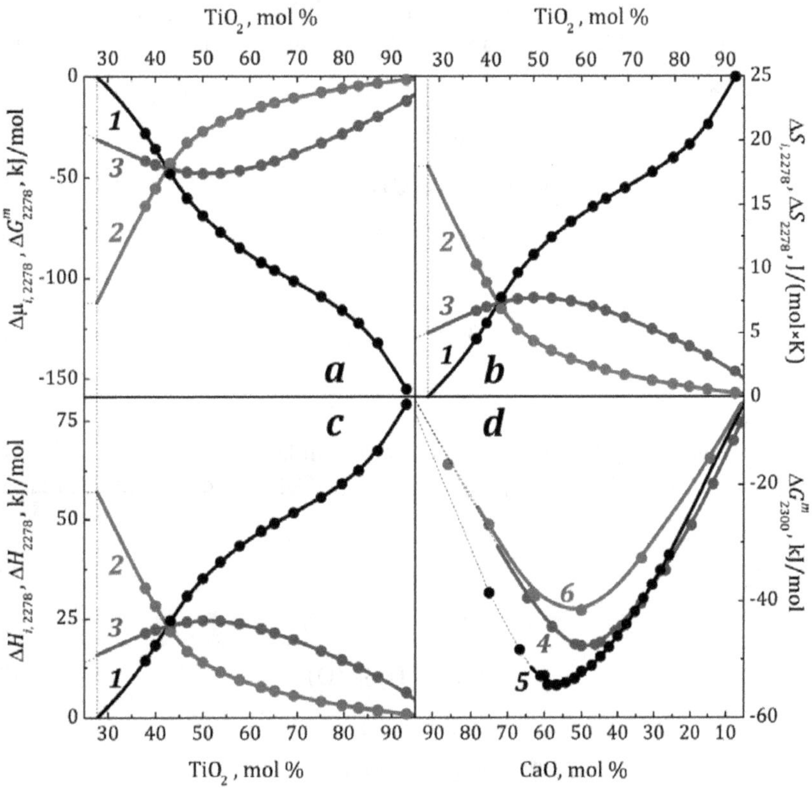

Figure 6.
The thermodynamic properties of the CaO-TiO_2 melts at 2278 K [64] (the chemical potentials of oxides and the mixing energy (a), the partial enthalpies of oxides and the enthalpy of formation (b), and the partial entropies of oxides and the entropy of formation (c)); symbols: (1) CaO, (2) TiO_2, (3) integral thermodynamic characteristics (mixing energy, enthalpy, and entropy of formation of the melts, respectively; the vertical dashed line marks the boundary of the "CaO + liquid" region and the melt) and the comparison of mixing energies (d) in the CaO-TiO_2 (4), CaO-SiO_2 (5), and CaO-Al_2O_3 (6) melts determined by the Knudsen effusion mass spectrometric method in [64, 68, 69], respectively (the dashed lines correspond to heterogeneous areas).

perovskite (**Figure 5**, lines 5 and 6). Differences with values obtained by Stolyarova et al. [63] (**Figure 5**, points 1), are caused, apparently, by the low accuracy of the latter.

The partial and integral thermodynamic regularities presented in **Figure 6** characterizing the CaO-TiO_2 melts are symbate. The enthalpy and entropy of melt formation are positive. The extreme values of the integral thermodynamic properties of the melts are in the concentration ranges close to perovskite, which confirms its stability in the melt. Some displacement of the extremum of integral thermodynamic functions can be caused by the presence of oxide compounds with a large amount of CaO in comparison with perovskite $CaTiO_3$ in the melt. A comparison of mixing energies in the CaO-TiO_2melts at 2300 K with those for the CaO-SiO_2 [68] and CaO-Al_2O_3 [69] melts (**Figure 6d**) indicates a stronger chemical interaction in the CaO-TiO_2 melts than the CaO-Al_2O_3 melts, but smaller than in the CaO-SiO_2 melts. It manifests in more positive values of the mixing energy of the melts.

5. The gas phase over perovskite

The evaporation processes and the thermodynamic properties of simple oxides CaO and TiO_2 were considered in detail in reference books [23, 24, 57, 70, 71].

The gas phase over calcium oxide consists of the molecular components (O), (O_2), (O_3), (O_4), (Ca), (Ca_2), and (CaO) possibly formed by the following reactions:

$$[\mathrm{CaO}] = (\mathrm{CaO}) \tag{17}$$

$$(\mathrm{CaO}) = (\mathrm{Ca}) + (\mathrm{O}) \tag{18}$$

$$2(\mathrm{Ca}) = (\mathrm{Ca_2}) \tag{19}$$

$$2(\mathrm{O}) = (\mathrm{O_2}) \tag{20}$$

$$3(\mathrm{O}) = (\mathrm{O_3}) \tag{21}$$

$$4(\mathrm{O}) = (\mathrm{O_4}). \tag{22}$$

The gas phase over titanium oxide contains similar vapor molecular forms (O), (O_2), (O_3), (O_4), (Ti), (Ti_2), (Ti_3), (TiO), and (TiO_2) formed by similar reactions:

$$[\mathrm{TiO_2}] = (\mathrm{TiO_2}) \tag{23}$$

$$(\mathrm{TiO_2}) = (\mathrm{TiO}) + (\mathrm{O}) \tag{24}$$

$$(\mathrm{TiO}) = (\mathrm{Ti}) + (\mathrm{O}) \tag{25}$$

$$2(\mathrm{Ti}) = (\mathrm{Ti_2}) \tag{26}$$

$$3(\mathrm{Ti}) = (\mathrm{Ti_3}). \tag{27}$$

Balducci et al. [72] detected (Ti_2O_3) and (Ti_2O_4) molecules in the gas phase over cobalt titanate $CoTiO_3$ at 2210–2393 K by the Knudsen effusion mass spectrometric method, which can be involved in the following equilibria:

$$2(\mathrm{TiO_2}) = (\mathrm{Ti_2O_4}), \tag{28}$$

$$(\mathrm{Ti_2O_4}) = (\mathrm{Ti_2O_3}) + (\mathrm{O}). \tag{29}$$

Note that the predominant components of the gas phase over these oxides are (Ca), (CaO), (TiO), (TiO_2), (O), and (O_2); the content of other vapor species does not exceed 1% of the total concentration at 1700–2200 K.

The properties of the gas phase over perovskite were studied in less detail. The experimental conditions and results of high-temperature studies of perovskite evaporation we will consider below.

Zakharov and Protas [73] studied ion emission from the perovskite surface under the action of laser radiation and identified the ion of a complex molecule ($CaTiO_3$) in addition to the ions of simple oxides (O^+, Ca^+, CaO^+, Ti^+, TiO^+) in the mass spectra of the vapor. They explained the presence of this ion by the similarity of - high temperature evaporation of alkaline-earth oxide titanates, which is confirmed by the composition of the observed condensates ($BaTiO_3$, $SrTiO_3$, and $CaTiO_3$) formed under similar conditions [74, 75]. The intensity ratio of ion currents in the mass spectra of vapor over perovskite obtained at laser pulse duration of 800–1000 μs at a wavelength of 6943 Å and energy of 3–5 J was as follows: I_O:I_{Ca}:I_{CaO}:I_{Ti}:I_{TiO}:I_{CaTiO_3} = 0.41:100:0.17:4.32:0.54:0.05. Note that the TiO_2^+ ion was not observed in the mass spectrum of vapor over both perovskite $CaTiO_3$ and rutile TiO_2. This is explained by the peculiarity of the mass spectrometric experiment using a laser, in which the easily ionizable molecular species dominate the mass spectra of vapor and the not readily ionizable molecules are discriminated. This selectivity of detected ions in the mass spectra of vapor significantly limits the accuracy and applicability of this method [76]. According to the data of [57], the estimated temperature of heating of perovskite under the action of laser radiation based on the possible equilibrium in the gas phase.

$$(Ca) + (TiO) = (CaO) + (Ti) \tag{30}$$

is 4890 ± 70 K, which is approximate, but does not contradict the conditions of similar laser-impact mass spectrometry experiments in the range 4000–6000 K.

Banon et al. [52] studied the evaporation of the $CaTiO_3$-Ti_2O_3-TiO_2 melts from molybdenum Knudsen effusion cells at 1900–2200 K by differential mass spectrometry. The mass spectra were recorded at a low ionizing voltage of 13 eV in order to avoid possible fragmentation of the TiO_2^+ molecular ion into Ti^+ and TiO^+ fragmentation ions, which were also molecular ions.

Atomic calcium was the dominant component of the gas phase over the composites. The complex gaseous oxide ($CaTiO_3$) was not detected. The partial pressures of the (Ca), (TiO), (TiO_2), and (O) vapor over perovskite at 2150 K were calculated using the thermochemical data of [57] and are shown in **Figure 7** as a function of the inverse temperature (for easily understanding, the temperature scale was scaled appropriately).

Gaseous perovskite was also not detected in the mass spectrometric studies of high-temperature evaporation of various compositions of the CaO-TiO_2-SiO_2 system from molybdenum and tungsten Knudsen effusion cells at 1700–2500 K [53, 63, 82, 83] presumably because the sensitivity of the equipment used in [63, 83] was insufficient for determining the $CaTiO_3^+$ ion or because this was not the purpose of the study [53, 82].

Lopatin and Semenov [84] studied the evaporation of a mixture of calcium carbonate and titanium dioxide from tungsten cells by the Knudsen effusion mass spectrometry method in the temperature range 2100–2500 K. The following ions were detected in the mass spectra of vapor over the mixture: Ca^+, CaO^+, Ti^+, TiO^+, TiO_2^+, and $CaTiO_3^+$. The energies of ion appearance in the mass spectra allowed the authors to determine the molecular origin of the Ca^+, CaO^+, TiO^+, TiO_2^+, and $CaTiO_3^+$ ions. The TiO^+ion also contained a fragment component of the TiO_2^+ion,

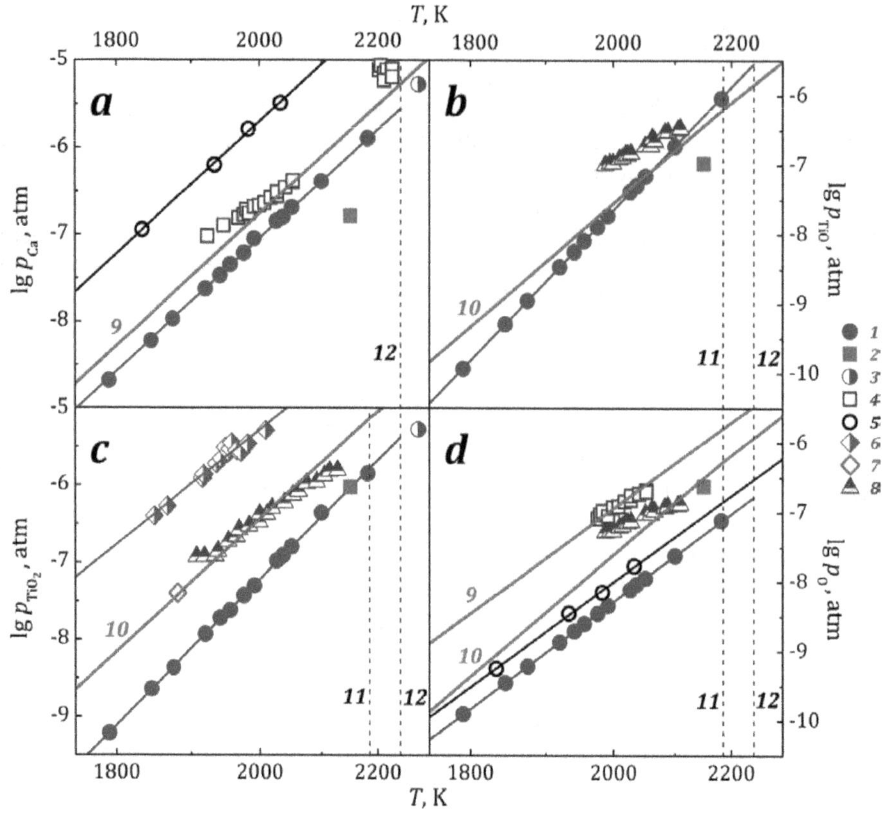

Figure 7.
The partial pressure of Ca (a), TiO (b), TiO_2 (c), and O (d) over perovskite (1–3) and oxides of calcium (4, 5, 9) and titanium (6–8, 10) vs. the inverse temperature, determined via Knudsen mass spectrometry: (1) in [77], (2) in [52], (4) in [78], (5) in [79], (6) in [80], (7) in [81], and (8) in [71]; using the vacuum furnace (according to Langmuir), (3) in [4]; and calculated, (9) and (10), according to the thermochemical data [57]; the vertical dashed lines (11) and (12) indicate melting points of titanium oxide and perovskite, respectively.

and the Ti^+ ion was completely fragmentary. The energy of appearance of the $CaTiO_3^+$ molecular ion was determined to be 9 ± 1 eV (the energy of appearance of the gold ion was used as a standard). The partial pressures of vapor species (p_i) were calculated by comparison with the accepted partial vapor pressures of gold taken as standard pressures (p_s) by the equation:

$$p_i = \frac{I_i T_i}{I_s T_s} p_s \times \frac{\sigma_s \gamma_s \eta_s}{\sigma_i \gamma_i \eta_i}, \tag{31}$$

where I_i (I_s) is the intensity of the ion current of the ith component of vapor (standard substance) recorded at a temperature T_i (T_s). The calculation should also include the ratios of the effective ionization cross sections of the ith molecular form and the standard substance (σ_i/σ_s), isotope distributions (η_i/η_s), and individual ion efficiencies (γ_i/γ_s), which depend on various parameters of ion current recording devices. The pressures p_{CaO}, p_{TiO_2}, and p_{CaTiO_3} calculated by (31) were used to determine the temperature dependence of the equilibrium constant in the gas phase:

$$(CaTiO_3) = (CaO) + (TiO_2) \tag{32}$$

Reaction	T, K	$\Delta_r H_{298}$, kJ/mol	$\Delta_r H_T$, kJ/mol	$\Delta_r S_T$, J/(mol K)	Refs.
$(CaTiO_3) = (CaO) + (TiO_2)$	2287–2466	545 ± 8	284 ± 44	28 ± 19	[84]
	2000	—	298 ± 30	—	[85]
	1956–2182	—	287 ± 12	18 ± 6	[77]
$[Ca] + [Ti] + 3/2 (O_2) = (CaTiO_3)$	2287–2466	−826 ± 26	—	—	[84]
	1956–2182	—	−760 ± 10	−242 ± 5	[77]
$(CaTiO_3) = (Ca) + (Ti) + 3(O)$	2287–2466	2225 ± 26	—	—	[84]
	2000	—	1983 ± 81	—	[85]
	1956–2182	—	1993 ± 15	396 ± 7	[77]
$[CaTiO_3] = (CaTiO_3)$	2000	—	1030 ± 22	—	[85]
	1956–2182	—	1027 ± 10	297 ± 5	[77]

Table 3.
Enthalpies and entropies of the reactions involving the $CaTiO_3$ molecule.

and subsequently calculate the enthalpies of formation ($\Delta_f H_{298}$) and atomization ($\Delta_{at} H_{298}$) of the $CaTiO_3$ molecule (**Table 3**).

Zhang et al. [4] studied the isotope fractionation of calcium and titanium during the evaporation of a perovskite melt suspended on an iridium wire in a vacuum furnace at a temperature of 2278 K (according to Langmuir method). The change in the composition of the residual perovskite melt during evaporation suggested that the component that evaporated predominantly from the melt was its calcium component. The total vapor pressure over perovskite could be evaluated from the data obtained (**Figure 7**).

Shornikov [64, 77] investigated the evaporation of perovskite at 1791–2182 K and its melts at 2241–2441 K from molybdenum Knudsen effusion cells by high-temperature mass spectrometric method.

The TiO_2^+, Ca^+, TiO^+, and O^+ ions prevailed in the mass spectra of vapor over perovskite and its melts at the ionizing electron energy of 20 eV, as well as other ions characteristic of the mass spectra over individual oxides [57, 58, 71]. A small amount of $CaTiO_3^+$ ion was observed, which was fragmented into $CaTi^+$, $CaTiO^+$, and $CaTiO_2^+$ ions ($I_{CaTi}:I_{CaTiO}:I_{CaTiO2}:I_{CaTiO3}$ = 6:10:13:100). The ratio of the ion current intensities in the mass spectra of vapor over perovskite at 2182 K was the following: $I_{Ca}:I_{CaO}:I_{Ti}:I_{TiO}:I_{TiO2}:I_{CaTiO3}:I_O:I_{O2}$ = 80:0.04:0.1:75:100:0.02:40:0.3. It corresponded to that observed by Samoilova and Kazenas [78] in the same temperature range at evaporation of CaO from alundum cell and by Semenov [86] at evaporation of TiO_2 from a tungsten cell.

The ratio of the ion current intensities in the mass spectra of vapor over perovskite melt containing 57.81 ± 0.15 mol% TiO_2 at 2278 K was the following: $I_{Ca}:I_{CaO}:I_{Ti}:I_{TiO}:I_{TiO2}:I_{CaTiO3}:I_O:I_{O2}$ = 25:0.02:0.1:56:100:0.13:0.44:0.012, which is different from that for the case of perovskite [77].

The presence of MoO_i^+($i = 0-3$) ions in the mass spectra of vapor over perovskite was due to the evaporation of molybdenum cell at high temperature:

$$[Mo] = (Mo), \tag{33}$$

as well as the interaction of perovskite with the cell material ($I_{TiO2}:I_{Mo}:I_{MoO}:I_{MoO2}:I_{MoO3}$ = 100:0.8:2.6:7.3:0.6) that was detected according to the following equilibria:

$$(\mathrm{Mo}) + (\mathrm{O}) = (\mathrm{MoO}) \tag{34}$$

$$(\mathrm{MoO}) + (\mathrm{O}) = (\mathrm{MoO_2}) \tag{35}$$

$$(\mathrm{MoO_2}) + (\mathrm{O}) = (\mathrm{MoO_3}). \tag{36}$$

Note that Berkowitz et al. [81] found that during evaporation of titanium oxide from a molybdenum liner inserted into a tantalum crucible at 1881 K, p_{MoO2} was initially 10–10^2 times higher than p_{TiO2}. The p_{MoO2} value gradually decreased and became comparable with the p_{TiO2} value, which is significantly different from the other results [52, 53, 56, 64, 77, 82]. The high p_{MoO2} observed in [81] probably was due to poor quality of the molybdenum liner material (or its alloy).

Possibly it was made using powder technology from MoO_3 reduced to metal molybdenum at ~1300 K. It could lead to such an excess of partial pressure of (MoO_3) and its decrease as it evaporates from the surface layers of the liner material.

The appearance energies of ions in the mass spectra of vapor over perovskite were determined by the Warren method [87] and corresponded to the accepted values of the ionization energies of atoms and molecules [88]. The appearance energy of $CaTiO_3^+$ ion in the mass spectra of vapor over perovskite was equal to 8.5 ± 0.6 eV (the appearance energy of silver ion was used as a standard) and corresponded to obtained by Lopatin and Semenov [84].

The established molecular composition of the gas phase over perovskite allowed us to draw a conclusion o n the predominant evaporation of perovskite according to the reactions (17), (18), (20), (23), (24), and (25), typical for evaporation of simple oxides [57, 71, 78, 86]. The presence of a small amount of ($CaTiO_3$) molecules in the gas phase over perovskite is probably due to the reaction:

$$[\mathrm{CaTiO_3}] = (\mathrm{CaTiO_3}). \tag{37}$$

The partial pressure values of vapor species in the gas phase over perovskite were determined by the Hertz-Knudsen equation, written in the following form [89]:

$$p_i = K_\alpha \frac{q_i}{s_{or} C_{or} t} \sqrt{\frac{2\pi RT}{M_i}}, \tag{38}$$

where q_i is the amount of ith substance component evaporated from the effusion cell, M_i is the molecular weight, t is time of evaporation,T is temperature, C_{or} is the Clausing coefficient characterized the effusion hole, and s_{or} is the hole area.

The K_α constant value was calculated taking into account the evaporation coefficient (α_i) of substance component associated with the molecule changing during its transition to the gas phase from the surface with an S_v area, using the Komlev equation [90]:

$$K_\alpha = \frac{1}{C_{or}} + s_{or} \frac{1 - C_c \alpha_i}{S_v \alpha_i C_c}, \tag{39}$$

where C_c is the Clausing coefficient characterized effusion cell.

The Clausing coefficient is associated with the collision of vapor species inside the effusion orifice channel of effusion cell and their reverse reflection from the channel walls. Its value does not exceed 1 and depends on the ratio of the diameter of the effusion hole to its thickness.

Taking into account predominance of typical for CaO and TiO_2 vapor species in the gas phase over perovskite and small amounts of $CaTiO_3$, the α_i values were used from [91].

The partial pressures of vapor species over perovskite at 1791–2182 K and its melts at 2278 K calculated using the relationships (38) and (39) with an error not exceeding 8% are shown in **Figure 8**.

The partial pressure of atomic oxygen determined using the relationships (38) and (39) agrees satisfactorily with those calculated using the thermochemical data [57] on $K_r(T)$ equilibrium constants of possible reactions (18), (20), (24), (25), and (36) in the gas phase over perovskite in the following relations:

$$p_{\mathrm{O}} = \frac{p_{\mathrm{CaO}}}{p_{\mathrm{Ca}}} K_{18}(T) \tag{40}$$

$$p_{\mathrm{O}} = \frac{p_{\mathrm{TiO_2}}}{p_{\mathrm{TiO}}} K_{24}(T) \tag{41}$$

$$p_{\mathrm{O}} = \frac{p_{\mathrm{TiO}}}{p_{\mathrm{Ti}}} K_{25}(T) \tag{42}$$

$$p_{\mathrm{O}} = \sqrt{\frac{p_{\mathrm{O_2}}^2}{K_{20}(T)}} \tag{43}$$

$$p_{\mathrm{O}} = \frac{p_{\mathrm{MoO_3}}}{p_{\mathrm{MoO_2}}} K_{36}(T). \tag{44}$$

It should be noted that the p_{O} values calculated according to the independent reactions in the gas phase over perovskite (40)–(45) were the same. It confirmed the assumption about the molecular origin of the identified ions in the mass spectrum of vapor over perovskite.

As it follows from **Figure 8a**, the defined partial pressures of vapor species over perovskite can be represented as linear logarithmic dependence vs. the inverse temperature:

$$\lg p_i = a_i/T + b_i. \tag{45}$$

Note that the relationship (45) is the same as the expression for the reaction constant [57]:

$$R \ln K_r(T) = -\frac{\Delta_r G_T}{T} = -\frac{\Delta_r H_T}{T} + \Delta_r S_T, \tag{46}$$

which allows to determine the enthalpy ($\Delta_r H_T$) and entropy ($\Delta_r S_T$) of a reaction.

The partial pressures of the predominant vapor species of the gas phase over perovskite (Ca, TiO, TiO_2 and O) are compared in **Figure 7** with the results on evaporation of simple oxides (CaO and TiO_2) under similar redox conditions caused by the interaction of oxygen with molybdenum [79, 81], tungsten [71], and tantalum [80] effusion cells or in chemically neutral conditions (in the absence of this interaction) for alundum cell [78].

We used the TiO and TiO_2 activities as well as the Gibbs energy of perovskite obtained by Banon et al. [52] and thermochemical data [52] on equilibriums (17), (18), (23), and (24) to estimate the partial pressure of vapor species over the perovskite at 2150 K. Therefore, the obtained values characterized by the evaporation of perovskite were not under reducing conditions (from molybdenum cell),

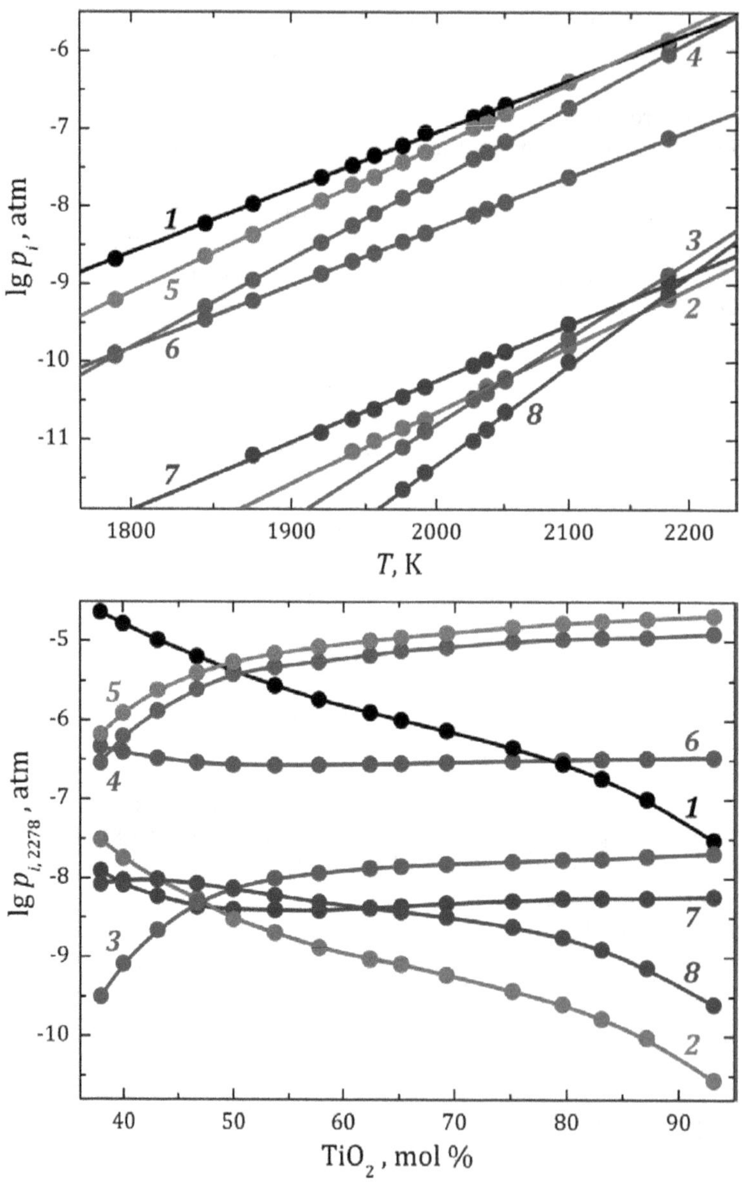

Figure 8.
The partial pressure values of vapor species over perovskite (a) [77] and over the CaO-TiO_2 melts at 2278 K (b) [64]: (1) Ca, (2) CaO, (3) Ti, (4) TiO, (5) TiO_2, (6) O, (7) O_2, and (8) $CaTiO_3$.

but, in contrary, under chemically neutral conditions (in the absence of interaction of perovskite with the cell material).

Figure 7 also shows the partial pressures of vapor species over calcium and titanium oxides calculated using thermochemical data [57]. By comparison of the experimental data obtained in [71, 77–81] and the calculated results, we can see the effect of reducing properties of cell materials on gas phase composition: tantalum [80], molybdenum [79, 81], tungsten [71], and alundum [78]. As we noted earlier [85], the greatest effect of cell materials on the vapor composition are observed with the oxygen-"deficient" species such as atomic calcium (**Figure 7a**), titanium monoxide (**Figure 7b**), and atomic oxygen itself (**Figure 7d**). There are no differences in the partial pressure of gaseous titanium dioxide (**Figure 7c**) determined in evaporation experiments using molybdenum [81] and tungsten [71] cells.

The total vapor pressure over the perovskite melt at 2278 K obtained by Zhang et al. [4] is consistent with the extrapolated values of partial pressures of the predominant vapor species of the gas phase—atomic calcium and titanium dioxide (**Figure 7a** and **c**) found in [77].

Similar slopes of lg p_{Ca} and lg p_O vs. the inverse temperature for calcium oxide and perovskite in **Figure 7a** and **d** (lines 1, 5 and 9, respectively) indicate a predominant effect of calcium component on evaporation of calcium from perovskite. This also can explain the difference in the slope of lg p_{TiO} in **Figure 7b** in the case of perovskite (line 1) and rutile (line 10) according to the equilibriums (18) and (24).

The enthalpy and entropy of reactions involving $CaTiO_3$ gaseous complex oxide calculated by the relationship (46) are given in **Table 3**. They are in a good agreement with those found by Lopatin and Semenov [84] and our earlier estimates [85].

The concentration dependences of partial pressures of vapor species over perovskite melts show a sharp decrease in p_{Ca} and p_{CaO} (**Figure 8b**, lines 1 and 2) with increasing of TiO_2 content in the melt. The vapor species containing titanium—(Ti), (TiO), and (TiO_2)—are increased with increasing TiO_2 concentration up to 65–70 mol% TiO_2, and further they are almost constant in the region of 75–100 mol% TiO_2 (**Figure 8b**, lines 3–5), which may indicate the immiscibility of the melt observed by Banon et al. [52]. The partial pressures of (O) and (O_2) slightly vary throughout the concentration range under consideration, showing a minimum in the perovskite concentration (**Figure 8b**, lines 6 and 7). The ($CaTiO_3$) partial pressures have maximum values in the melt region with a high calcium content compared to the perovskite concentration (**Figure 8b**, line 8), as it was noted earlier.

6. Conclusions

The thermodynamic properties of perovskite determined by different calorimetric approaches and EMF method agree with the results obtained via Knudsen effusion mass spectrometry at high temperatures. The resulting values of oxide activities in perovskite, as well as the Gibbs energy, the entropy and enthalpy of the formation of perovskite from simple oxides, and the melting enthalpy of perovskite are consistent with each other. The enthalpy of perovskite formation is constant throughout the temperature range, and the entropy of perovskite formation tends to increase slightly.

The oxide activities in perovskite melts were determined by mass spectrometric Knudsen effusion method. The thermodynamic properties of melts (chemical potentials of oxides and mixing energies, as well as partial and integral enthalpies and entropies of melt's formation) were calculated based on the experimental data. The obtained experimental information testifies to the symbate behavior of ther-modynamic functions characterizing the melts. The extreme values of the integral thermodynamic properties of melts are in the concentration region close to perovskite, which confirms its stability in the melt. The displacement of the extremum of the integral thermodynamic functions in the CaO-TiO_2 melts can be caused by the presence in the melt of oxide compounds with a large amount of CaO compared to perovskite. A comparison of mixing energies in the CaO-TiO_2 melts with those for the CaO-SiO_2 and CaO-Al_2O_3 melts indicates a stronger chemical interaction in the CaO-TiO_2 melts than the similar CaO-Al_2O_3 melts, but smaller than in the CaO-SiO_2 melts.

The evaporation of perovskite and its melts from a molybdenum cell at high temperature was studied by the Knudsen effusion mass spectrometric method. The molecular components typical of simple oxides and the ($CaTiO_3$) gaseous complex oxide were identified in the gas phase over perovskite. The partial vapor pressures of the molecular components of the gas phase over perovskite were determined. A comparison of these values with the available experimental data and with the values corresponding to simple oxides showed that the character of perovskite evaporation is mainly affected by the calcium component of perovskite. The observed concentration dependences of the partial pressures of vapor species over the perovskite melts correspond to those characterizing the condensed phase.

Acknowledgements

This study was financially supported by the Presidium of the Russian Academy of Sciences (Program No. 7 "Experimental and Theoretical Studies of Solar System and Star Planetary System Objects. Transition Processes in Astrophysics") and by the Russian Foundation for Basic Research (Grant No. 19-05-00801A "Thermodynamics of Formation Processes of Substance of Refractory Inclusions in Chondrites").

Author details

Sergey Shornikov
Vernadsky Institute of Geochemistry and Analytical Chemistry of Russian Academy of Sciences, Moscow, Russia

*Address all correspondence to: sergey.shornikov@gmail.com

References

[1] Rose G. Uber einige neue mineralien des Urals. Journal für Praktische Chemie. 1840;**19**:459-468

[2] Stefanovsky SV, Yudintsev SV. Titanates, zirconates, aluminates and ferrites as waste forms for actinide immobilization. Russian Chemical Reviews. 2016;**85**:962-994

[3] Wark D, Boynton WV. The formation of rims on calcium-aluminum-rich inclusions: Step I—Flash heating. Meteoritics & Planetary Science. 2001;**36**:1135-1166

[4] Zhang J, Huang S, Davis AM, Dauphas N, Hashimoto A, Jacobsen SB. Calcium and titanium isotopic fractionations during evaporation. Geochimica et Cosmochimica Acta. 2014;**140**:365-380

[5] Ivanova MA. Ca-Al-rich inclusions in carbonaceous chondrites: The oldest solar system objects. Geochemistry International. 2016;**54**: 387-402

[6] Parga Pondal I, Bergt K. Sobre la combinacion de la cal en los sistemas $CaO–TiO_2$ y $CaO–SiO_2–TiO_2$. Anales de la Sociedad Española de Física y Química. 1933;**31**:623-637

[7] Roth RS. Revision of the phase equilibrium diagram of the binary system calcia–titania, showing the compound $Ca_4Ti_3O_{10}$. Journal of Research of NBS. 1958;**61**:437-440

[8] Lazaro SR. Estudo teorico-experimental do titanato de calcio—$CaTiO_3$ [thesis]. Sao Carlos: Universidade Estadual Paulista, Intituto de Quimica; 2002. p. 57

[9] Wartenberg HV, Reusch HJ, Saran E. Schmelzpunktsdiagramme hochstfeuerfester oxyde. VII. Systeme mit CaO und BeO. Zeitschrift für Anorganische und Allgemeine Chemie. 1937;**230**:257-276

[10] Pontes FM, Pinheiro CD, Longo E, Leite ER, Lazaro SR, Varela JA, et al. The role of network modifiers in the creation of photoluminescence in $CaTiO_3$. Materials Chemistry and Physics. 2003;**78**:227-233

[11] Tulgar HE. Solid state relationships in the system calcium oxide—Titanium dioxide. Istanbul Teknik Üniversitesi Bülteni. 1976;**29**:111-129

[12] Kisel NG, Limar TF, Cherednichenko IF. On calcium dititanate. Russian Journal of Inorganic Chemistry. 1972;**8**: 1782-1785

[13] Pfaff G. Peroxide route to synthesize calcium titanate powders of different composition. Journal of the European Ceramic Society. 1992;**9**:293-299

[14] Seko A, Hayashi H, Kashima H, Tanaka I. Matrix- and tensor-based recommender systems for the discovery of currently unknown inorganic compounds. Physical Review Materials. 2018;**2**:13805-1-13805-9

[15] Savenko VG, Sakharov VV. Formation of calcium titanate $Ca_2Ti_5O_{12}$ on thermolysis of mixed titanium and calcium hydroxides. Russian Journal of Inorganic Chemistry. 1979;**24**:1389-1391

[16] Limar TF, Kisel NG, Cherednichenko IF, Savos'kina AI. On calcium tetratitanate. Russian Journal of Inorganic Chemistry. 1972;**17**:559-561

[17] Oganov AR, Ma Y, Lyakhov AO, Valle M, Gatti C. Evolutionary crystal structure prediction as a method for the discovery of minerals and materials. Reviews in Mineralogy and Geochemistry. 2010;**71**:271-298

[18] DeVries RC, Roy R, Osborn EF. Phase equilibria in the system CaO–TiO_2. The Journal of Physical Chemistry. 1954;**58**:1069-1073

[19] Gong W, Wu L, Navrotsky A. Combined experimental and computational investigation of thermodynamics and phase equilibria in the CaO–TiO_2 system. Journal of the American Ceramic Society. 2018;**101**: 1361-1370

[20] Kelley KK, Mah AD. Metallurgical thermochemistry of titanium. U. S. Bur. Min. Repts. 1959. No. 5490. 48 p

[21] Reznitskii LA, Guzei AS. Thermodynamic properties of alkaline earth titanates, zirconates, and hafnates. Russian Chemical Reviews. 1978;**47**: 99-119

[22] Robie RA, Hemingway BS. Thermodynamic properties of mineral and related substances at 298.15 K and 1 bar (105 Pascals) pressure and high temperatures. U. S. Geol. Surv. Bull. 1995. No. 2131. p. 461

[23] Barin I. Thermochemical Data of Pure Substances. VCH: Weinheim; 1995. p. 2003

[24] Bale CW, Belisle E, Chartrand P, Decterov SA, Eriksson G, Gheribi AE, et al. FactSage thermochemical software and databases, 2010–2016. Calphad. 2016;**54**:35-53

[25] Shomate CH. Heat capacities at low temperatures of the metatitanates of iron, calcium and magnesium. Journal of the American Chemical Society. 1946; **68**:964-966

[26] Woodfield BF, Shapiro JL, Stevens R, Boerio-Goates J, Putnam RL, Helean KB, et al. Molar heat capacity and thermodynamic functions for $CaTiO_3$. The Journal of Chemical Thermodynamics. 1999;**31**:1573-1583

[27] Buykx WJ. Specific heat, thermal diffusivity and thermal conductivity of synroc, perovskite, zirconolite and barium hollandite. Journal of Nuclear Materials. 1982;**107**:78-82

[28] Sato T, Yamazaki S, Yamashita T, Matsui T, Nagasaki T. Enthalpy and heat capacity of $(Ca_{1-x}Pu_x)TiO_3$ (x = 0 and 0.20). Journal of Nuclear Materials. 2001;**294**:135-140

[29] Naylor BF, Cook OA. High-temperature contents of the metatitanates of calcium, iron and magnesium. Journal of the American Chemical Society. 1946;**68**:1003-1005

[30] Guyot F, Richet P, Courtial P, Gillet P. High-temperature heat capacity and phase transitions of $CaTiO_3$ perovskite. Physics and Chemistry of Minerals. 1993;**20**:141-146

[31] Yashima M, Ali R. Structural phase transition and octahedral tilting in the calcium titanate perovskite $CaTiO_3$. Solid State Ionics. 2009;**180**:120-126

[32] Panfilov BI, Fedos'ev NN. Heat of formation of metatitanate calcium, strontium and barium. Russian Journal of Inorganic Chemistry. 1964;**9**: 2685-2692

[33] Kelley KK, Todd SS, King EG. Heat and free energy data for titanates of iron and the alkaline-earth metals. U. S. Bur. Min. Repts. 1954. No. 5059. p. 37

[34] Feng D, Shivaramaiah R, Navrotsky A. Rare-earth perovskites along the $CaTiO_3$–$Na_{0.5}La_{0.5}TiO_3$ join: Phase transitions, formation enthalpies, and implications for loparite minerals. American Mineralogist. 2016;**101**: 2051-2056

[35] Rezukhina TN, Levitskii VA, Frenkel' MY. Thermodynamic properties of barium and calcium tungstates. Russian Inorganic Materials. 1966;**2**:325-331

[36] Jacob KT, Abraham KP. Thermodynamic properties of calcium titanates: $CaTiO_3$, $Ca_4Ti_3O_{10}$, and $Ca_3Ti_2O_7$. The Journal of Chemical Thermodynamics. 2009;**41**:816-820

[37] Putnam RL, Navrotsky A, Woodfield BF, Boerio-Goates J, Shapiro JL. Thermodynamics of formation for zirconolite ($CaZrTi_2O_7$) from T = 298.15 K to T = 1500 K. The Journal of Chemical Thermodynamics. 1999;**31**:229-243

[38] Navi NU, Shneck RZ, Shvareva TY, Kimmel G, Zabicky J, Mintz MH, et al. Thermochemistry of $(Ca_xSr_{1-x})TiO_3$, $(Ba_xSr_{1-x})TiO_3$, and $(Ba_xCa_{1-x})TiO_3$ perovskite solid solutions. Journal of the American Ceramic Society. 2012;**95**: 1717-1726

[39] Sahu SK, Maram PS, Navrotsky A. Thermodynamics of nanoscale calcium and strontium titanate perovskites. Journal of the American Ceramic Society. 2013;**96**:3670-3676

[40] Helean KB, Navrotsky A, Vance ER, Carter ML, Ebbinghaus B, Krikorian O, et al. Enthalpies of formation of Ce-pyrochlore, $Ca_{0.93}Ce_{1.00}Ti_{2.035}O_{7.00}$, U-pyrochlore, $Ca_{1.46}U^{4+}_{0.23}U^{6+}_{0.46}Ti_{1.85}O_{7.00}$ and Gd-pyrochlore, $Gd_2Ti_2O_7$: Three materials relevant to the proposed waste form for excess weapons plutonium. Journal of Nuclear Materials. 2002;**303**: 226-239

[41] Takayama-Muromachi E, Navrotsky A. Energetics of compounds $(A^{2+}B^{4+}O_3)$ with the perovskite structure. Journal of Solid State Chemistry. 1988;**72**:244-256

[42] Prasanna TRS, Navrotsky A. Energetics in the brownmillerite-perovskite pseudobinary $Ca_2Fe_2O_5$–$CaTiO_3$. Journal of Materials Research. 1994;**9**:3121-3124

[43] Linton J, Navrotsky A, Fei Y. The thermodynamics of ordered perovskites on the $CaTiO_3$–$FeTiO_3$ join. Physics and Chemistry of Minerals. 1998;**25**: 591-596

[44] Koito S, Akaogi M, Kubota O, Suzuki T. Calorimetric measurements of perovskites in the system $CaTiO_3$–$CaSiO_3$ and experimental and calculated phase equilibria for high-pressure dissociation of diopside. Physics of the Earth and Planetary Interiors. 2000;**120**: 1-10

[45] Golubenko AN, Rezukhina TN. Thermodynamic properties of calcium titanate from electrochemical measurements at elevated temperatures. Russian Journal of Physical Chemistry. 1964;**38**:2920-2923

[46] Gillet P, Guyot F, Price GD, Tournerie B, LeCleach A. Phase changes and thermodynamic properties of $CaTiO_3$. Spectroscopic data, vibrational modelling and some insights on the properties of $MgSiO_3$ perovskite. Physics and Chemistry of Minerals. 1993;**20**:159-170

[47] Klimm D, Schmidt M, Wolff N, Guguschev C, Ganschow S. On melt solutions for the growth of $CaTiO_3$ crystals. Journal of Crystal Growth. 2018;**486**:117-121

[48] Shornikov SI. High-temperature mass spectrometric study of the thermodynamic properties of $CaTiO_3$ perovskite. Russian Journal of Physical Chemistry A. 2019;**93**:1428-1434

[49] Topor ND, Suponitskii YL. The high-temperature microcalorimetry of inorganic substances. Russian Chemical Reviews. 1984;**53**:827-850

[50] Aronson S. Estimation of the heat of formation of refractory mixed oxides. Journal of Nuclear Materials. 1982;**107**: 343-346

[51] Taylor RW, Schmalzried H. The free energy of formation of some titanates,

silicates and magnesium aluminate from measurements made with galvanic cells involving solid electrolyte. The Journal of Physical Chemistry. 1964;**68**:2444-2449

[52] Banon S, Chatillon C, Allibert M. Free energy of mixing in $CaTiO_3$–Ti_2O_3–TiO_2 melts by mass spectrometry. Canadian Metallurgical Quarterly. 1981; **20**:79-84

[53] Shornikov SI, Archakov IY. Mass spectrometric study of thermodynamic properties of the SiO_2–$CaTiO_3$ melts. In: Kudin L, Butman M, Smirnov A, editors. Proc. II Intern. Symp. on High Temperature Mass Spectrometry. Ivanovo: ISUCST; 2003. pp. 112-116

[54] Cho SW, Suito H. Thermodynamics of oxygen and nitrogen in liquid nickel equilibrated with CaO–TiO_x and CaO–TiO_x–Al_2O_3 melts. Metallurgical and Materials Transactions A. 1994;**25**:5-13

[55] Kishi M, Suito H. Thermodynamics of oxygen, nitrogen and sulfur in liquid iron equilibrated with CaO–TiO_x and CaO–TiO_x–Al_2O_3 melts. Steel Research. 1994;**65**:261-266

[56] Shornikov SI, Archakov IY, Shultz MM. Thermodynamic properties of the melts, containing titanium dioxide. In: Gorynin IV, Ushkov SS, editors. Proc. 9th World Conf. on Titanium. Vol. 3. Saint-Petersburg: Prometey; 2000. pp. 1469-1473

[57] Glushko VP, Gurvich LV, Bergman GA, Veitz IV, Medvedev VA, Khachkuruzov GA, et al. Thermodynamic Properties of Individual Substances. Moscow: Nauka; 1978–1982 [in Russian]

[58] Chase MW. NIST-JANAF themochemical tables. Journal of Physical and Chemical Reference Data. 1998. p. 1951

[59] Nerad I, Danek V. Thermodynamic analysis of pseudobinary subsystems of the system CaO–TiO_2–SiO_2. Chemical Papers. 2002;**56**:77-83

[60] Seo W-G, Zhou D, Tsukihashi F. Calculation of thermodynamic properties and phase diagrams for the CaO–CaF_2, BaO–CaO and BaO–CaF_2 systems by molecular dynamics simulation. Materials Transactions. 2005;**46**:643-650

[61] Bottinga Y, Richet P. Thermodynamics of liquid silicates, a preliminary report. Earth and Planetary Science Letters. 1978;**40**:382-400

[62] Kulikov IS. Thermal Dissociation of Compounds. 2nd ed. Metallurgiya: Moscow; 1969. p. 576 [in Russian]

[63] Stolyarova VL, Zhegalin DO, Stolyar SV. Mass spectrometric study of the thermodynamic properties of melts in the CaO–TiO_2–SiO_2 system. Glass Physics and Chemistry. 2004;**30**: 142-150

[64] Shornikov SI. Thermodynamic properties of CaO–TiO_2 system: Experimental data and calculations. In: Gladyshev PP, editor. Physical and Analytical Chemistry of Natural and Technogenic Systems, New Technologies and Materials—Khodakovsky's Readings. Dubna: Dubna State University; 2019. pp. 185-190 [in Russian]

[65] Lewis GN, Randall M. Thermodynamics and the Free Energy of Chemical Substances. N. Y. & London: McGrow-Hill Book Comp.; 1923. p. 676

[66] Belton GR, Fruehan RJ. The determination of activities of mass spectrometry: Some additional methods. Metallurgical and Materials Transactions B. 1971;**2**:291-296

[67] Prigogine I, Defay R. Chemical Thermodynamics. London: Longman; 1954. p. 543

[68] Shornikov SI, Archakov IY. Mass spectrometric study of phase relations and vaporization processes in the CaO–SiO_2 system. Glass Science and Technology: International Journal of the German Society of Glass Technology. 2000;**73C2**:51-57

[69] Shornikov SI, Stolyarova VL, Shultz MM. A mass-spectrometric study of vapor composition and thermodynamic properties of CaO–Al_2O_3 melts. Russian Journal of Physical Chemistry. 1997;**71**:19-22

[70] Kulikov IS. Thermodynamics of Oxides. Metallurgiya: Moscow; 1986. p. 344 [in Russian]

[71] Kazenas EK, Tsvetkov YV. Evaporation of Oxides. Moscow: Nauka; 1997. p. 543 [in Russian]

[72] Balducci G, Gigli G, Guido M. Identification and stability determinations for the gaseous titanium oxide molecules Ti_2O_3 and Ti_2O_4. The Journal of Chemical Physics. 1985;**83**:1913-1916

[73] Zakharov VP, Protas IM. Mass spectrometric study of ion emission during evaporation of complex substances under the influence of laser radiation. Izvestiya Akademii Nauk SSSR. Seriya Fizicheskaya (Bulletin of the Russian Academy of Sciences: Physics) 1974;**38**:238-243

[74] Schwarz H, Tourltellotte HA. Vacuum deposition by high-energy laser with emphasis on barium titanate films. Journal of Vacuum Science and Technology B. 1969;**6**:373-378

[75] Hao J, Si W, Xia XX, Guo R, Bhalla AS, Cross LE. Dielectric properties of pulsed-laser-deposited calcium titanate thin films. Applied Physics Letters. 2000;**76**:3100-3102

[76] Knox BE. Laser ion source analysis of solids. In: Ahearn AJ, editor. Trace Anal ysis by Mass Spectrometry. N. Y. & London: Academic Press; 1972. pp. 423-444

[77] Shornikov SI. Mass-spectrometric study of perovskite evaporation. Russian Journal of Physical Chemistry A. 2019;**93**:866-873

[78] Samoilova IO, Kazenas EK. Thermodynamics of dissociation and sublimation of calcium oxide. Russel Metals. 1995:33-35

[79] Shornikov SI. Evaporation processes and thermodynamic properties of the CaO–Al_2O_3–SiO_2 system and the materials based on this system [thesis]. Institute of Silicate Chemistry of RAS: Saint-Petersburg; 1993. p. 21 [in Russian]

[80] Groves WO, Hoch M, Johnston HL. Vapor–solid equilibria in the titanium–oxygen system. The Journal of Physical Chemistry. 1955;**59**:127-131

[81] Berkowitz J, Chupka WA, Inghram MG. Thermodynamics of the Ti–Ti_2O_3 system and the dissociation energy of TiO and TiO_2. The Journal of Physical Chemistry. 1957;**61**:1569-1572

[82] Archakov IY, Shornikov SI, Tchemekova TY, Shultz MM. The behavior of titanium dioxide in the slag melts. In: Gorynin IV, Ushkov SS, editors. Proc. 9th World Conf. on Titanium. Vol. 3. Saint-Petersburg: Prometey; 2000. pp. 1464-1468

[83] Ostrovski O, Tranell G, Stolyarova VL, Shultz MM, Shornikov SI, Ishkildin AI. High-temperature mass spectrometric study of the CaO–TiO_2–SiO_2 system. High Temperature Materials and Processes. 2000;**19**:345-356

[84] Lopatin SI, Semenov GA. Thermochemical study of gaseous salts of oxygen-containing acids: XI. Alkaline-earth metal titanates. Russ. J. General Chemistry. 2001;**71**:1522-1526

[85] Shornikov SI, Yakovlev OI. Study of complex molecular species in the gas phase over the $CaO–MgO–Al_2O_3–TiO_2–SiO_2$ system. Geochemistry International. 2015;**53**:690-699

[86] Semenov GA. The evaporation of titanium dioxide. Russian Inorganic Materials. 1969;**5**:67-73

[87] Warren JW. Measurement of appearance potentials of ions produced by electron impact, using a mass spectrometer. Nature. 1950;**165**:810-811

[88] Gurvich LV, Karachevtsev GV, Kondratev VN, Lebedev YA, Medvedev VA, Potapov VK, et al. Bond Dissociation Energies, Ionization Potentials, and Electron Affinity. Moscow: Nauka; 1974. p. 351 [in Russian]

[89] Shornikov SI. Thermodynamic study of the mullite solid solution region in the $Al_2O_3–SiO_2$ system by mass spectrometric techniques. Geochemistry International. 2002;**40**:S46-S60

[90] Komlev GA. On determination of saturated steam pressure by effusion method. Russian Journal of Physical Chemistry. 1964;**38**:2747-2748

[91] Shornikov SI. Vaporization coefficients of oxides contained in the melts of Ca-Al-inclusions in chondrites. Geochemistry International. 2015;**53**: 1080-1089

3

Significant Role of Perovskite Materials for Degradation of Organic Pollutants

Someshwar Pola and Ramesh Gade

Abstract

The advancement and the use of visible energy in ecological reparation and photodegradation of organic pollutants are being extensively investigated worldwide. Through the last two decades, great exertions have been dedicated to emerging innocuous, economical, well-organized and photostable photocatalysts for ecofriendly reparation. So far, many photocatalysts mostly based on ternary metal oxides and doped with nonmetals and metals with various systems and structures have been described. Among them, perovskite materials and their analogs (layer-type perovskites) include an emerged as semiconductor-based photocatalysts due to their flexibility and simple synthesis processes. This book chapter precisely concentrates on the overall of related perovskite materials and their associated systems; precisely on the current progress of perovskites that acts as photocatalysts and ecofriendly reparation; explores the synthesis methods and morphologies of perovskite materials; and reveals the significant tasks and outlooks on the investiga-tion of perovskite photocatalytic applications.

Keywords: layered-type perovskite materials, photocatalysis, photodegradation, organic pollutants

1. Introduction

Solar energy is one of the primary sources in the field of green and pure energy that points to the power predicament and climate change task. Solar energy consumption is an ecological reconciliation, and then, the chemical change in solar is presence exhaustive, considered throughout global [1, 2]. In general, solar energy is renewed into a wide range of developments, such as degradation of organic pollutants as photocatalysis, splitting of water molecules for producing clean energy, and reduction of CO_2 gas [3, 4]. Consuming a similar perception, metal-oxide photocatalysis has also been widely examined for possible exertions in ecological restitution as well as the photodegradation and elimination of organic toxins in the aquatic system [5, 6], decrease of bacterial inactivation [7–9], and heavy metal ions [10–12]. Throughout the earlier few years, excellent applications have been dedicated to evolving well-organized, less expensive, and substantial photocatalysts, particularly those that can become active under visible light such as $NaLaTiO_6$, $Ag_3PO_4/BaTiO_3$, $Pt/SrTiO_3$, $SrTiO_3$-TiN, noble-metal-$SrTiO_3$ composites, $GdCoO_3$, orthorhombic perovskites $LnVO_3$ and $Ln_{1-x}Ti_xVO_3$ (Ln = Ce, Pr, and Nd), $Ca_{0.6}Ho_{0.4}MnO_3$, Ce-doped $BaTiO_3$, fluorinated Bi_2WO_6, graphitic carbon nitride -Bi_2WO_6, $BaZrO_3$ $_{\delta}$, $CaCu_3Ti_4O_{12}$, [13–24], graphene-doped perovskite materials, and nonmetal-doped perovskites [25]. Furthermore, directed to years extended exhaustive investigation exertions on the pursuit of innovative photocatalytic systems, particularly that can produce the overall spectrum of visible-light. Out of a vast assemblage of photocatalysts, perovskite or layered-type perovskite systems and its analogs include a better candidate for capable semi-conductor-based photocatalysts due to their framework easiness and versatility, excellent photostability, and systematic photocatalytic nature. In general, the ideal perovskite structure is cubic, and the formula is ABO_3. Where A is different metal cations having charge +1 or +2 or +3 nature and B site occupies with tri or tetra and pentavalent nature, which covers the whole family of perovskite oxide materials by sensibly various metal ions at A and B locations [26], aside from a perfect cubic perovskite system, basic alteration perhaps persuaded by several cations exchange. Such framework alteration could undoubtedly vary the photophysical, optical, and photocatalytic activities of primary oxides.

Moreover, a sequence of layered-type perovskite materials contains many 2D blocks of the ABO_3 framework, which are parted by fixed blocks. The scope of formulating multicomponent perovskite systems by whichever fractional change of cations in A and B or both positions or injecting perovskite oxides into a layered-type framework agrees scientists investigate and control the framework of crystals and the correlated electronic and photocatalytic activities of the perovskites. So far, hundreds of various types of perovskite or perovskite-based catalysts have been published, and more outstandingly, some ABO_3-related materials became renowned with "referred" accomplishment for catalytic activities. Thus, these systems (perovskite materials) have exposed highly capable of upcoming applications on the source of applying more attempts to them. While several outstanding reviews mean that explained that perovskites performed as photocatalyst for degradation of organic pollutants [27–30], only an insufficient of them content consideration to inorganic perovskite (mostly ABO_3-related) photocatalysts [31–33]. A wide range of tagging and complete attention of perovskite materials, for example, layered-type perovskite acting as photocatalysts, is relatively deficient. The purpose of this book chapter is to precise the current progress of perovskite-based photocatalysts for ecological reparation, deliberate current results, and development on perovskite oxides as catalysts, and allow a view on the upcoming investigation of perovskite materials. After a short outline on the wide-ranging structure of perovskite oxides, it was stated that perovskites act as a photocatalyst that are incorporated, arranged and explored based on preparation methods [29, 34], photophysical properties based on bandgap energies, morphology-based framework and the photocatalytic activities depends on either UV or visible light energy of the semiconducting materials. Finally, this chapter is based on the current advancement and expansion of perovskite photocatalytic applications under solar energy consumption. The potential utilization, new tasks, and the research pathway will be accounted for the final part of the chapter [35].

2. Results and discussion

2.1 Details of perovskite oxide materials

2.1.1 Perovskite frameworks

The standard system of perovskite-based materials could be designated as ABO_3, where the A and B are cations with 12-fold coordinated and 6-fold coordinated to concerning oxygen anions. **Figure 1a** describes the typically coordinated basic of the ABO_3 system, which consists of a 3D system, BO_6 octahedra as located at corner, and at the center, A cation are occupied. Within the ABO_3 system, the A cation usually is group I and II or a lanthanide metal, whereas the B is commonly a transition metal ion. The tolerance factor $(t) = 1$ calculated by using an equation $t = (rA + rO)/\sqrt{2}\,(rB + rO)$, where r_O, r_A, and r_B are the radii of respective ions A and B and oxygen elements for a cubic crystal structure ABO_3 perovskite system [36].

For constituting a stable perovskite, it is typically the range of t value present in between 0.75 and 1.0. The lower value of t builds a somewhat slanted perovskite framework with rhombohedral or orthorhombic symmetry. In the case of t, it is approximately 1; then, perovskite structure is an ideal cubic system at high temperatures. Even though the value of t, obtained by the size of metal ion, is a significant guide for the permanency of perovskite systems, the factor of octa-hedral (u) $u = r_B/r_O$ and the role of the metal ions composition of A and B atoms and the coordination number of respective metals are considered [37]. Given the account of those manipulating factors and the electro-neutrality, the ABO_3 perovskite can hold a broad variety of sets of A and B by equal or dissimilar oxidation states and ionic radii. Moreover, the replacement of A or B as well as both the cations could be partly by the doping of various elements, to range the ABO_3 perovskite into a wide-ranging family of $A_m{}^1 A_{1-}{}^1{}_m\ B^1{}_n B_{1-}{}^1{}_n\ O_{3\pm\delta}$ [38]. The replacement of several cations into the either A or B positions could modify the structure of the original system and therefore improve the photocatalytic activities [23]. After various metal ions in perovskite oxide are doped, the optical and electronic band positions, which influence the high impact on the photocatalytic process, are modified [24].

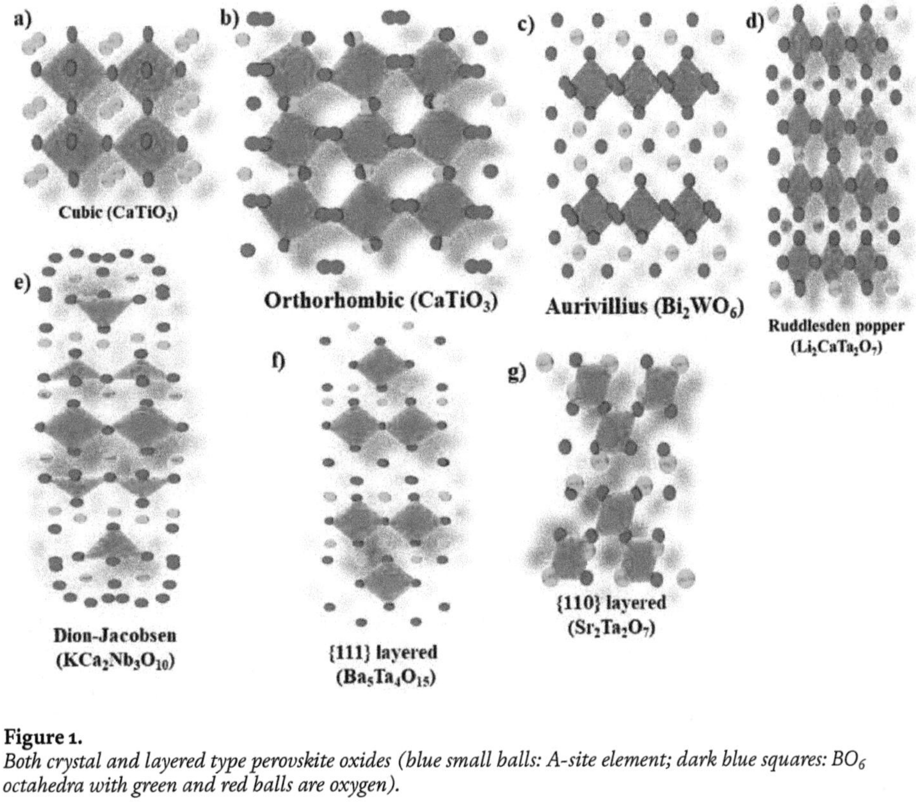

Figure 1.
Both crystal and layered type perovskite oxides (blue small balls: A-site element; dark blue squares: BO_6 octahedra with green and red balls are oxygen).

2.1.2 Layered perovskite-related systems

Moreover, to the overall ABO_3 system, further characteristic polymorphs of the perovskite system are Brownmillerite (BM) ($A_2B_2O_5$) framework [39]. BM is a type of oxygen-deficient perovskite, in which the unit cell is a system of well-organized BO_4 and BO_6 units. The coordination number of cations occupied by A-site was decreased to eight because of the oxygen deficiency. Perovskite (ABO_3) oxides have three dissimilar ionic groups, construction for varied and possibly useful imperfection chemistry. Moreover, the partial replacement of A and B ions is permitted even though conserving the perovskite system and shortages of cations at the A-site or of oxygen anions are common [40]. The Ion-exchange method is used for the replacement of existing metal ions with similar sized or dissimilar oxidation states; then, imperfections can be announced into the system. The imperfection concentrations of perovskites could be led by doping of different cations [24]. Oxygen ion interstitials or vacancies could be formed by the replacement of B-position cations with higher or lower valence, respectively, fabricating new compounds of $AB_{(1-m)}B_m{}^IO_{3-\delta}$ [41]. A typical oxygen-deficient perovskite system is Brownmillerite ($A_2B_2O_5$), in which one part of six oxygen atoms is eliminated. Moreover, the replacement of exciting a site cation to new cation with higher oxidation state metal ions then the formed new materials with new framework with different stoichiometry is $A_{1-m}A_m{}^IBO_3$ [41]. In the case of the replacement of A-site ions with smaller oxidation state cations, consequences in oxygen-deficient materials with new framework such as $A_{1-m}A_m{}^IBO_{3-x}$ are developed.
Thermodynamically, the replacement of B-position vacancies in perovskite systems is not preferable due to the compact size and the high charge of B cations [42]. A-position vacancies are more detected due to the BO_3 range in perovskite system forms a stable network [43]; the 12 coordinated sites can be partly absent due to bigger-size A cations. Lately, presenting suitable imperfections on top of the surface of perovskite oxides has been thoroughly examined as a means of varying the bands' position and optical properties of the starting materials. For this reason, perovskite materials afford a tremendous objective for imperfection originating to vary the photocatalytic activity of perovskite material-based photocatalysts [44].

The typical formula for the furthermost recognized layered perovskite materials is designated as $A_{n+1}B_nO_{3n+1}$ or $A_2{}^IA_{n-1}BnO_{3n+1}$ (Ruddlesden-Popper (RP) phase), $A^I[A_{n-1}B_nO_{3n+1}]$ (Dion-Jacobson (DJ) phase) for {100} series, ($A_nB_nO_{3n+2}$) for {110} series and ($A_{n+1}B_nO_{3n+3}$) for {111}, and $(Bi_2O_2)(A_{n-1}B_nO_{3n+1})$ (Aurivillius phase) series. In these systems, n represents the number of BO_6 octahedra that duration a layer, which describes the width of the layer. Typical samples of these layered sys-tems are revealed in **Figure 1c–g**. For RP phases, their frameworks consist of A^IO as the spacing layer for the intergrowth ABO_3 system. These materials hold fascinating properties such as ferroelectricity, superconductivity, magnetoresistance, and pho-tocatalytic activity. Sr_2SnO_4 and $Li_2CaTa_2O_7$ systems are materials of simple RP kind photocatalysts. A common formula for DJ phase is $A^I[A_{n-1}B_nO_{3n+1}]$ ($n > 1$), where A^I splits the perovskite-type slabs and is characteristically a monovalent alkali cation. The typical DJ kind photocatalysts are $RbLnTa_2O_7$ ($n = 2$) and $KCa_2Nb_3O_{10}$ ($n = 3$). Associates of the $A_nB_nO_{3n+2}$ and $A_{n+1}B_nO_{3n+3}$ structural sequences with dissimilar layered alignments have also been recognized in some photocatalysts like $Sr_2Ta_2O_7$ and $Sr_5Ta_4O_{15}$ ($n = 4$). For Aurivillius phases, their frameworks are con-structed by one after another fluctuating layers of $[Bi_2O_2]^{2+}$ and virtual perovskite blocks. Bi_2WO_6 and $BiMoO_6$ ($n = 1$), found as the primary ferroelectric nature for Aurivillius materials, lately have been extensively investigated as visible light photocatalysts.

2.2 Perovskite systems for photocatalysis

A broad array of perovskite photocatalysts have been advanced for organic pollutant degradation in the presence of ultraviolet or visible-light-driven through the last two decades [45]. These typical examples and brief investigational consequences on perovskites are concise giving to their systems, then perovskite materials categorized into six groups. Precisely, ABO_3-type perovskites, AA^IBO_3, A^IABO_3, ABB^IO_3 and $AB(ON)_3$-type perovskites, and $AA^IBB^{II}O_3$-type perovskites are listed in **Table 1**.

2.3 Photocatalytic properties perovskite oxides

$NaTaO_3$ has been a standard perovskite material for a well-organized UV-light photocatalyst for degradation of organic pollutants and production of H_2 and O_2 through water splitting [46–57]. It can be prepared by various methods such as solid-state [46–48, 53, 56], hydrothermal [49, 52, 54, 55], molten salt [57] and sol-gel [50, 51] and with wide bandgap of 4.0 eV. In order to enhance the surface area of $NaTaO_3$ bulk material, many investigators tried to use further synthetic ways to make nanosized particles as an additional study on the $NaTaO_3$ photocatalyst for degradation of organic pollutants. Kondo et al. prepared a colloidal range of $NaTaO_3$ nanoparticles consuming three-dimensional mesoporous carbon as a pattern, which was pretend by the colloidal arrangement of silica nanospheres. After calcining the mesoporous carbon matrix, a colloidal arrangement of $NaTaO_3$ nanoparticles with a range of 20 nm and a surface area of 34 $m^2\ g^{-1}$ was attained. C-doped $NaTaO_3$ material was tested for degradation of NO_x under UV light [36]. Several titanates such as $BaTiO_3$ [58–60], Rh or Fe-doped $BaTiO_3$ [61, 62], $CaTiO_3$ [63, 64] and Cu [65], Rh [66], Ag and La-doped $CaTiO_3$ [67], and $PbTiO_3$ [68, 69] were also described as UV or visible light photocatalysts. Magnetic $BiFeO_3$, recognized as the one of the multiferric perovskite materials in magnetoelectric properties, was also examined as a visible light photocatalyst for photodegradation of organic pollutants because of small bandgap energy (2.2 eV) [70–79]. In a previous account, $BiFeO_3$ with a bandgap of around 2.18 eV produced by a citric acid-supported sol-gel technique has revealed its visible-light-driven photocatalytic study by the disintegration of methyl orange dye [70]. The subsequent investigations on $BiFeO_3$ are primarily concentrated on the synthesis of new framework $BiFeO_3$ with various morphologies. For instance, Lin and Nan et al. prepared $BiFeO_3$ unvarying microspheres and microcubes by a using hydrothermal technique as revealed in **Figure 2** [73].

The bandgap energies of $BiFeO_3$ compounds were found to be about 1.82 eV for $BiFeO_3$ microspheres and 2.12–2.27 eV for microcubes. This indicated that the absorption edge was moved toward the longer wavelength that is influenced by the crystal-field strength, particle size, and morphology. The microcube material showed the maximum photocatalytic degradation performance of congo red dye under visible-light irradiation due to the quite low bandgap energy. Further, a simplistic aerosol-spraying method was established for the synthesis of mesoporous $BiFeO_3$ hollow spheres with improved activity for the photodegradation of RhB dye and 4-chlorophenol, because of improved light absorbance ensuing from various light reflections in a hollow chamber and a very high surface area [71]. Moreover, a unusually improved water oxidation property on Au nanoparticle-filled $BiFeO_3$ nanowires under visible-light-driven was described [77]. The Au-$BiFeO_3$ hybrid system was encouraged by the electrostatic contact of negatively charged Au nanoparticles and positively charged $BiFeO_3$ nanowires at pH 6.0 giving to their various isoelectric points. An improved absorbance between 500 and 600 nm was found for Au/$BiFeO_3$systems because of the characteristic Au surface plasmon band existing visible light region then which greater influenced in the photodegradation of organic pollutants. Also, the study of photoluminescence supported improvement of the photocatalytic property due to the effective charge transfer from $BiFeO_3$ to Au. Even though Ba, Ca, Mn, and Gd-doped $BiFeO_3$ nanomaterials have exhibited noticeable photocatalytic property for the degradation of dyes [80–84], several nano-based $LaFeO_3$ with various morphologies such as nanoparticles, nanorods, nanotubes, nanosheets, and nanospheres have also been synthesized for visible light photocatalysts for degradation of organic dyes [85–93]. Sodium bismuth titanate ($Bi_{0.5}Na_{0.5}TiO_3$) has been extensively used for ferroelectric and piezoelectric devices. It was also investigated as a UV-light photocatalyst with a bandgap energy of 3.0 eV [94–97]. Hierarchical micro/nanostructured $Bi_{0.5}Na_{0.5}TiO_3$ was produced by in situ self-assembly of $Bi_{0.5}Na_{0.5}TiO_3$ nanocrystals under precise hydrothermal conditions, through the evolution mechanism was examined in aspect means that during which the growth mechanism was studied [95]. It was anticipated that the hierarchical nanostructure was assembled through a method of nucleation and growth and accumulation of nanoparticles and following in situ dissolution-recrystallization of the microsphere type nanoparticles with extended heating period and enhanced temperature or basic settings. The 3D hierarchical $Bi_{0.5}Na_{0.5}TiO_3$ showed very high photocatalytic activity for the decomposition of methyl orange dye because of the adsorption of dye molecules and bigger surface area. The properties of $Bi_{0.5}Na_{0.5}TiO_3$ were also assessed by photocatalytic degradation of nitric oxide in the gas phase [95]. $La_{0.7}Sr_{0.3}MnO_3$, acting as a photocatalyst, was examined for solar light-based photocatalytic decomposition of methyl orange [96–98]. In addition, $La_{0.5}Ca_{0.5}NiO_3$ [99], $La_{0.5}Ca_{0.5}CoO_{3-\delta}$ [100], and $Sr_{1-x}Ba_xSnO_3$ (x = 0–1) [101] nanoparticles were synthesized for revealing improved photocatalytic degradation of dyes. A-site strontium-based perovskites such as $SrTi_{1-x}Fe_xO_{3-\delta}$, $SrTi_{0.1}Fe_{0.9}O_{3-\delta}$, $SrNb_{0.5}Fe_{0.5}O_3$, and $SrCo_{0.5}Fe_{0.5}O_{3-\delta}$ compounds were prepared through solid-state reaction and sol-gel approaches, and were examined for the degradation of organic pollutants under visible light irradiation [102–105]. Also, some other researchers modified A-site with lanthanum-based perovskites such as $LaNi_{1-x}Cu_xO_3$ and $LaFe_{0.5}Ti_{0.5}O_3$ were confirmed as effective visible light photocatalysts for the photodegradation of p-chlorophenol [91, 106, 107]. The other ABB^IO_3 kind photocatalysts with $Ca(TiZr)O_3$ [108], $Ba(ZrSn)O_3$ [109], $Na(BiTa)O_3$ [110], $Na(TiCu)O_3$ [111], $Bi(MgFeTi)O_3$ [112], and $Ag(TaNb)O_3$ [113] have also been studied. Related to AA^IBO_3-type perovskites, the ABB^IO_3 kind system means that BI-site substitution by a different cation is another option for tuning the physicochemical or photocatalytic properties of perovskites materials as photocatalyst, due to typically the B-position cations in ABO_3 mostly regulate the position of the conduction band, moreover to construct the structure of perovskite system with oxygen atoms. The band positions of photocatalyst can be magnificently modified by sensibly coalescing dual or ternary metal cations at the B-position, or changing the ratio of several cations, which has been fine verified by the various materials as mentioned above. More studies on ABB^IO_3 kind of photocatalysts are projected to show their new exhilarating photocatalytic efficiency.

The mesoporous nature of $LaTiO_2N$ of photocatalyst attended due to thermal ammonolysis process of $La_2Ti_2O_7$ precursor from polymer complex obtained from the solid-state reaction. The oxynitride analysis revealed that the pore size and shape, lattice defects and local defects, and oxidation states' local analysis related between morphology and photocatalytic activity were reported by Pokrant et al. [114]. Due to the high capability of accommodating an extensive array of cations and valences at both A- and B-sites, ABO_3-kind perovskite materials are capable materials for fabricating solid-solution photocatalysts. On the other hand, equally the A and B cations can be changed by corresponding cations subsequent in a perovskite with the formula of $(ABO_3)_x(A^IB^IO_3)_{1-x}$. Additional solid solution examples with $CaZrO_3$–$CaTaO_2N$

[115], $SrTiO_3$–$LaTiO_2N$ [116], $La_{0.8}Ba_{0.2}Fe_{0.9}Mn_{0.1}O_{3-x}$ [117], $Na_{1-x}La_xFe_{1-x}Ta_xO_3$ [118], $Na_{0.5}La_{0.5}TiO_3$–$LaCrO_3$ [119], Cu-$(Sr_{1-y}Na_y)$-$(Ti_{1-x}Mo_x)O_3$ [120], $Na_{1-x}La_xTa_{1-x}Cr_xO_3$ [121], $BiFeO_3$–$(Na_{0.5}Bi_{0.5})TiO_3$ [122], and $Sr_{1-x}Bi_xTi_{1-x}Cr_xO_3$ [123] have been used as photocatalysts for splitting of water molecules under visible light.

2.4 Photocatalytic activity of layered perovskite materials

In the general formula of the RP phase, $A_{n-1}A_2{}^{I}B_nO_{3n+1}$, A and A^I are alkali, alkaline earth, or rare earth metals, respectively, while B states to transition met-als. A and A^I cations are placed in the perovskite layer and boundary with 12-fold cuboctahedral and 9-fold coordination to the anions, respectively, whereas B cations are sited inside the perovskite system with anionic squares, octahedra, and pyramids. The tantalum-based RP phase materials have been examined as photocatalysts for degradation of organic pollutants under UV light irradiation conditions; such materials are $K_2Sr_{1.5}Ta_3O_{10}$ [124], $Li_2CaTa_2O_7$ [125], $H_{1.81}Sr_{0.81}Bi_{0.19}Ta_2O_7$ [126], and N-alkyl chain inserted $H_2CaTa_2O_7$ [127]. A series of various metals and N-doped perovskite materials were synthesized, such as Sn, Cr, Zn, V, Fe, Ni, W, and N-doped $K_2La_2Ti_3O_{10}$, for photocatalysis studies under UV and visible light irradiation [128–133]. Still, only Sn-doping efficiently decreased the bandgap energy of $K_2La_2Ti_3O_{10}$ from 3.6 eV to 2.7 eV. The bandgap energy of N-doped $K_2La_2Ti_3O_{10}$ was measured to be around 3.4 eV. Additional RP phase kind titanates like Sr_2SnO_4 [134], $Sr_3Ti_2O_7$ [135], Cr-doped Sr_2TiO_4 [136], $Sr_4Ti_3O_{10}$ [137], $Na_2Ca_2Nb_4O_{13}$ [138], and Rh- and Ln-doped $Ca_3Ti_2O_7$ [139] have also been examined. Bi_2WO_6 (2.8 eV) shows very high oxygen evolution efficacy than Bi_2MoO_6 (3.0 eV) from aqueous $AgNO_3$ solution under visible-light-driven. Because of the appropriate bandgap energy, comparatively elevated photocatalytic performance, and good constancy, Bi_2MO_6 materials have been thoroughly examined as the Aurivillius phase kind that acts as photocatalysts under visible light. In this connection, hundreds of publications associated to the Bi_2MoO_6 and Bi_2WO_6 act as photocatalysts so far reported. Most of the investigations in the reports are concentrated on the synthesis of various nanostructured Bi_2MoO_6 and Bi_2WO_6 as well as nanofibers, nanosheets, ordered arrays, hollow spheres, hierarchical architectures, inverse opals, and nanoplates, etc., by various synthesis techniques like solvothermal, hydrothermal, electrospinning, molten salt, thermal evaporation deposition, and microwave. All these methods of hydrothermal process have been frequently working for the controlled sizes, shapes, and morphologies of the particles. The photocatalytic properties of these perovskite materials are mostly examined by the photodegradation of organic pollutants. Moreover, the investigations on the simple Bi_2MoO_6 and Bi_2WO_6, doped with various metals and nonmetals such as Zn, Er, Mo, Zr, Gd, W, F, and N, into Bi_2MoO_6 and Bi_2WO_6 was studied for increasing the photocatalytic performance under visible light. Therefore, these Bi_2MO_6-based photocatalysts is not specified here, due to further full deliberations that can be shown in many reviews [140–142].

$ABi_2Nb_2O_9$ where A is Ca, Sr, Ba and Pb is other type of the AL-like layered perovskite material [143–150]. The bandgap energy of $PbBi_2Nb_2O_9$ is 2.88 eV and originally described as an undoped with single-phase layered-type perovskite material used as photocatalyst employed under visible light irradiation [144]. $Bi_5FeTi_3O_{15}$ is also Aurivillius (AL) type multi-layered nanostructured perovskite material with a low bandgap energy (2.1 eV) and also shows photocatalytic activity under visible light [151, 152]. Mostly, these materials were synthesized using the hydrothermal method that has been frequently working for the controlled shapes such as flower-like hierarchical morphology, nanoplate-based, and the complete advance process from nanonet-based to nanoplate-based micro-flowers was shown. The photocatalytic activity of $Bi_5FeTi_3O_{15}$ was studied by the degradation of rhodamine

B and acetaldehyde under visible light [151]. The La substituted $Bi_{5-x}La_xTi_3FeO_{15}$ (x = 1, 2) Al-type layered materials were synthesized through hydrothermal method and these materials were used for photodegradation of rhodamine B under solar-light irradiation [153]. Among all AL-type perovskite materials, only $PbBi_2Nb_2O_9$, Bi_2MO_6 (M = W or Mo), and $Bi_5Ti_3FeO_{15}$ are very high photocatalytic active under visible-light-driven due to low bandgap energy and photostability. Another type of layered perovskite material is Dion-Jacobson phase (DJ), a simple example is $CsBa_2M_3O_{10}$ (M = Ta, Nb) and oxynitride crystals used for degradation of caffeine from wastewater under UVA- and visible-light-driven [154]. Similarly, another DJ phase material such means Dion–Jacobsen (DJ) as $CsM_2Nb_3O_{10}$ (M = Ba and Sr) and also doped with nitrogen used for photocatalysts for degradation of methylene blue [155]. Zhu et al. prepared tantalum-based {111}-layered type of perovskite material such as $Ba_5Ta_4O_{15}$ from hydrothermal method, which has been frequently employed for the controlled shape like hexagonal structure with nanosheets and used as pho-tocatalyst for photodegradation of rhodamine B and gaseous formaldehyde [156]. Pola et al. synthesized a layered-type perovskite material constructed on $A^IA^{II}Ti_2O_6$ (A^I = Na or Ag or Cu and A^{II} = La) structure for the photodegradation of several organic pollutants and industrial wastewater under visible-light-driven [157–162].

Perovskite system	Synthesis process	Light source	Pollutants	References
$NaTaO_3$	HT	UV	CH_3CHO	[163]
La-doped $NaTaO_3$	SG	UV	MB	[164]
La-doped $NaTaO_3$	HT	UV	MB	[165]
Cr-doped $NaTaO_3$	HT	UV	MB	[166]
Eu-doped $NaTaO_3$	SS	UV	MB	[167]
Bi-doped $NaTaO_3$	SS	UV	MB	[168]
N-doped $NaTaO_3$	SS	UV	MB	[169]
C-doped $NaTaO_3$	HT	Visible	NO_x	[36]
N/F co-doped $NaTaO_3$	HT	UV	RhB	[170]
$SrTiO_3$	HT	UV	RhB	[42, 43, 171]
Fe-doped $SrTiO_3$	SG	Visible	RhB	[172]
N-doped $SrTiO_3$	HT	Visible	MB, RhB, MO	[173]
F-doped $SrTiO_3$	BM	Visible	NO	[174]
Ni/La-doped $SrTiO_3$	SG	Visible	MG	[175]
S/C co-doped $SrTiO_3$	SS	Visible	2-Propanol	[176]
N/La-doped $SrTiO_3$	SG	Visible	2-Propanol	[177]
Fe-doped $SrTiO_3$	ST	Visible light	TC	[178]
$SrTiO_3/Fe_2O_3$	HT	Visible	TC	[179]
$BaTiO_3$	SG	UV	Pesticide	[36]
$BaTiO_3$	SG	UV	Aromatics	[58]
$BaTiO_3$	HT	UV	MO	[58]
$KNbO_3$	HT	Visible	RhB	[180]
$KNbO_3$	HT	UV	RhB	[181]
$KNbO_3$	HT	Visible	MB	[182]
$NaNbO_3$	SS	UV	RhB	[183]
$NaNbO_3$	Imp.	UV	2-Propanol	[184]
$NaNbO_3$	SS	UV	MB	[185]

Perovskite system	Synthesis process	Light source	Pollutants	References
N-doped $NaNbO_3$	SS	UV	2-Propanol	[186–188]
Ru-doped $NaNbO_3$	HT	Visible	Phenol	[189]
$AgNbO_3$	SS	UV	MB	[190]
La-doped $AgNbO_3$	SS	Visible	2-Propanol	[191]
$BiFeO_3$	SG	UV-Vis	MO, RhB, 4-CP	[69–77]
Ba-doped $BiFeO_3$	ES	Visible	CR	[79]
Ca-doped $BiFeO_3$	ES	Visible	CR	[80]
Ba or Mn-doped $BiFeO_3$	ES	Visible	CR	[82]
Ca or Mn-doped $BiFeO_3$	HT	UV-Visible	RhB	[82]
Gd-doped $BiFeO_3$	SG	Visible	RhB	[83]
$LaFeO_3$	Comb.	UV	Methyl phenol	[84]
$LaFeO_3$	SG	Visible	RhB	[85]
$LaFeO_3$	HT	Visible	RhB, MB, chlorophenol	[86, 90, 91, 192]
Ca-doped $LaFeO_3$	SS	Visible	MB	[92]
$LnFeO_3$ (Pr,Y)	SG	Visible	RhB	[193]
$SrFeO_{3-x}$	US	Visible	Phenol	[194]
$SrFeO_3$	SS	Visible	MB	[195]
$BaZrO_3$	SG	UV	MB	[196]
$BaZrO_3$	HT	UV	MO	[197]
$ATiO_3$ (A = Fe, Pb) and $AFeO_3$ (A = Bi, La, Y)	SG	Visible	MB	[198]
$Zn_{0.9}Mg_{0.1}TiO_3$	SG	Visible	MB	[199]
$SrTiO_3$ nanocube-coated CdS microspheres	HT	Visible	Antibiotic pollutants	[200]
$Ag/AgCl/CaTiO_3$	HT	Visible	RhB	[201]
TiO_2-coupled $NiTiO_3$	SS	Visible	MB	[202]
$ZnTiO_3$	HT	UV	MO and PCP	[203]
Mg-doped $BaZrO_3$	SS	UV	MB	[204]
$SrSnO_3$	MW	UV	MO	[205]
$LaCoO_3$	MW	Visible	MO	[206]
$LaCoO_3$	Ads.	UV	MB, MO	[207]
$LaCoO_3$	ES	UV	RhB	[208]
$LaNiO_3$	SG	Visible	MO	[209]
$Bi_{0.5}Na_{0.5}TiO_3$	HT	UV	MO	[93]
$La_{0.7}Sr_{0.3}MnO_3$	SG	Solar light	MO	[97]
$La_{0.5}Ca_{0.5}NiO_3$	SG	UV	RB5	[98]
$La_{0.5}Ca_{0.5}CoO_3$	SG	UV	CR	[99]
$Sr_{1-x}Ba_xSnO_3$	SS	UV	Azo-dye	[100]
$BaCo_{1/2}Nb_{1/2}O_3$	SG	Visible	MB	[210]
$Ba(In_{1/3}Pb_{1/3}M_{1/3})O_3$ (M = Nb and Ta)	SS	Visible	MB, 4-CP	[211]
$A(In_{1/3}Nb_{1/3}B_{1/3})O_3$ (A = Sr, Ba; B = Sn, Pb)	SS	Visible	MB, 4-CP	[212]
$SrTi_{1-x}Fe_xO_{3-\delta}$	SS	Visible	MB	[102]

Perovskite system	Synthesis process	Light source	Pollutants	References
$SrTi_{0.1}Fe_{0.9}O_{3-\delta}$	SG	Solar light	MO	[103]
$SrFe_{0.5}Co_{0.5}O_{3-\delta}$	SG	Solar light	CR	[213]
$LaFe_{0.5}Ti_{0.5}O_3$	SG	UV	Phenol	[90]
$Bi(Mg_{3/8}Fe_{2/8}Ti_{3/8})O_3$	MS	Visible	MO	[110]
$LaTi(ON)_3$	SG	Visible	Acetone	[214]
$(Ag_{0.75}Sr_{0.25})(Nb_{0.75}Ti_{0.25})O_3$	SS	Visible	CH_3CHO	[215]
$La_{0.8}Ba_{0.2}Fe_{0.9}Mn_{0.1}O_{3-x}$	SG	Solar light	MO	[115]
Cu-$(Sr_{1-y}Na_y)(Ti_{1-x}Mo_x)O_3$	HT	Visible	Propanol	[118]
$BiFeO_3$–$(Na_{0.5}Bi_{0.5})TiO_3$	SG	Visible	RhB	[120]
$SrBi_2Nb_2O_9$	SG SS	UV	Aniline, RhB	[145, 146]
$ABi_2Nb_2O_9$ (A = Sr, Ba)	SG	UV	MO	[147]
$Bi_5Ti_3FeO_{15}$	HT SS	Visible	RhB, CH_3CHO IPA	[149, 150]
$Bi_{5-x}La_xTi_3FeO_{15}$	SS	Solar light	RhB	[151]
$Bi_3SrTi_2TaO_{12}$ $Bi_2LaSrTi_2TaO_{12}$	SS	UV	RhB	[216]
$Ba_5Ta_4O_{15}$	HT	UV	RhB	[154]
N-doped $Ln_2Ti_2O_7$ (Ln = La, Pr, Nd)	HT	Visible	MO	[217]
$CdS/Ag/Bi_2MoO_6$	SG	Visible	RhB	[218]

SS: solid state; HT: hydrothermal; SG: sol-gel; BM: ball-milling; ES: electronspun; MW: microwave; Comb.: combustion; US: ultrasonic; MS: molten salt; Imp.: impregnation; Ads.: adsorption; ST: solvothermal; RhB: rhodamine B; MO: methyl orange; MB: methylene blue; 4-cp: 4-chlorophenol; MG: malachite green; CR: congo red; NO: nitrogen monoxide; PA: isopropyl alcohol; TC: tetracycline; and PCP: pentachlorophenol.

Table 1.
Perovskite materials used as photocatalysts (ABO_3, AA^IBO_3, AA^IBO_3, ABB^IO_3, $AB(ON)_3$, and $AA^IBB^{II}O_3$) for degradation of pollutants.

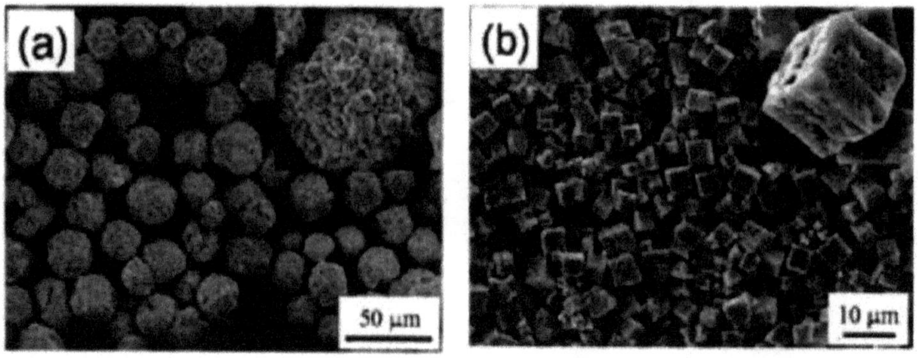

Figure 2.
SEM patterns of $BiFeO_3$: (a) microspheres and (b) microcubes. The intensified pictures are revealed in the upper part inserts.

Acknowledgements

Authors would like to thank DST-FIST schemes and CSIR, New Delhi. One of us (Ramesh Gade) thanks Council of Scientific & Industrial Research (CSIR), New Delhi, for the award of Junior Research Fellowship.

Author details

Someshwar Pola* and Ramesh Gade
Department of Chemistry, University College of Science, Osmania University, Hyderabad, Telangana, India

*Address all correspondence to: somesh.pola@gmail.com

References

[1] Barbara DA, Jared JS, Tyson AB, William WA. Insights from placing photosynthetic light harvesting into context. Journal of Physical Chemistry Letters. 2014;**5**:2880-2889

[2] von Erika S, Paul SM, James DA, Lori B, Piers F, David F, et al. Chemistry and the linkages between air quality and climate change. Chemical Reviews. 2015;**115**:3856-3897

[3] Rowan JB, Andrew TH. Review of major design and scale-up considerations for solar photocatalytic reactors. Industrial and Engineering Chemistry Research. 2009;**48**:8890-8905

[4] Michael GW, Emily LW, James RM, Shannon WB, Qixi M, Elizabeth AS, et al. Solar water splitting cells. Chemical Reviews. 2010;**110**:6446-6473

[5] Manvendra P, Rahul K, Kamal K, Todd M, Charles UP Jr, Dinesh M. Pharmaceuticals of emerging concern in aquatic systems: Chemistry, occurrence, effects, and removal methods. Chemical Reviews. 2019;**119**:3510-3673

[6] Marta M, Jolanta K, Iseult L, Marianne M, Jan K, Steve B, et al. Changing environments and biomolecule coronas: Consequences and challenges for the design of environmentally acceptable engineered nanoparticles. Green Chemistry. 2018;**20**:4133-4168

[7] Wanjun W, Guiying L, Dehua X, Taicheng A, Huijun Z, Po KW. Photocatalytic nanomaterials for solar-driven bacterial inactivation: recent progress and challenges. Environmental Science: Nano. 2017;**4**:782-799

[8] Wanjun WYY, Taicheng A, Guiying L, Ho YY, Jimmy CY, Po KW. Visible-light-driven photocatalytic inactivation of *E. coli* K-12 by bismuth vanadate nanotubes: bactericidal performance and mechanism. Environmental Science & Technology. 2012;**46**:4599-4606

[9] Nadine C, Stefanie I, Elisabeth S, Marjan V, Karolin K, David D, et al. Inactivation of antibiotic resistant bacteria and resistance genes by ozone: From laboratory experiments to full-scale wastewater treatment. Environmental Science & Technology. 2016;**50**:11862-11871

[10] Yidong Z, Xiangxue W, Ayub K, Pengyi W, Yunhai L, Ahmed A, et al. Environmental remediation and application of nanoscale zero-valent iron and its composites for the removal of heavy metal ions: A review. Environmental Science & Technology. 2016;**50**:7290-7304

[11] Wenya H, Kelong A, Xiaoyan R, Shengyan W, Lehui L. Inorganic layered ion-exchangers for decontamination of toxic metal ions in aquatic systems. Journal of Materials Chemistry A. 2017;**5**:19593-19606

[12] Manolis JM, Mercouri GK. Metal sulfide ion exchangers: Superior sorbents for the capture of toxic and nuclear waste-related metal ions. Chemical Science. 2016;**7**:4804-4824

[13] Chunli K, Kunkun X, Yuhan W, Dongmei H, Ling Z, Fang L, et al. Synthesis of $SrTiO_3$–TiN nanocomposites with enhanced photocatalytic activity under simulated solar irradiation. Industrial and Engineering Chemistry Research. 2018;**57**:11526-11534

[14] Vaidyanathan S, Ryan KR, Eduardo EW. Synthesis and UV- visible-light photoactivity of noble-metal-$SrTiO_3$ composites. Industrial and Engineering Chemistry Research. 2006;**45**:2187-2193

[15] Partha M, Aarthi T, Giridhar M, Srinivasan N. Photocatalytic Degradation of Dyes and Organics with Nanosized $GdCoO_3$. Journal of Physical Chemistry C. 2007;**111**(4):1665-1674

[16] Dipankar S, Sudarshan M, Guru Row TN, Giridhar M. Synthesis, structure, and photocatalytic activity in orthorhombic perovskites $LnVO_3$ and $Ln_{1-x}Ti_xVO_3$ (Ln = Ce, Pr, and Nd). Industrial and Engineering Chemistry Research. 2009;**48**:7489-7497

[17] Barrocas B, Sério S, Rovisco A, Melo Jorge ME. Visible-light photocatalysis in $Ca_{0.6}Ho_{0.4}MnO_3$ films deposited by RF-magnetron sputtering using nanosized powder compacted target. Journal of Physical Chemistry C. 2014;**118**:590-597

[18] Tokeer A, Umar F, Ruby P. Fabrication and photocatalytic applications of perovskite materials with special emphasis on alkali-metal-based niobates and tantalates. Industrial and Engineering Chemistry Research. 2018;**57**:18-41

[19] Senthilkumar P, Arockiya Jency D, Kavinkumar T, Dhayanithi D, Dhanuskodi S, Umadevi M, et al. Built-in electric field assisted photocatalytic dye degradation and photoelectrochemical water splitting of ferroelectric Ce doped $BaTiO_3$ nanoassemblies. ACS Sustainable Chemistry & Engineering. 2019;**7**:12032-12043

[20] Hongbo F, Shicheng Z, Tongguang X, Yongfa Z, Jianmin C. Photocatalytic degradation of RhB by fluorinated Bi_2WO_6 and distributions of the intermediate products. Environmental Science & Technology. 2008;**42**:2085-2091

[21] Sandeep K, Sunita K, Meganathan T, Ashok KG. Achieving enhanced visible-light-driven photocatalysis using type-II $NaNbO_3$/CdS core/shell heterostructures. ACS Applied Materials & Interfaces. 2014;**6**:13221-13233

[22] Yanlong T, Binbin C, Jiangli L, Jie F, Fengna X, Xiaoping D. Hydrothermal synthesis of graphitic carbon nitride–Bi_2WO_6 heterojunctions with enhanced visible light photocatalytic activities. ACS Applied Materials & Interfaces. 2013;**5**:7079-7085

[23] Anindya SP, Gaurangi G, Mohammad Q. Ordered–disordered $BaZrO_{3-\delta}$ hollow nanosphere/carbon dot hybrid nanocomposite: A new visible-light-driven efficient composite photocatalyst for hydrogen production and dye degradation. ACS Omega. 2018;**3**:10980-10991

[24] Joanna HC, Matthew SD, Robert GP, Christopher PI, James RD, John BC, et al. Visible light photo-oxidation of model pollutants using $CaCu_3Ti_4O_{12}$: An experimental and theoretical study of optical properties, electronic structure, and selectivity. Journal of the American Chemical Society. 2011;**133**:1016-1032

[25] Rasel D, Chad DV, Agnes S, Bin C, Ahmad FI, Xianbo L, et al. Recent advances in nanomaterials for water protection and monitoring. Chemical Society Reviews. 2017;**46**:6946-7020

[26] Pena MA, Fierro JLG. Chemical structures and performance of perovskite oxides. Chemical Reviews. 2001;**101**:1981-2018

[27] Anna K, Marcos FG, Gerardo C. Advanced nanoarchitectures for solar photocatalytic applications. Chemical Reviews. 2012;**112**:1555-1614

[28] Wei W, Moses OT, Zongping S. Research progress of perovskite materials in photocatalysis- and photovoltaics-related energy conversion and environmental treatment. Chemical Society Reviews. 2015;**44**:5371-5408

[29] Guan Z, Gang L, Lianzhou W, John TSI. Inorganic perovskite photocatalysts for solar energy utilization. Chemical Society Reviews. 2016;**45**:5951-5984

[30] Xiubing H, Guixia Z, Ge W, John TSI. Synthesis and applications of nanoporous perovskite metal oxides. Chemical Science. 2018;**9**:3623-3637

[31] Beata B, Ewa K, Joanna N, Zhishun W, Maya E, Bunsho O, et al. Preparation of CdS and Bi_2S_3 quantum dots co-decorated perovskite-type $KNbO_3$ ternary heterostructure with improved visible light photocatalytic activity and stability for phenol degradation. Dalton Transactions. 2018;**47**:15232-15245

[32] Hamidreza A, Yuan W, Hongyu S, Mehran R, Hongxing D. Ordered meso- and macroporous perovskite oxide catalysts for emerging applications. Chemical Communications. 2018;**54**:6484-6502

[33] Bin L, Gang L, Lianzhou W. Recent advances in 2D materials for photocatalysis. Nanoscale. 2016;**8**:6904-6920

[34] Chunping X, Prasaanth Ravi A, Cyril A, Rafael L, Samuel M. Nanostructured materials for photocatalysis. Chemical Society Reviews. 2019;**48**:3868-3902

[35] Xiaoyun C, Jun X, Yueshan X, Feng L, Yaping D. Rare earth double perovskites: a fertile soil in the field of perovskite oxides. Inorganic Chemistry Frontiers. 2019;**6**:2226-2238

[36] Xiaoyong W, Shu Y, Qiang D, Tsugio S. Preparation and visible light induced photocatalytic activity of C-$NaTaO_3$ and C-$NaTaO_3$–Cl-TiO_2 composite. Physical Chemistry Chemical Physics. 2013;**15**:20633-20640

[37] Shaodong S, Xiaojing Y, Qing Y, Zhimao Y, Shuhua L. Mesocrystals for photocatalysis: A comprehensive review on synthesis engineering and functional modifications. Nanoscale Advances. 2019;**1**:34-63

[38] Ming Y, Xian LH, Shicheng Y, Zhaosheng L, Tao Y, Zhigang Z. Improved hydrogen evolution activities under visible light irradiation over $NaTaO_3$ codoped with lanthanum and chromium. Materials Chemistry and Physics. 2010;**121**:506-510

[39] Shin S, Yonemura M. Order-disorder transition of Sr2Fe_2O_5 from brownmillerite to perovskite structure at an elevated temperature. Materials Research Bulletin. 1978;**13**:1017-1021

[40] Wagner FT, Somorjai GA. Photocatalytic and photoelectrochemical hydrogen production on strontium titanate single crystals. Journal of the American Chemical Society. 1980;**102**:5494-5502

[41] Guowei L, Graeme RB, Thomas TMP. Vacancies in functional materials for clean energy storage and harvesting: The perfect imperfection. Chemical Society Reviews. 2017;**46**:1693-1706

[42] Xiao W, Gang X, Zhaohui R, Chunxiao X, Ge S, Gaorong H. PVA-assisted hydrothermal synthesis of $SrTiO_3$ nanoparticles with enhanced photocatalytic activity for degradation of RhB. Journal of the American Ceramic Society. 2008;**91**:3795-3799

[43] Xiao W, Gang X, Zhaohui R, Chunxiao X, Wenjian W, Ge S, et al. Single-crystal-like mesoporous $SrTiO_3$ spheres with enhanced photocatalytic performance. Journal of the American Ceramic Society. 2010;**93**:1297-1305

[44] Bo W, Ming-Yu Q, Chuang H, Zi-Rong T, Yi-Jun X. Photocorrosion inhibition of semiconductor-based photocatalysts: Basic principle, current development, and future perspective. ACS Catalysis. 2019;**9**:4642-4687

[45] Natalita MN, Xingdong W, Rachel AC. High-throughput synthesis and screening of titania-based photocatalysts. ACS Combinatorial Science. 2015;**17**:548-569

[46] Hideki K, Akihiko K. Highly efficient decomposition of pure water into H_2 and O_2 over $NaTaO_3$ photocatalysts. Catalysis Letters. 1999;**58**:153-155

[47] Hideki K, Akihiko K. Water splitting into H_2 and O_2 on alkali tantalate photocatalysts $ATaO_3$ (A = Li, Na, and K). Journal of Physical Chemistry B. 2001;**105**(19):4285-4292

[48] Hideki K, Akihiko K. Photocatalytic water splitting into H_2 and O_2 over various tantalate photocatalysts. Catalysis Today. 2003;**78**:561-569

[49] Liu JW, Chen G, Li ZH, Zhang ZG. Hydrothermal synthesis and photocatalytic properties of $ATaO_3$ and $ANbO_3$ (A = Na and K). International Journal of Hydrogen Energy. 2007;**32**:2269-2272

[50] Che-Chia H, Hsisheng T. Influence of structural features on the photocatalytic activity of $NaTaO_3$ powders from different synthesis methods. Applied Catalysis A. 2007;**331**:44-50

[51] Che-Chia H, Chien-Cheng T, Hsisheng T. Structure characterization and tuning of perovskite-like $NaTaO_3$ for applications in photoluminescence and photocatalysis. Journal of the American Ceramic Society. 2009;**92**:460-466

[52] Xia L, Jinling Z. Facile hydrothermal synthesis of sodium tantalate ($NaTaO_3$) nanocubes and high photocatalytic properties. Journal of Physical Chemistry C. 2009;**113**:19411-19418

[53] Fu X, Wang X, Leung DYC, Xue W, Ding Z, Huang H, et al. Photocatalytic reforming of glucose over La doped alkali tantalate photocatalysts for H_2 production. Catalysis Communications. 2010;**12**:184-187

[54] Yokoi T, Sakuma J, Maeda K, Domen K, Tatsumi T, Kondo JN. Preparation of a colloidal array of $NaTaO_3$ nanoparticles via a confined space synthesis route and its photocatalytic application. Physical Chemistry Chemical Physics. 2011;**13**:2563-2570

[55] Shi J, Liu G, Wang N, Li C. Microwave-assisted hydrothermal synthesis of perovskite $NaTaO_3$ nanocrystals and their photocatalytic properties. Journal of Materials Chemistry. 2012;**22**:18808-18813

[56] Meyer T, Priebe JB, Silva RO, Peppel T, Junge H, Beller M, et al. Advanced charge utilization from $NaTaO_3$ photocatalysts by multilayer reduced graphene oxide. Chemistry of Materials. 2014;**26**:4705-4711

[57] Li Y, Gou H, Lu J, Wang C. A two-step synthesis of $NaTaO_3$ microspheres for photocatalytic water splitting. International Journal of Hydrogen Energy. 2014;**39**:13481-13485

[58] Gomathi Devi L, Krishnamurthy G. $TiO_2/BaTiO_3$-assisted photocatalytic mineralization of diclofop-methyl on UV-light irradiation in the presence of oxidizing agents. Journal of Hazardous Materials. 2009;**162**:899-905

[59] Gomathi Devi L, Krishnamurthy G. TiO_2- and $BaTiO_3$-assisted photocatalytic degradation of selected chloroorganic compounds in aqueous medium: Correlation of reactivity/ orientation effects of substituent groups of the pollutant molecule on the degradation rate. The Journal of Physical Chemistry. A. 2011;**115**:460-469

[60] Liu J, Sun Y, Li Z. Ag loaded flower-like $BaTiO_3$ nanotube arrays: Fabrication

and enhanced photocatalytic property. CrystEngComm. 2012;**14**:1473-1478

[61] Maeda K. Rhodium-doped barium titanate perovskite as a stable p-type semiconductor photocatalyst for hydrogen evolution under visible light. ACS Applied Materials & Interfaces. 2014;**6**:2167-2173

[62] Upadhyay S, Shrivastava J, Solanki A, Choudhary S, Sharma V, Kumar P, et al. Enhanced photoelectro-chemical response of $BaTiO_3$ with Fe doping: Experiments and first-principles analysis. Journal of Physical Chemistry C. 2011;**115**:24373-24380

[63] Mizoguchi H, Ueda K, Orita M, Moon S-C, Kajihara K, Hirano M, et al. Decomposition of water by a $CaTiO_3$ photocatalyst under UV light irradiation. Materials Research Bulletin. 2002;**37**:2401-2406

[64] Shimura K, Yoshida H. Hydrogen production from water and methane over Pt-loaded calcium titanate photocatalyst. Energy & Environmental Science. 2010;**3**:615-617

[65] Zhang H, Chen G, Li Y, Teng Y. Electronic structure and photocatalytic properties of copper-doped $CaTiO_3$. International Journal of Hydrogen Energy. 2010;**35**:2713-2716

[66] Nishimoto S, Matsuda M, Miyake M. Photocatalytic activities of Rh-doped $CaTiO_3$ under visible light irradiation. Chemistry Letters. 2006;**35**:308-309

[67] Zhang H, Chen G, He X, Xu J. Electronic structure and photocatalytic properties of Ag–La codoped $CaTiO_3$. Journal of Alloys and Compounds. 2012;**516**:91-95

[68] Arney D, Watkins T, Maggard PA. Effects of particle surface areas and microstructures on photocatalytic H_2 and O_2 production over $PbTiO_3$. Journal of the American Ceramic Society. 2011;**94**:1483-1489

[69] Zhen C, Yu JC, Liu G, Cheng H-M. Selective deposition of redox co-catalyst(s) to improve the photocatalytic activity of single-domain ferroelectric $PbTiO_3$ nanoplates. Chemical Communications. 2014;**50**:10416-10419

[70] Gao F, Chen X, Yin K, Dong S, Ren Z, Yuan F, et al. Visible-Light photocatalytic properties of weak magnetic $BiFeO_3$ nanoparticles. Advanced Materials. 2007;**19**:2889-2892

[71] Huo Y, Miao M, Zhang Y, Zhu J, Li H. Aerosol-spraying preparation of a mesoporous hollow spherical $BiFeO_3$ visible photocatalyst with enhanced activity and durability. Chemical Communications. 2011;**47**:2089-2091

[72] Cho CM, Noh JH, Cho I-S, An J-S, Hong KS. Low-temperature hydrothermal synthesis of pure $BiFeO_3$ nanopowders using triethanolamine and their applications as visible-light photocatalysts. Journal of the American Ceramic Society. 2008;**91**:3753-3755

[73] Li S, Lin Y-H, Zhang B-P, Wang Y, Nan C-W. Controlled fabrication of $BiFeO_3$ uniform microcrystals and their magnetic and photocatalytic behaviors. Journal of Physical Chemistry C. 2010;**114**:2903-2908

[74] Xu X, Lin Y-H, Li P, Shu L, Nan C-W. Synthesis and photocatalytic behaviors of high surface area $BiFeO_3$ thin films. Journal of the American Ceramic Society. 2011;**94**:2296-2299

[75] Ji W, Yao K, Lim Y-F, Liang YC, Suwardi A. Large modulation of perpendicular magnetic anisotropy in a $BiFeO_3/Al_2O_3/Pt/Co/Pt$ multiferroic heterostructure via spontaneous polarizations. Applied Physics Letters. 2013;**103**:062901

[76] Chen XY, Yu T, Gao F, Zhang HT, Liu LF, Wang YM, et al. Application of weak ferromagnetic $BiFeO_3$ films as the photoelectrode material under visible-light irradiation. Applied Physics Letters. 2007;**91**:022114

[77] Li S, Zhang J, Kibria MG, Mi Z, Chaker M, Ma D, et al. Remarkably enhanced photocatalytic activity of laser ablated Au nanoparticle decorated $BiFeO_3$ nanowires under visible-light. Chemical Communications. 2013;**49**:5856-5858

[78] Huo Y, Jin Y, Zhang Y. Citric acid assisted solvothermal synthesis of $BiFeO_3$ microspheres with high visible-light photocatalytic activity. Journal of Molecular Catalysis A: Chemical. 2010;**331**:15-20

[79] Schultz AM, Zhang Y, Salvador PA, Rohrer GS. Effect of crystal and domain orientation on the visible-light photochemical reduction of Ag on $BiFeO_3$. ACS Applied Materials & Interfaces. 2011;**3**:1562-1567

[80] Feng YN, Wang HC, Shen Y, Lin YH, Nan CW. Magnetic and photocatalytic behaviors of Ba-doped $BiFeO_3$ nanofibers. International Journal of Applied Ceramic Technology. 2014;**11**:676-680

[81] Feng YN, Wang HC, Luo YD, Shen Y, Lin YH. Ferromagnetic and photocatalytic behaviors observed in Ca-doped $BiFeO_3$ nanofibres. Journal of Applied Physics. 2013;**113**:146101

[82] Wang HC, Lin YH, Feng YN, Shen Y. Photocatalytic behaviors observed in Ba and Mn doped $BiFeO_3$ nanofibers. Journal of Electroceramics. 2013;**31**:271

[83] Pei YL, Zhang C. Effect of ion doping in different sites on the morphology and photocatalytic activity of $BiFeO_3$ microcrystals. Journal of Alloys and Compounds. 2013;**570**:57-60

[84] Guo R, Fang L, Dong W, Zheng F, Shen M. Enhanced photocatalytic activity and ferromagnetism in Gd doped $BiFeO_3$ nanoparticles. Journal of Physical Chemistry C. 2010;**114**:21390-21396

[85] Deganello F, Tummino ML, Calabrese C, Testa ML, Avetta P, Fabbi D, et al. A new, sustainable $LaFeO_3$ material prepared from biowaste-sourced soluble substances. New Journal of Chemistry. 2015;**39**:877-885

[86] Li L, Wang X, Zhang Y. Enhanced visible light-responsive photocatalytic activity of $LnFeO_3$ (Ln = La, Sm) nanoparticles by synergistic catalysis. Materials Research Bulletin. 2014;**50**:18-22

[87] Thirumalairajan S, Girija K, Mastelaro VR, Ponpandian N. Photocatalytic degradation of organic dyes under visible light irradiation by floral-like $LaFeO_3$ nanostructures comprised of nanosheet petals. New Journal of Chemistry. 2014;**38**:5480-5490

[88] Parida KM, Reddy KH, Martha S, Das DP, Biswal N. Fabrication of nanocrystalline $LaFeO_3$: An efficient sol–gel auto-combustion assisted visible light responsive photocatalyst for water decomposition. International Journal of Hydrogen Energy. 2010;**35**:12161-12168

[89] Tijare SN, Joshi MV, Padole PS, Mangrulkar PA, Rayalu SS, Labhsetwar NK. Photocatalytic hydrogen generation through water splitting on nano-crystalline $LaFeO_3$ perovskite. International Journal of Hydrogen Energy. 2012;**37**:10451-10456

[90] Thirumalairajan S, Girija K, Ganesh I, Mangalaraj D, Viswanathan C, Balamurugan A. Controlled synthesis of perovskite $LaFeO_3$ microsphere composed of nanoparticles via self-assembly process and their associated photocatalytic activity. Chemical Engineering Journal. 2012;**209**:420-428

[91] Hu R, Li C, Wang X, Sun Y, Jia H, Su H, et al. Photocatalytic activities of $LaFeO_3$ and La_2FeTiO_6 in p-chlorophenol degradation under visible light. Catalysis Communications. 2012;**29**:35-39

[92] Thirumalairajan S, Girija K, Hebalkar NY, Mangalaraj D, Viswanathan C, Ponpandian N. Shape evolution of perovskite $LaFeO_3$ nanostructures: A systematic investigation of growth mechanism, properties and morphology dependent photocatalytic activities. RSC Advances. 2013;**3**:7549-7561

[93] Li FT, Liu Y, Liu RH, Sun ZM, Zhao DS, Kou CG. Preparation of Ca-doped $LaFeO_3$ nanopowders in a reverse microemulsion and their visible light photocatalytic activity. Materials Letters. 2010;**64**:223-225

[94] Invalid reference

[95] Li J, Wang G, Wang H, Tang C, Wang Y, Liang C, et al. In situ self-assembly synthesis and photocatalytic performance of hierarchical $Bi0.5Na0.5TiO_3$ micro/nanostructures. Journal of Materials Chemistry. 2009;**19**:2253-2258

[96] Wang L, Wang W. Photocatalytic hydrogen production from aqueous solutions over novel $Bi0.5Na0.5TiO_3$ microspheres. International Journal of Hydrogen Energy. 2012;**37**:3041-3047

[97] Ai Z, Lu G, Lee S. Efficient photocatalytic removal of nitric oxide with hydrothermal synthesized $Na0.5Bi0.5TiO_3$ nanotubes. Journal of Alloys and Compounds. 2014;**613**:260-266

[98] Hu CC, Lee YL, Teng H. Efficient water splitting over $Na_{1-x}K_xTaO_3$ photocatalysts with cubic perovskite structure. Journal of Materials Chemistry. 2011;**21**:3824-3830

[99] Ghiasi M, Malekzadeh MA. Solar photocatalytic degradation of methyl orange over $La_{0.7}Sr_{0.3}MnO_3$ nano-perovskite. Separation and Purification Technology. 2014;**134**:12-19

[100] Yazdanbakhsh M, Tavakkoli H, Hosseini SM. Characterization and evaluation catalytic efficiency of $La_{0.5}Ca_{0.5}NiO_3$ nanopowders in removal of reactive blue 5 from aqueous solution. Desalination. 2011;**281**:388-395

[101] Tavakkoli H, Beiknejad D, Tabari T. Fabrication of perovskite-type oxide $La_{0.5}Ca_{0.5}CoO_{3-\delta}$ nanoparticles and its dye removal performance. Desalination and Water Treatment. 2014;**52**:7377-7388

[102] Sales HB, Bouquet V, Deputier S, Ollivier S, Gouttefangeas F, Guilloux-Viry G, et al. $Sr_{1-x}Ba_xSnO_3$ system applied in the photocatalytic discoloration of an azo-dye. Solid State Sciences. 2014;**28**:67-73

[103] Bae SW, Borse PH, Lee JS. Dopant dependent band gap tailoring of hydrothermally prepared cubic $SrTi_xM_{1-x}O_3$ (M = Ru, Rh, Ir, Pt, Pd) nanoparticles as visible light photocatalysts. Applied Physics Letters. 2008;**92**:104107

[104] Ghaffari M, Huang H, Tan PY, Tan OK. Synthesis and visible light photocatalytic properties of $SrTi_{(1-x)}Fe_xO_{(3-\delta)}$ powder for indoor decontamination. Powder Technology. 2012;**225**:221-226

[105] Chen HX, Wei ZX, Wang Y, Zeng WW, Xiao CM. Preparation of $SrTi_{0.1}Fe_{0.9}O_{3-\delta}$ and its photocatalysis activity for degradation of methyl orange in water. Materials Chemistry and Physics. 2011;**130**:1387-1393

[106] Jeong ED, Yu SM, Yoon JY, Bae JS, Cho CR, Lim KT, et al. Efficient visible light photocatalysis in cubic Sr_2FeNbO_6. Journal of Ceramic Processing Research. 2012;**13**:305-309

[107] Li J, Jia L, Fang W, Zeng J. Enhancement of activity of $LaNi_{0.7}Cu_{0.3}O_3$ for photocatalytic water splitting by reduction treatment at moderate temperature. International Journal of Hydrogen Energy. 2010;**35**:5270-5275

[108] Li J, Zeng J, Jia L, Fang W. Investigations on the effect of Cu^{2+}/Cu^{1+} redox couples and oxygen vacancies on photocatalytic activity of treated $LaNi_{1-x}Cu_xO_3$ (x = 0.1, 0.4, 0.5). International Journal of Hydrogen Energy. 2010;**35**:12733-12740

[109] Li H, Liang Q, Gao LZ, Tang SH, Cheng ZY, Zhang BL, et al. Catalytic production of carbon nanotubes by decomposition of CH_4 over the pre-reduced catalysts $LaNiO_3$, $La_4Ni_3O_{10}$, $La_3Ni_2O_7$ and La_2NiO_4. Catalysis Letters. 2001;**74**:185-188

[110] Yuan Y, Zhao Z, Zheng J, Yang M, Qiu L, Li Z, et al. Polymerizable complex synthesis of $BaZr_{1-x}SnxO_3$ photocatalysts: Role of Sn^{4+} in the band structure and their photocatalytic water splitting activities. Journal of Materials Chemistry. 2010;**20**:6772-6779

[111] Kang HW, Lim SN, Park SB. Photocatalytic H_2 evolution under visible light from aqueous methanol solution on $NaBi_xTa_{1-x}O_3$ prepared by spray pyrolysis. International Journal of Hydrogen Energy. 2012;**37**:4026-4035

[112] Xu L, Li C, Shi W, Guan J, Sun Z. Visible light-response $NaTa_{1-x}Cu_xO_3$ photocatalysts for hydrogen production from methanol aqueous solution. Journal of Molecular Catalysis A: Chemical. 2012;**360**:42-47

[113] Zhang W, Chen J, An X, Wang Q, Fan L, Wang F, et al. Rapid synthesis, structure and photocatalysis of pure bismuth A-site perovskite of $Bi(Mg_{3/8}Fe_{2/8}Ti_{3/8})O_3$. Dalton Transactions. 2014;**43**:9255-9259

[114] Ni L, Tanabe M, Irie H. A visible-light-induced overall water-splitting photocatalyst: Conduction-band-controlled silver tantalite. Chemical Communications.2013;**49**:10094-10096

[115] Simone P, Marie CC, Stephan I, Alexandra EM, Rolf E. Mesoporosity in photocatalytically active oxynitride single crystals. Journal of Physical Chemistry C.2014;**118**(36):20940-20947

[116] Wu P, Shi J, Zhou Z, Tang W, Guo L. $CaTaO_2N–CaZrO_3$ solid solution: Band-structure engineering and visible-light-driven photocatalytic hydrogen production. International Journal of Hydrogen Energy. 2012;**37**:13704-13710

[117] Luo W, Li Z, Jiang X, Yu T, Liu L, Chen X, et al. Correlation between the band positions of $(SrTiO_3)_{1-x}\cdot(LaTiO_2N)_x$ solid solutions and photocatalytic properties under visible light irradiation. Physical Chemistry Chemical Physics. 2008;**10**:6717-6723

[118] Wei ZX, Xiao CM, Zeng WW, Liu JP. Magnetic properties and photocatalytic activity of $La_{0.8}Ba_{0.2}Fe_{0.9}Mn_{0.1}O_{3-\delta}$ and $LaFe_{0.9}Mn_{0.1}O_{3-\delta}$. Journal of Molecular Catalysis A: Chemical. 2013;**370**:35-43

[119] Kanhere P, Nisar P, Tang Y, Pathak B, Ahuja R, Zheng J, et al. Electronic structure, optical properties, and photocatalytic activities of $LaFeO_3–NaTaO_3$ solid solution. Journal of Physical Chemistry C. 2012;**116**:22767-22773

[120] Shi J, Ye J, Zhou Z, Li M, Guo L. Hydrothermal synthesis of $Na_{0.5}La_{0.5}TiO_3–LaCrO_3$ solid-solution single-crystal nanocubes for visible-light-driven photocatalytic H_2 evolution. Chemistry—A European Journal. 2011;**17**:7858-7867

[121] Qiu X, Miyauchi M, Yu H, Irie H, Hashimoto K. Visible-light-driven

Cu(II)-($Sr_{1-y}Na_y$)($Ti_{1-x}Mo_x$)O_3 photocatalysts based on conduction band control and surface ion modification. Journal of the American Chemical Society. 2010;**132**:15259-15267

[122] Yi ZG, Ye JH. Band gap tuning of $Na_{1-x}La_xTa_{1-x}Cr_xO_3$ for H_2 generation from water under visible light irradiation. Journal of Applied Physics. 2009;**106**:074910

[123] Liu H, Guo Y, Guo B, Dong W, Zhang D. $BiFeO_3$–($Na_{0.5}Bi_{0.5}$)TiO_3 butterfly wing scales: Synthesis, visible-light photocatalytic and magnetic properties. Journal of the European Ceramic Society. 2012;**32**:4335-4340

[124] Lv M, Xie Y, Wang Y, Sun X, Wu F, Chen H, et al. Bismuth and chromium co-doped strontium titanates and their photocatalytic properties under visible light irradiation. Physical Chemistry Chemical Physics. 2015;**17**:26320-26329

[125] Yao W, Ye J. Photocatalytic properties of a new photocatalyst $K_2Sr_{1.5}Ta_3O_{10}$. Chemical Physics Letters. 2007;**435**:96-99

[126] Liang Z, Tang K, Shao Q , Li G, Zeng S, Zheng H. Synthesis, crystal structure, and photocatalytic activity of a new two-layer Ruddlesden–Popper phase, $Li_2CaTa_2O_7$. Journal of Solid State Chemistry. 2008;**181**:964-970

[127] Li Y, Chen G, Zhou C, Li Z. Photocatalytic water splitting over a protonated layered perovskite tantalate $H_{1.81}Sr_{0.81}Bi_{0.19}Ta_2O_7$. Catalysis Letters. 2008;**123**:80-83

[128] Wang Y, Wang C, Wang L, Hao Q , Zhu X, Chen X, et al. Preparation of interlayer surface tailored protonated double-layered perovskite $H_2CaTa_2O_7$ with n-alcohols, and their photocatalytic activity. RSC Advances. 2014;**4**:4047-4054

[129] Kumar V, Uma GS. Investigation of cation (Sn^{2+}) and anion (N^{3-}) substitution in favor of visible light photocatalytic activity in the layered perovskite $K_2La_2Ti_3O_{10}$. Journal of Hazardous Materials. 2011;**189**:502-508

[130] Thaminimulla CTK, Takata T, Hara M, Knodo JN, Domen K. Effect of chromium addition for photocatalytic overall water splitting on Ni–$K_2La_2Ti_3O_{10}$. Journal of Catalysis. 2000;**196**:362-365

[131] Yang YH, Chen QY, Yin ZL, Li J. Study on the photocatalytic activity of $K_2La_2Ti_3O_{10}$ doped with zinc(Zn). Applied Surface Science. 2009;**255**:8419-8424

[132] Yang Y, Chen Q, Yin Z, Li J. Study on the photocatalytic activity of $K_2La_2Ti_3O_{10}$ doped with vanadium (V). Journal of Alloys and Compounds. 2009;**488**:364

[133] Huang Y, Wei Y, Cheng S, Fan L, Li Y, Lin J, et al. Photocatalytic property of nitrogen-doped layered perovskite K_2La_2Ti3O10. Solar Energy Materials and Solar Cells. 2010;**94**:761-766

[134] Wang B, Li C, Hirabayashi D, Suzuki K. Hydrogen evolution by photocatalytic decomposition of water under ultraviolet–visible irradiation over $K_2La_2Ti_{3-x}M_xO_{10+\delta}$ perovskite. International Journal of Hydrogen Energy. 2010;**35**:3306-3312

[135] Sato J, Saito N, Nishiyama H, Inoue Y. New photocatalyst group for water decomposition of RuO_2-loaded p-block metal (In, Sn, and Sb) oxides with d10 Configuration. The Journal of Physical Chemistry. B. 2001;**105**:6061-6063

[136] Jeong H, Kim T, Kim D, Kim K. Hydrogen production by the photocatalytic overall water splitting on NiO/$Sr_3Ti_2O_7$: Effect of preparation method. International Journal of Hydrogen Energy. 2006;**31**:1142-1146

[137] Sun X, Xie Y, Wu F, Chen H, Lv M, Ni S, et al. Photocatalytic hydrogen production over chromium doped layered perovskite Sr_2TiO_4. Inorganic Chemistry. 2015;**54**:7445-7453

[138] Ko YG, Lee WY. Effects of nickel-loading method on the water-splitting activity of a layered $NiO_x/Sr_4Ti_3O_{10}$ photocatalyst. Catalysis Letters. 2002;**83**:157-160

[139] Arney D, Fuoco L, Boltersdorf J, Maggard PA. Flux synthesis of $Na_2Ca_2Nb_4O_{13}$: The influence of particle shapes, surface features, and surface areas on photocatalytic hydrogen production. Journal of the American Ceramic Society. 2012;**96**:1158-1162

[140] Nishimoto S, Okazaki Y, Matsuda M, Miyake M. Photocatalytic H_2 evolution by layered perovskite $Ca_3Ti_2O_7$ codoped with Rh and Ln (Ln = La, Pr, Nd, Eu, Gd, Yb, and Y) under visible light irradiation. Journal of the Ceramic Society of Japan. 2009;**117**:1175-1179

[141] Zhang L, Wang H, Chen Z, Wong PK, Liu J. Bi_2WO_6 micro/nano-structures: Synthesis, modifications and visible-light-driven photocatalytic applications. Applied Catalysis B. 2011;**106**:1-13

[142] Zhang L, Zhu Y. A review of controllable synthesis and enhancement of performances of bismuth tungstate visible-light-driven photocatalysts. Catalysis Science & Technology. 2012;**2**:694-706

[143] Zhang N, Ciriminna R, Ragliaro M, Xu YJ. Nanochemistry-derived Bi_2WO_6 nanostructures: towards production of sustainable chemicals and fuels induced by visible light. Chemical Society Reviews. 2014;**43**:5276-5287

[144] Fu H, Pan C, Yao W, Zhu Y. Visible-light-induced degradation of rhodamine B by nanosized Bi_2WO_6. The Journal of Physical Chemistry. B. 2005;**109**:22432-22439

[145] Kim HG, Hwang DW, Lee JS. An undoped, single-phase oxide photocatalyst working under visible light. Journal of the American Chemical Society. 2004;**126**:8912-8913

[146] Kim HG, Becker OS, Jang JS, Ji SM, Borse PH, Lee JS. A generic method of visible light sensitization for perovskite-related layered oxides: Substitution effect of lead. Journal of Solid State Chemistry. 2006;**179**:1214-1218

[147] Kim HG, Borse PH, Jang JS, Jeong ED, Lee JS. Enhanced photochemical properties of electron rich W-doped $PbBi_2Nb_2O_9$ layered perovskite material under visible-light irradiation. Materials Letters. 2008;**62**:1427-1430

[148] Wu W, Liang S, Chen Y, Shen L, Zheng H, Wu L. High efficient photocatalytic reduction of 4-nitroaniline to p-phenylenediamine over microcrystalline $SrBi_2Nb_2O_9$. Catalysis Communications. 2012;**17**:39-42

[149] Kim JH, Hwang KT, Kim US, Kang YM. Photocatalytic characteristics of immobilized $SrBi_2Nb_2O_9$ film for degradation of organic pollutants. Ceramics International. 2012;**38**:3901-3906

[150] Wu W, Liang S, Wang X, Bi J, Liu P, Wu L. Synthesis, structures and photocatalytic activities of microcrystalline $ABi_2Nb_2O_9$ (A = Sr, Ba) powders. Journal of Solid State Chemistry. 2011;**184**:81-88

[151] Li Y, Chen G, Zhang H, Lv G. Band structure and photocatalytic activities for H_2 production of $ABi_2Nb_2O_9$ (A = Ca, Sr, Ba). International Journal of Hydrogen Energy. 2010;**35**:2652-2656

[152] Sun S, Wang W, Xu H, Zhou L, Shang M, Zhang L. $Bi_5FeTi_3O_{15}$

hierarchical microflowers: Hydrothermal synthesis, growth mechanism, and associated visible-light-driven photocatalysis. Journal of Physical Chemistry C. 2008;**112**:17835-17843

[153] Jang JS, Yoon SS, Borse PH, Lim KT, Hong TE, Jeong ED, et al. Synthesis and characterization of aurivilius phase $Bi_5Ti_3FeO_{15}$layered perovskite for Visible light photocatalysis. Journal of the Ceramic Society of Japan. 2009;**117**:1268-1272

[154] Naresh G, Mandal TK. Excellent sun-light-driven photocatalytic activity by aurivillius layered perovskites, $Bi_{5-x}La_xTi_3FeO_{15}$ (x = 1, 2). ACS Applied Materials & Interfaces. 2014;**6**:21000-21010

[155] Bożena C, Mirabbos H. UVA- and visible-light-driven photocatalytic activity of three-layer perovskite Dion-Jacobson phase $CsBa_2M_3O_{10}$ (M = Ta, Nb) and oxynitride crystals in the removal of caffeine from model wastewater. Journal of Photochemistry and Photobiology A: Chemistry. 2016;**324**:70-80

[156] Reddy JR, Kurra S, Guje R, Palla S, Veldurthi NK, Ravi G, et al. Photocatalytic degradation of methylene blue on nitrogen doped layered perovskites, $CsM_2Nb_3O_{10}$ (M = Ba and Sr). Ceramics International. 2015;**41**:2869-2875

[157] Xu TG, Zhang C, Shao X, Wu K, Zhu YF. Monomolecular-layer $Ba_5Ta_4O_{15}$ nanosheets: Synthesis and investigation of photocatalytic properties. Advanced Functional Materials. 2006;**16**:1599-1607

[158] Ramesh G, Jakeer A, Kalyana LY, Seid YA, Tao YT, Someshwar P. Photodegradation of organic dyes and industrial wastewater in the presence of layer-type perovskite materials under visible light irradiation. Journal of Environmental Chemical Engineering. 2018;**6**:4504-4513

[159] Friedmann D, Mendice C, Bahnemann D. TiO_2 for water treatment: Parameters affecting the kinetics and mechanisms of photocatalysis. Applied Catalysis B: Environmental. 2010;**99**:398-406

[160] Rizzo L, Meric S, Kassino D, Guida M, Russo F, Belgiorno V. Degradation of diclofenac by TiO_2 photocatalysis: UV absorbance kinetics and process evaluation through a set of toxicity bioassays. Water Research. 2009;**43**:979-988

[161] Konstantinou T, Albanis A. TiO_2-assisted photocatalytic degradation of azo dyes in aqueous solution: Kinetic and mechanistic investigations. A review. Applied Catalysis B: Environmental. 2004;**49**:1-14

[162] Kim SH, Ngo HH, Shon HK, Vigneswaran S. Adsorption and photocatalysis kinetics of herbicide onto titanium oxide and powdered activated carbon. Separation and Purification Technology. 2008;**58**:335-342

[163] Liu JW, Chen G, Li ZH, Zhang ZG. Electronic structure and visible light photocatalysis water splitting property of chromium-doped $SrTiO_3$. Journal of Solid State Chemistry. 2006;**179**:3704-3708

[164] Torres-Martínez LM, Cruz-López A, Juárez-Ramírez I, Elena Meza-de la Rosa M. Methylene blue degradation by $NaTaO_3$ sol–gel doped with Sm and La. Journal of Hazardous Materials. 2009;**165**:774-779

[165] Li X, Zang J. Hydrothermal synthesis and characterization of lanthanum-doped $NaTaO_3$ with high photocatalytic activity. Catalysis Communications. 2011;**12**:1380-1383

[166] Yiguo S, Wang S, Meng Y, Wang HHX. Dual substitutions of

single dopant Cr^{3+} in perovskite $NaTaO_3$: synthesis, structure, and photocatalytic performance. RSC Advances. 2012;**2**:12932-12939

[167] Wang B, Kanhere PD, Chen Z, Nisar J, Biswarup PA. Anion-doped natao3 for visible light photocatalysis. Journal of Physical Chemistry C. 2014;**118**:10728-10739

[168] Kanhere PD, Zheng J, Chen Z. Site specific optical and photocatalytic properties of Bi-doped $NaTaO_3$. Journal of Physical Chemistry C. 2011;**115**:11846-11853

[169] Liu D-R, Wei C-D, Xue B, Zhang X-G, Jiang Y-S. Synthesis and photocatalytic activity of N-doped $NaTaO_3$ compounds calcined at low temperature. Journal of Hazardous Materials. 2010;**182**:50-54

[170] Zhang J, Jiang Y, Gao W, Hao H. Synthesis and visible photocatalytic activity of new photocatalyst MBI_2O_4 (M = Cu,Zn). Journal of Materials Science: Materials in Electronics. 2014;**25**:3807-3815

[171] Da Silva LF, Avansi W, Andres J, Ribeiro C. Long-range and short-range structures of cube-like shape $SrTiO_3$ powders: Microwave-assisted hydrothermal synthesis and photocatalytic activity. Physical Chemistry Chemical Physics. 2013;**15**:12386-12393

[172] Xie T-H, Sun X, Lin J. Enhanced photocatalytic degradation of RhB driven by visible light-induced MMCT of Ti(IV)-O-Fe(II)formed in Fe-doped $SrTiO_3$. Journal of Physical Chemistry C. 2008;**112**:9753-9759

[173] Zou F, Zheng J, Qin X, Zhao Y, Jiang L, Zhi J, et al. Chemical Communications. 2012;**48**:8514-8516

[174] Wang J, Yin S, Zhang Q, Saito F, Sato T. Mechanochemical synthesis of $SrTiO_{3-x}F_x$ with high visible light photocatalytic activities for nitrogen monoxide destruction. Journal of Materials Chemistry. 2003;**13**:2348-2352

[175] Jia A, Liang X, Su Z, Zhu T, Liu S. Synthesis and effect of calcinations temperature on the physical-chemical properties and photocatalytic activities of Ni, La coded $SrTiO_3$. Journal of Hazardous Materials. 2010;**178**:233-242

[176] Ohno T, Tsubota T, Nakamura Y, Sayama K. Preparation of S, C cation-codoped $SrTiO_3$ and its photocatalytic activity under visible light. Applied Catalysis A. 2005;**288**:74-79

[177] Miyauchi M, Takashio M, Tobimastu H. Photocatalytic activity of $SrTiO_3$ codoped with nitrogen and lanthanum under visible light illumination. Langmuir. 2004;**20**:232-236

[178] Li P, Liu C, Wu G, Yang H, Lin S, Ren A, et al. Solvothermal synthesis and visible light-driven photocatalytic degradation for tetracycline of Fe-doped $SrTiO_3$. RSC Advances. 2014;**4**:47615-47624

[179] Liu C, Wu G, Chen J, Hang K, Shi W. Fabrication of a visible-light-driven photocatalyst and degradation of tetracycline based on the photoinduced interfacial charge transfer of $SrTiO_3$/ Fe_2O_3 nanowires. New Journal of Chemistry. 2016;**40**:5198-5208

[180] Lan J, Zhou X, Liu G, Yu J, Zhang J, Zhi L, et al. Enhancing photocatalytic activity of one-dimensional $KNbO_3$ nanowires by Au nanoparticles under ultraviolet and visible-light. Nanoscale. 2011;**3**:5161-5167

[181] Jiang L, Qui Y, Yi Z. Potassium niobate nanostructures: controllable morphology, growth mechanism, and photocatalytic activity. Journal of Materials Chemistry A. 2013;**1**:2878-2885

[182] Yan L, Zhang T, Lei W, Xu Q, Zhou X, Xu P, et al. Catalytic activity of gold nanoparticles supported on $KNbO_3$ microcubes. Catalysis Today. 2014;**224**:140-146

[183] Li G, Yi Z, Bai Y, Zhang W, Zhang H. Anisotropy in photocatalytic oxidization activity of $NaNbO_3$ photocatalyst. Dalton Transactions. 2012;**41**:10194-10198

[184] Hard template synthesis of nanocrystalline $NaNbO_3$ with enhanced photocatalytic performance. Catalysis Letters. 2012;**142**:901-906

[185] Katsumata K-I, Cordonier CEJ, Shichi T, FuJishima A. Photocatalytic activity of $NaNbO_3$ thin films. Journal of the American Chemical Society. 2009;**131**:3856-3857

[186] Yu J, Yu Y, Zhou P, Xiao W, Cheng B. Morphology-dependent photocatalytic-production activity of CDS. Applied Catalysis B. 2014:156-157

[187] Pai MR, Majeed J, Banerjee AM, Arya A, Bhattacharya S, Rao R, et al. Role of Nd3+ ions in modifying the band structure and photocatalytic properties of substituted indium titanates, $In_{2(1-x)}Nd_{2x}TiO_5$ oxides. Journal of Physical Chemistry C. 2012;**116**:1458-1471

[188] Haifeng S, Xiukai L, Hideol W, Zhingang Z, Jinhua Y. 3- Propanol photodegradation over nitrogen-doped $NaNbO_3$ powders under visible-light irradiation. Journal of Physics and Chemistry of Solids. 2009;**70**:931-935

[189] Paul B, Choo KH. Visible light active Ru-doped sodium niobate perovskite decorated with platinum nanoparticles via surface capping. Catalysis Today. 2014;**230**:138-144

[190] Shu H, Xie J, Xu H, Li H, Gu Z, Sun G, et al. Structural characterization and photocatalytic activity of NiO/ $AgNbO_3$. Journal of Alloys and Compounds. 2010;**496**:633-637

[191] Li G, Kako T, Wamg D, Zou Z, Ye Y. Enhanced photocatalytic activity of La-doped $AgNbO_3$ under visible light irradiation. Dalton Transactions. 2009:2423-2427

[192] Wang Y, Gao Y, Chen L, Zhang H. Geothite as an efficient heterogeneous fenton catalyst for the degradation of methyl orange. Chemical Engineering Journal. 2012;**239**:322-331

[193] Li L, Zhang M, Tan P, Gu W, Wang X. Synthetic photocatalytic activity of $LnFeO_3$ (Ln=Pr, Y) perovskites under visible-light illumination. Ceramics International. 2014;**40**:13813-13817

[194] Jia L, Ding T, Li Q, Tang Y. Study of photocatalytic performance of $SrFeO_{3-x}$ by ultrasonic radiation. Catalysis Communications. 2007;**8**:963-966

[195] Gaffari M, Tan PY, Oruc ME, Tan OK, Tse MS. Effect of ball milling on the characteristics of nano structure $SrFeO_3$ powder for photocatalytic degradation of methylene blue under visible light irradiation and its reaction kinetics. Catalysis Today. 2011;**161**:70-77

[196] Ye T-N, Xu M, Fu W, Cai Y-Y, Wei X, Wang K-X, et al. The crystallinity effect of mesocrystalline $BaZrO_3$ hollow nanosheres on charge separation for photocatalysis. Chemical Communications. 2014;**50**:3021-3023

[197] Prastomo N, Zakaria NHB, Kawamura G, Muto H, Matsuda A. High surface area $BaZrO_3$ photocatalyst prepared by base-hot-water treatment. Journal of the European Ceramic Society. 2011;**31**:2699-2705

[198] Li L, Liu X, Zhang Y, Nuhfer NT, Barmak K, Salvador PA, et al. Visible-light photochemical activity of

heterostructured core-shell materials composed of selected ternary titanates and ferrites coated by TiO_2. ACS Applied Materials & Interfaces. 2013;**5**:5064-5071

[199] Cai Z, Zhou H, Song J, Zhao F, Li J. Preparation and characterization of $Zn_{0.9}Mg_{0.1}TiO_3$ via electrospinning. Dalton Transactions. 2011;**40**:8335-8339

[200] Wu G, Xiao L, Gu W, Shi W, Jiang D, Liu C. Fabrication and excellent visible-light-driven photodegradation activity for antibiotics of $SrTiO_3$ nanocube coated CdS microsphere heterojunctions. RSC Advances. 2016;**6**:19878-19886

[201] Wang Y, Niu C-G, Liang W, Wang Y, Zhang X-G, Zeng G-M. Synthesis of fern-like Ag/AgCl/$CaTiO_3$ plasmonic photocatalysts and their enhanced visible-light photocatalytic properties. RSC Advances. 2016;**6**:47873-47882

[202] Shu X, He J, Chen D. Visible-light-induced photocatalyst based on nickel titanate nanoparticles. Industrial and Engineering Chemistry Research. 2008;**47**:4750-4753

[203] Ding N, Chen X, Wu C-ML, Li H. Adsorption of nucleobase pairs on hexagonal boron nitride sheet: hydrogen bonding versus stacking. Physical Chemistry Chemical Physics. 2013;**15**:20203-20209

[204] Ma X, Zhang J, Li H, Duan B, Guo L, Que M, et al. Violet blue long-lasting phosphorescence properties of Mg-doped $BaZrO_3$ and its ability to assist photocatalysis. Journal of Alloys and Compounds. 2013;**580**:564-569

[205] Alammar T, Hamm I, Grasmik V, Wark M, Mudring A-V. Microwave-assisted synthesis of perovskite $SrSnO_3$ nanocrystals in ionic liquids for photocatalytic applications. Inorganic Chemistry. 2017;**56**(12):6920-6932

[206] Jung WY, Hong S-S. Synthesis of $LaCoO_3$ nanoparticles by microwave process and their photocatalytic activity under visible light irradiation. Journal of Industrial and Engineering Chemistry. 2013;**19**:157-160

[207] Dong B, Li Z, Li Z, Xu X, Song M, Zheng W, et al. Highly efficient $LaCoO_3$ nanofibers catalysts for photocatalytic degradation of rhodamine B. Journal of the American Ceramic Society. 2010;**93**:3587-3590

[208] Shasha F, Niu H, Tao Z, Song J, Mao C, Zhang S, et al. Low temperature synthesis and photocatalytic property of perovskite-type La-CoO_3 hollow spheres. Journal of Alloys and Compounds. 2013;**576**:5-12

[209] Lia Y, Yaoa S, Wenb W, Xuea L, Yana Y. Sol-gel combustion synthesis and visible-light-driven photocatalytic property of perovskite $LaNiO_3$. Journal of Alloys and Compounds. 2010;**491**:560-564

[210] Capon RJ, Rooney F, Murray LM, Collins E, Sim ATR, Rostas JAP, et al. Dragmacidins: New Protein Phosphatase Inhibitors from a Southern Australian Deep-Water Marine Sponge, Spongosorites sp. Materials Letters. 2007;**61**:3959-3962

[211] Hwang DW, Kim HG, Lee JS, Kim J, Li W, Se Hyuk O. Photocatalytic hydrogen production from water over M-Doped $La_2Ti_2O_7$ (M = Cr, Fe) under visible light irradiation (λ > 420 nm). The Journal of Physical Chemistry. B. 2005;**109**:15001-15007

[212] Hur SG, Kim TW, Hwang S-J, Choy J-H. Influences of A- and B-site cations on the physicochemical properties of perovskite-structured A(In1/3Nb1/3B1/3)O3 (A = Sr, Ba; B = Sn, Pb) photocatalysts. Journal of Photochemistry and Photobiology A. 2006;**183**:176-181

[213] Thirunavukkarasu SK, Abdul AR, Chidambaram J, Govindasamy R, Sampath M, Arivarasan VK, et al. Green synthesis of titanium dioxide nanoparticles using *Psidium guajava* extract and its antibacterial and antioxidant properties. Advances in Materials Research. 2014;**58**:968-976

[214] Aguiar R, Kalytta A, Reller A, Weidenkaff A, Ebbinghausc SG. Photocatalytic decomposition of acetone using LaTi(O,N)3 nanoparticles under visible light irradiation. Journal of Materials Chemistry. 2008;**18**:4260-4265

[215] Wang D, Kako T, Ye J. Efficient photocatalytic decomposition of acetaldehyde over a solid-solution perovskite $(Ag_{0.75}Sr_{0.25})(Nb_{0.75}Ti_{0.25})O_3$ under visible-light irradiation. Journal of the American Chemical Society. 2008;**130**:2724-2725

[216] Dong W, Kaibin T, Zhenhua L, Huagui Z. Synthesis, crystal structure, and photocatalytic activity of the new three-layer aurivillius phases, $Bi_2ASrTi_2TaO_{12}$ (A = Bi, La). Journal of Solid State Chemistry.2010;**183**:361-366

[217] Fanke M, Zhanglian H, James A, Ming L, Mingjia Z, Feng Y, et al. Visible light photocatalytic activity of nitrogen-doped $La_2Ti_2O_7$ nanosheets originating from band gap narrowing. Nano Research. 2012;**5**:213-221

[218] Danjun W, Huidong S, Li G, Feng F, Yucang L. Design and construction of the sandwich-like z-scheme multicomponent cds/ag/ bi2moo6 heterostructure with enhanced photocatalytic performance in rhb photodegradation. New Journal of Chemistry. 2016;**40**:8614-8624

4

Synthesis and Investigation of the Physical Properties of Lead-Free BCZT Ceramics

Dang Anh Tuan, Vo Thanh Tung, Le Tran Uyen Tu and Truong Van Chuong

Abstract

This work presents the structure, microstructure, and physical properties of low sintering temperature lead-free ceramics $0.52(Ba_{0.7}Ca_{0.3})TiO_3$-$0.48Ba(Zr_{0.2}Ti_{0.8})O_3$ doped with nano-sized ZnO particles (noted as BCZT/x, x is the content of ZnO nanoparticles in wt.%, x = 0.00, 0.05, 0.10, 0.15, 0.20, and 0.25). The obtained results of Raman scattering and dielectric measurements have confirmed that Zn^{2+} has occupied B-site, to cause a deformation in the ABO_3-type lattice of the BCZT/x specimens. The 0.15 wt.% ZnO-modified ceramic sintered at 1350°C exhibited excellent piezoelectric parameters: d_{33} = 420 pC/N, d_{31} = −174 pC/N, k_p = 0.483, k_t = 0.423, and k_{33} = 0.571. The obtained results indicate that the high-quality lead-free BCZT ceramic could be successfully synthesized at a low sintering temperature of 1350°C with an addition of appropriated amount of ZnO nanoparticles. This work also reports the influence of the sintering temperature on structure, micro-structure, and piezoelectric properties of BCZT/0.15 compound. By rising sintering temperature, the piezoelectric behaviors were improved and rose up to the best parameters at a sintering temperature of 1450°C (d_{33} = 576 pC/N and k_p = 0.55). The corresponding properties of undoped BCZT ceramics were investigated as a comparison. It also presented that the sintering behavior and piezo-parameters of doped BCZT samples are better than the undoped BCZT samples at each sintering temperature.

Keywords: lead-free, BZT-BCT, ceramics, nanoparticle, ZnO

1. Introduction

Perovskite ABO_3-type compounds with high flexibility in symmetry play an important role in materials science. Typical materials such as lead zirconate titanate (PZT) based on family $BaTiO_3$ have received a lot of attention due to their outstand-ing dielectric, ferroelectric, and piezoelectric performance.

Nevertheless, PZT systems are globally restricted due to evaporating toxic lead oxide to the environment during preparation. With the recent growing demand of global environmental and human health protection, many non-lead materials have been systematically studied to replace the lead-based ceramics [1, 2].

In 2009, based on alternating with A- and/or B-sites in perovskite $BaTiO_3$, Liu and Ren established a new lead-free ferroelectric system Ba $(Zr_{0.2}Ti_{0.8})O_3$-x$(Ba_{0.7}Ca_{0.3})$

TiO_3 (abbreviated as BZT-BCT) that has excellent piezoelectricity (d_{33} = 620 pC/N at x = 50, i.e., morphotropic phase boundary or MPB composition) [3]. After that the BZT-BCT materials have been widely studied [4–6]. It is noted that based $BaTiO_3$ ceramics have been usually sintered at a very high temperature to obtain the desired properties [7–9] which causes many difficulties in the preparation and application of these materials. It is well-known that sintering behavior can be improved by using nanostructured raw materials and/or sintering aids [10, 11]. Among commonly used dopants, ZnO (in nano- or microscale) is known as an effective sintering aid for enhancing density and electric properties of piezoceramics [12–14].

In this work, the effects of ZnO nanoparticles as well as sintering temperature on structure, microstructure, and some electric properties of $0.48Ba(Zr_{0.2}Ti_{0.8})O_3$-$0.52(Ba_{0.7}Ca_{0.3})TiO_3$ or BCZT composition were detailedly presented.

2. Experimental produces

2.1 Preparing ZnO nanoparticles

ZnO nanoparticles were synthesized using zinc acetate and ammonia solutions as initial materials. Accordingly, zinc acetate was dissolved in distilled water to form solution. Then NH_4OH solution was gradually dropped into zinc acetate solution and stirred until a white precipitate was received. The amount of NH_4OH was enough so that the overall reaction to form ZnO is as follows [15]:

$$(CH_3COO)_2Zn + 2NH_4OH \rightarrow ZnO + 2(CH_3COO)NH_4 + H_2O.$$

The obtained white precipitate was filtered and washed several times with the aid of vacuum filter machine and then annealed at a temperature of 250°C for 1 h to remove unwanted products.

Figure 1 shows XRD and microstructure image (measured by D8 Advance, Bruker AXS, and Nova NanoSEM 450-FEI, respectively) of the obtained ZnO powder after annealing at 250°C for 1 h.

Detailed structural characterization demonstrated that the synthesized product possesses pure hexagonal symmetry [16]. The obtained ZnO particles are spherical in shape with their average diameter of 59 nm (according to Scherrer equation). The nanostructured ZnO powder was used as a sintering aid in fabrication of BCZT ceramics.

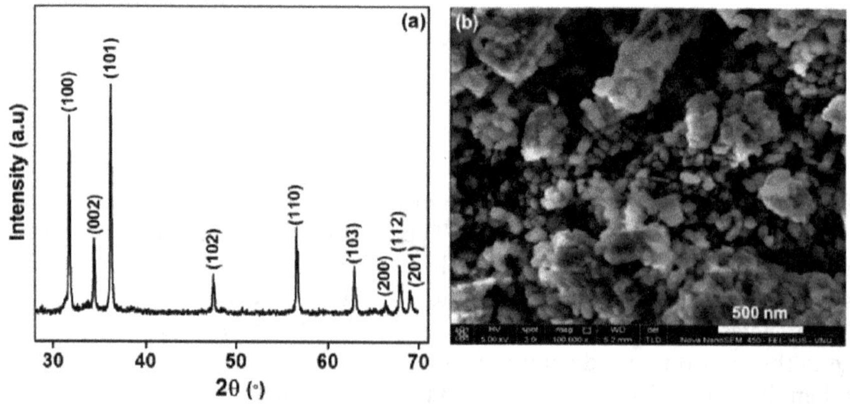

Figure 1.
(a) X-ray diffraction (XRD) and (b) micrograph of as-prepared ZnO powder.

2.2 Fabrication of ZnO nanoparticles doped BCZT ceramics at low sintering temperature

A conventional ceramic fabrication technique was used to prepare lead-free ceramics $0.52(Ba_{0.7}Ca_{0.3})TiO_3$-$0.48Ba(Zr_{0.2}Ti_{0.8})O_3$ doped with ZnO nanoparticles (abbreviated as BCZT/x, x is the content of ZnO in wt.%, x = 0.00, 0.05,0.10, 0.15, 0.20, 0.25). The raw materials with high purity (>99%) of $BaCO_3$, $CaCO_3$, ZrO_2, and TiO_2 (Merck) were weighed and mixed in a planetary milling machine (PM400/2-MA-Type) using ethanol as a medium for 20 h. The obtained powders were calcined at 1250°C for 3 h. The calcined powder was milled again in ethanol for 20 h; after that the x wt.% of ZnO nanoparticles were added, finely mixed, and then pressed into desired-shape specimens by uniaxial pressing with a pressure of 100 MPa. Sintering was carried out at various temperatures for 4 h. The X-ray diffraction patterns were recorded at room temperature by a D8 Advance, Bruker AXS. The tetragonal and rhombohedral volume fractions, τ_T and τ_R, were, respectively, evaluated using the equations below [17]:

$$\tau_T = \frac{I^T_{200} + I^T_{002}}{I^T_{200} + I^R_{200} + I^T_{002}}, \quad (1)$$

$$\tau_R = \frac{I^R_{200}}{I^T_{200} + I^R_{200} + I^T_{002}}, \quad (2)$$

where $I^{T/R}_{200/002}$ are the corresponding tetragonal (T) and rhombohedral (R) peak intensities.

The crystalline structure and lattice parameters of all samples were estimated from fitting results of the XRD data by using the Powder Cell software [18]. The surface of the sintered samples was processed and cleaned by an ultrasonic cleaner and then observed by scanning electron microscopy (SEM, Nova NanoSEM 450-FEI). Particle size is the mean linear intercept length that was determined using an intercept method with the assistance of Lince software [19]. The silver pastes were fired at 450°C for 30 minutes on both sides of these sintered bulks as electrodes for electrical measurements. Dielectric properties of the materials were determined together using an impedance analyzer (Agilent 4396B, Agilent Technologies, America, HIOKI3532) by measuring the capacitance of the specimens from room temperature to 120°C. Raman scattering spectra was measured using LabRAM-1B (Horiba Jobin Yvon). Ferroelectricity was studied by using the Sawyer-Tower circuit method. In order to study piezoelectric properties, the samples were polled in silicon oil bath by applying the DC electric field of 2 kV/mm for 60 minutes at room temperature. The main piezoelectric parameters were calculated when using a resonance method (Agilent 4396B, HP4193A) and all formulas in the IEEE standard for piezoelectric ceramic characterization [20].

3. Results and discussion

3.1 Structure, microstructure, and electric properties of BCZT/x ceramics sintered at temperature of 1350°C

Figure 2(a) shows X-ray diffraction patterns of the BCZT/x ceramics at various contents of ZnO nanoparticles measured at room temperature. All the compositions have demonstrated pure perovskite phases, and no trace of secondary phase was detected in the investigated region. **Figure 2(b)** plots the enlarged XRD patterns in the range of (44–46)° of BCZT/x ceramics. As shown, all BCZT/x ceramics have tetragonal symmetry (space group P4mm) characterized by splitting (002)/(200) peaks at around 2θ of 45° with the intensity changing between samples. Moreover, the position of these diffraction peaks shifted to lower angles as increasing x.

Figure 3 illustrates the variation in the lattice parameters, (a, c), and tetragonality, c/a, as a function of the addition of ZnO nanoparticles, x, for BCZT/x ceramics sintered at 1350°C. As increasing x, constant (a) increases significantly, whereas constant (c) and tetragonality (c/a) reach their maximum values at x = 0.15 wt.%. It likely indicated that Zn^{2+}ions were incorporated into the BCZT lattices, and a stable solid solution was formed in the ceramics. However,Zn^{2+}ions did not change crystal symmetry of the materials that only varied the size of unit cells. Considering the radii of Ba^{2+}, Ca^{2+}, Ti^{4+}, Zr^{4+}, O^{2-}, and Zn^{2+} of 1.44, 1.34, 0.605, 0.79, 1.4, and 0.74 Å, respectively, Zn^{2+} is possibly substituted for the B-site at (Ti^{4+}, Zr^{4+}) positions within ABO_3perovskite structure that induced a lattice distortion in the BCZT/x ceramics. This result would cause the diffuse transition behavior in the materials and will be detailedly discussed in following section.

Surface morphologies of the BCZT/x ceramics sintered at 1350°C are shown in **Figure 4**.

It is evident that addition of nanostructured ZnO has strongly influenced the microstructure. Clean surfaces were observed for BCZT/ *x* samples with

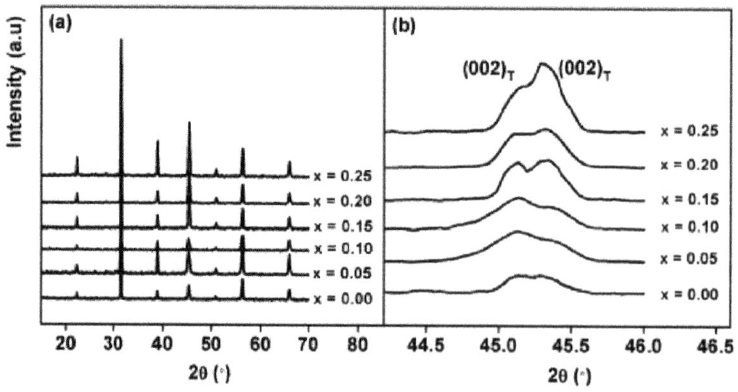

Figure 2.
XRD patterns (a) and expanded XRD patterns in the 2θ range of (44–46)° and (b) of the BCZT/x ceramics sintered at temperature of 1350°C.

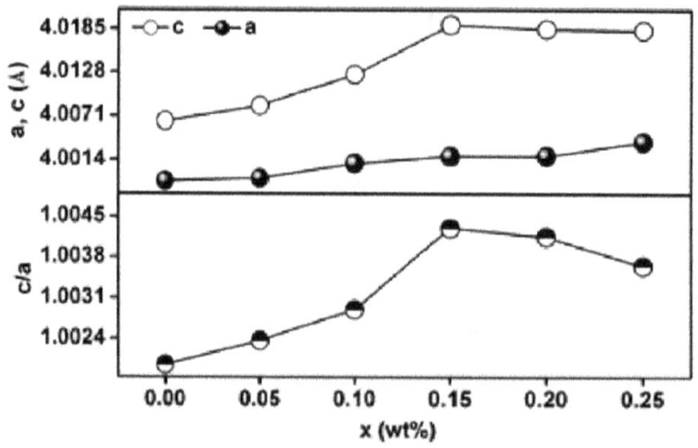

Figure 3.
Lattice parameters and tetragonality for the BCZT/x ceramics.

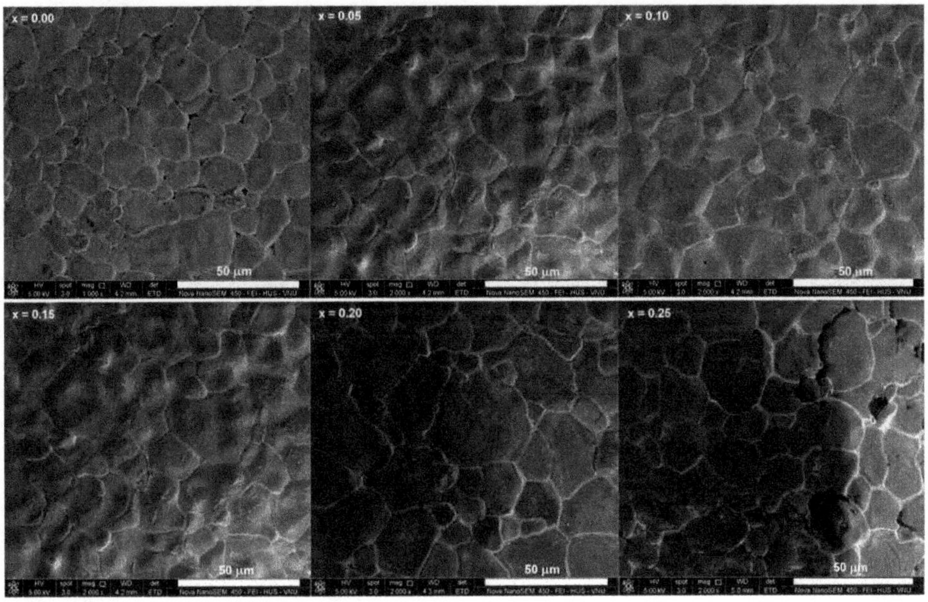

Figure 4.
SEM images of the BCZT/x ceramics sintered at 1350°C.

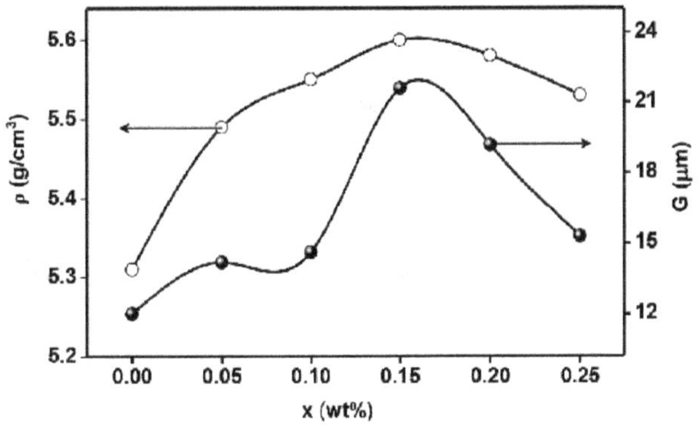

Figure 5.
Dependence of grain size, G, and density, ρ, on the ZnO nanoparticle content.

x = 0.00–0.15, and the grain size raised and reached the maximum value of 21.6μm at x = 0.15 compound. The liquid phase, however, appeared on the grain surface and boundary as x > 0.15. It may be an excess amount of nano-sized ZnO particles during sintering, accumulating at the surface and grain boundary to restrict grain size evolution. Therefore, the experimental results indicate that solubility limit of ZnO nanoparticles in BCZT substrate is below 0.15 wt.% at a sintered temperature of 1350°C. Dependence of grain size and density on the ZnO nanoparticle content is in the same manner (**Figure 5**). Thus, the BCZT/0.15 composition was expected to possess excellent electric properties.

Figure 6 illustrates temperature dependences of permittivity, ε (T), of the BCZT/x ceramics measured with frequency of 1 kHz at room temperature. It is clear to see that the samples with ZnO addition exhibit a typical temperature dependence of permittivity. A wide cubic-tetragonal phase transition at the temperature around 70°C was observed for all samples. Furthermore, another phase transition was found around 40°C for BCZT/0.00 composition (without ZnO nanoparticles) that is in the MPB region of BCZT and seems to be related to a tetragonal-rhombohedral

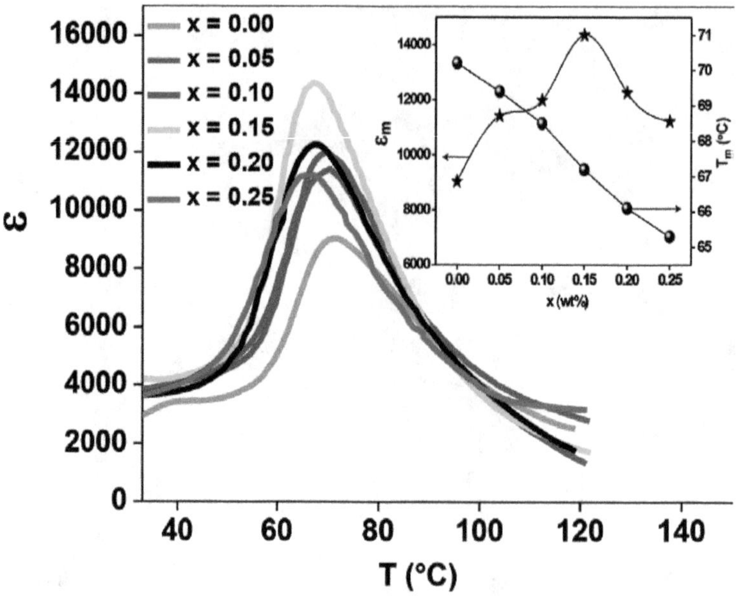

Figure 6.
Plot of permittivity versus temperature measured at 1 kHz for BCZT/x systems.

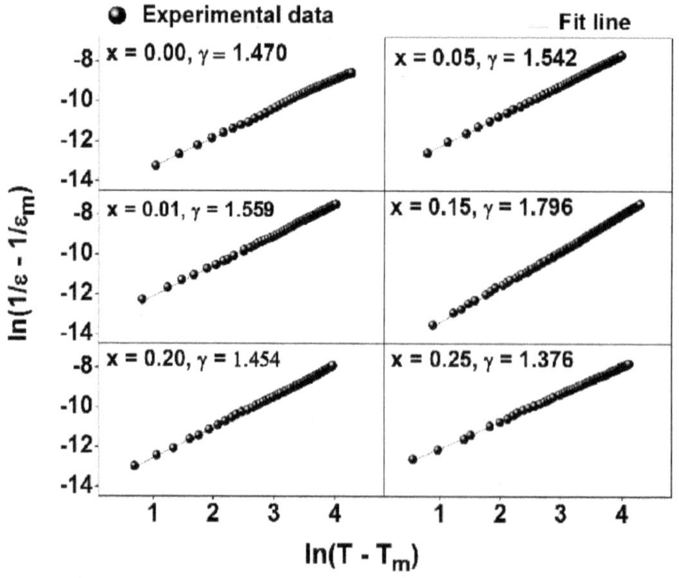

Figure 7.
$ln(1/\varepsilon - 1/\varepsilon_m)$ versus $ln(T - T_m)$ for BCZT/x systems.

phase transition as reported in the literatures [3, 21]. It is supposed that a part of the material has changed into rhombohedral phase with small amount so that this phase was not identified in X-ray patterns but can be observed in ε(T) curve. The mentioned phase transition was disappeared as raising content of ZnO nanoparticles. It may be shifted to lower temperature. It can be seen that ZnO nanoparticles have strongly affected dielectric properties. First, the shape of permittivity-temperature curves of the ZnO-added samples is broadened and shifted toward the lower-region. This is due to the lattice distortion and indicates the ferroelectric diffuse transitions as reported in the literature [22]. The highest permittivity of the samples, ε_m, nonlinearly depends on ZnO content. It increases with increasing ZnO content, reaches a maximal value of 14,361 for the BCZT/0.15 composition, and decreases monotonously after that. The temperature T_m corresponding to the maximum permittivity, ε_m, reduces with the increase of ZnO concentration due to the lattice distortion, as shown in XRD patterns. **Figure 7** presents the plots of $\ln(1/\varepsilon - 1/\varepsilon_m)$ versus $\ln(T - T_m)$ measured at 1 kHz for the BCZT/x ceramic that was fitted with modified Curie-Weiss law to obtain diffuseness degree parameter, γ. It can be observed that γ was changed as a function of x and reached the highest value of 1.796 at x = 0.15 composition.

As mentioned above, the diffuse characterization of BCZT/x ceramics may be a result of replacing Zn^{2+} for B-site ion (Ti^{4+}, Zr^{4+}). To put it more clearly, the room temperature Raman spectrum of the BCZT/x ceramics was recorded and analyzed (**Figure 8**). As shown, Raman modes of $BaTiO_3$-based systems are named as A_1(TO1), A_1(TO2), E(TO2), A_1(TO3), and A_1(LO3)/E(LO3) in the range of 150–1000 cm^{-1} [6]. The position and half-width of these modes were determined by fitting Raman data with Lorentzian function. E(TO2) vibration mode that has been associated with the tetragonal-cubic phase transition shifted to a lower wavenumber (**Figure 9(a)**). It means that substitution for B-site by Zn^{2+} results in reducing average B-O bonding energies. Thus, tetragonal-cubic

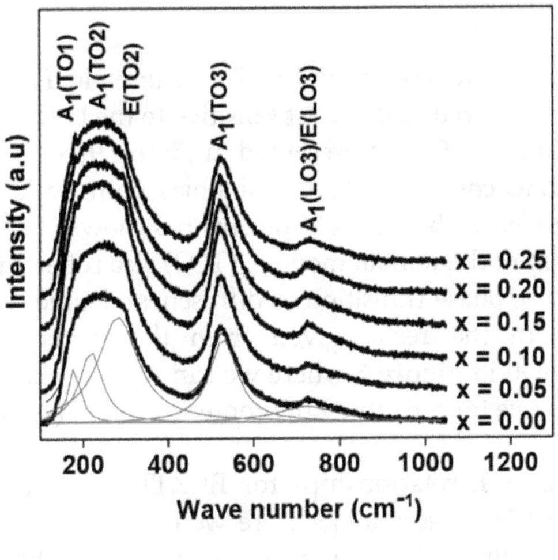

Figure 8.
Raman spectrum of BCZT/x systems recorded at room temperature.

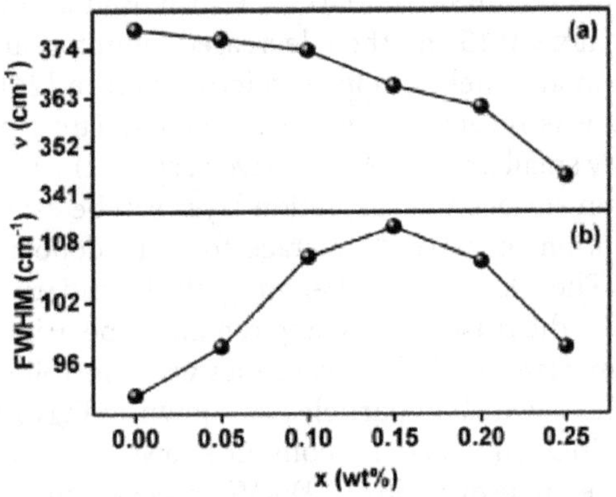

Figure 9.
(a) Raman shift, ν, of E(TO2) and (b) half-width, FWHM, of A_1(TO2) as a function of ZnO nanoparticles, x.

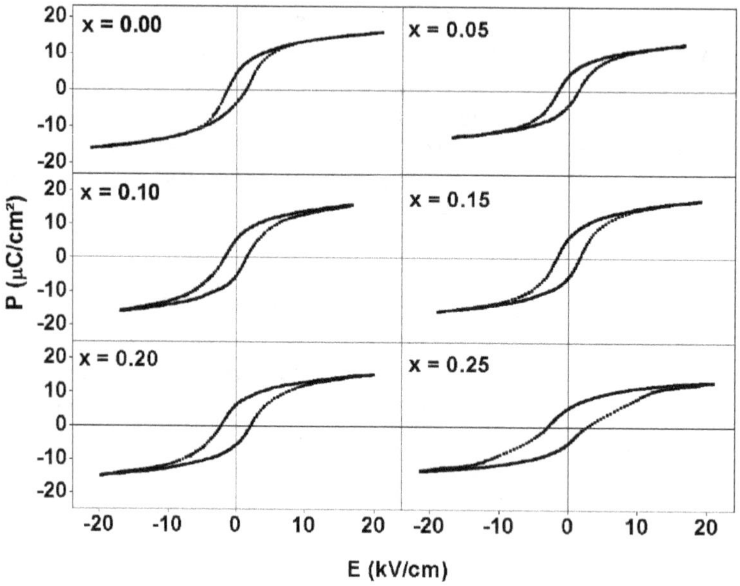

Figure 10.
Ferroelectric (P-E) hysteresis loops of BCZT/x systems.

phase transition temperature was diminished (see inset in **Figure 6**). The A_1(TO2) vibration mode was considered as the most sensitive to the lattice distortion that can be evaluated by FWHM of A_1(TO2) as presented in **Figure 9(b)**. It has been shown that FWHM depends on ZnO content and reached a maximum value at x = 0.15. It means the material sample added with 0.15 wt.% of ZnO has shown the highest disorder [23]. In other words, FWHM of the Raman mode A_1(TO2) also reflects the diffuseness of the ferroelectric-paraelectric phase transition as diffuseness degree. This obtained result is well appropriated with the result given from the temperature dependence of permittivity as presented in **Figure 6** where we can see that the diffuseness degree, γ, reached a maximum value for the same ZnO content x = 0.15 (see **Figure 7**).

Figure 10 presents P-E relationships for BCZT/x ceramics measured at room temperature. Received hysteresis loops were well saturated and fairly slim for all samples that assert again diffuse ferroelectric nature in BCZT/x ceramics. The characterized values of remanent polarization, P_r, and coercive field, E_c, depended on ZnO nanoparticles concentration as illustrated in **Figure 11**. As shown, when x content varied in the region of 0.00–0.25, P_r increased and reached a maximum value of 6.19 μC/cm^2 at x = 0.15 and then decreased monotonously. This result could be explained based on an amelioration in microstructure [24]. According to that, poor ferroelectricity was received at grain boundary. Thus, polarization of grain boundary may be very small or zero. Alternatively, space charges eliminate polarization charge from grain surface that depletion layer can be established. That caused polarization interruption on particle surface to form depolarization field which lowers polarization. The reduced number of grain boundary is due to increasing grain size that could be the reason for raising remanent polarization and vice versa. In this study, the grain size, G, of these ceramics was controlled by varying doping concentration of nano-sized ZnO particles as shown in **Figure 5**. The dependence of coercive field on ZnO nanoparticles content shows that the parameter continuously intensified in the range of (1.36–2.72) kV/cm as raising x. In other words, ZnO nanoparticles made the ceramics harder. The enhancement of $E_c$value could be due to the increase of charged oxygen vacancies as doping that pinned to the movement

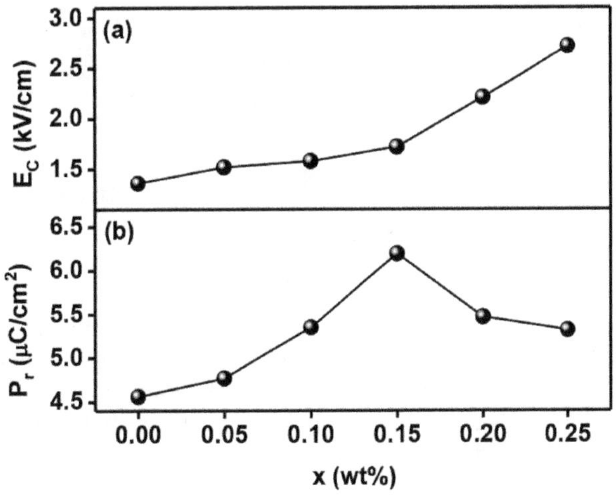

Figure 11.
Values of E_c and P_r as a function of x.

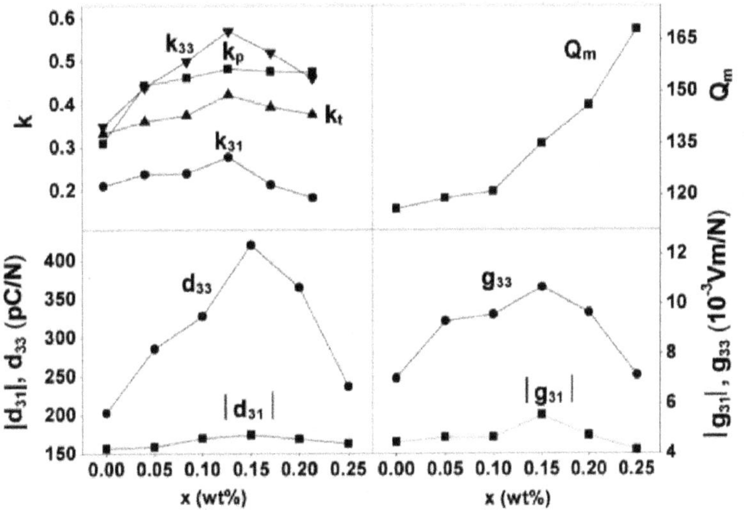

Figure 12.
Nano-sized ZnO content dependence of some piezo-parameters for BCZT/x ceramics.

of ferroelectric domain walls. The obtained values of E_c demonstrated that the BCZT/x materials are typically soft compared to electric properties.

Figure 12 displays the electromechanical coupling factor (k), piezoelectric constant (d_{33}, d_{31}, g_{33}, g_{31}), and mechanical quality factor (Q_m) with various amounts of ZnO nanoparticles.

It can be observed that k, g_{33}, g_{31}, d_{33}, and d_{31} curves possess maximum values for the BCZT/0.15 compound as presented in **Table 1**.

As mentioned above, the comprehensive analysis of X-ray diffraction, SEM images, and dielectric properties have proven the ZnO addition induced lattice distortion. The degree of this local lattice distortion increased up to the maximal value as ZnO concentration raised up to 0.15 wt.%. It is supposed the spontaneous polarization in each nano-domain has contributed to overall spontaneous polarization that enhanced piezoelectric characteristics of the material samples. Beyond the value of 0.15, the piezoelectric parameters decrease due to residual amount of ZnO nanoparticles agglomerating at the surface and grain boundary restricting

k_p	k_t	k_{31}	k_{33}	d_{33} (pC/N)	d_{31} (pC/N)	g_{33} (10^{-3} V m/N)	g_{31} (10^{-3} V m/N)
0.483	0.423	0.278	0.571	420	−174	10.68	−5.55

Table 1.
Values of coupling factors (k_p, k_t, k_{31}, k_{33}) and piezoelectric constants (d_{31}, d_{33}, g_{31}, g_{33}) of BCZT/0.15 sample sintered at 1350°C.

grain size growth. According to Ying-Chieh Lee et al., $Zn^{2+}$is substituted into the B-site to generate a doubly oxygen vacancy for compensation [12]. The presence of charged oxygen vacancies would be pinned by the movement of ferroelectric domain walls and consequently to enhance the $Q_m$value. The result is similar to that reported by Jiagang Wu et al., who used micro-size ZnO as an acceptor for the $Ba_{0.85}Ca_{0.15}Ti_{0.90}Zr_{0.10}O_3$ ceramic sintered at 1450°C [25].

3.2 Influence of sintering temperature on structure, microstructure, and piezoelectric properties of doped BCZT ceramics

In this section, effects of sintering temperature on the structure, microstructure, and piezoelectric properties of 0.15 wt.% modified BCZT or BCZT/0.15 compound are presented.

Figure 13 shows XRD patterns of BCZT/0.15 ceramic sintered at various temperatures of 1300, 1350, 1400, and 1450°C. All samples exhibited a pure perovskite phase, and no second-phase trace was detected in the investigated region. To study the effect of the sintering temperatures on the structure of BCZT/0.15 material, enlarged XRD patterns in the range of (44–46)° corresponded to each sintering temperature were analyzed by fitting XRD data with Gaussian function (inset in **Figure 13**). As shown, there were splitting (200)/(002) diffraction peaks at around 2θ of 45° for specimens sintered at 1300, 1350, and 1400°C, which means these samples have tetragonal symmetry. However, coexistence of rhombohedral and tetragonal phases was observed for the sample sintered at 1450°C in which tetragonal volume fraction, calculated by using Eq. (1), was 67.3%. In other words, BCZT/0.15 sample sintered at 1450°C could be MPB composition. Mixture of tetragonal-rhombohedral phases was also observed for BCZT/0.00 specimen (without ZnO nanoparticles) sintered at 1450°C (**Figure 14**) in which tetragonal volume

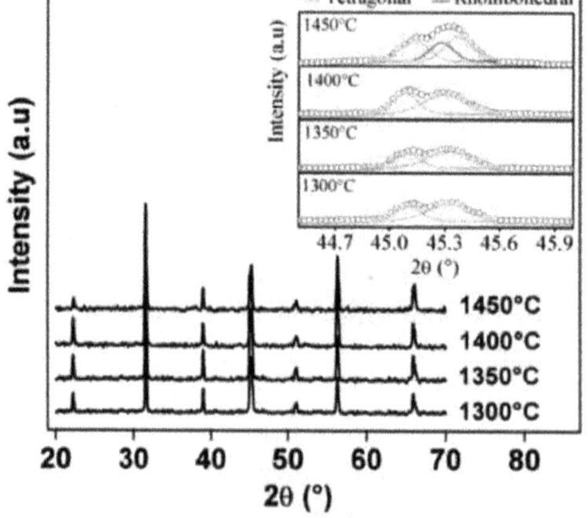

Figure 13.
XRD patterns of BCZT/0.15 sintered at different temperatures.

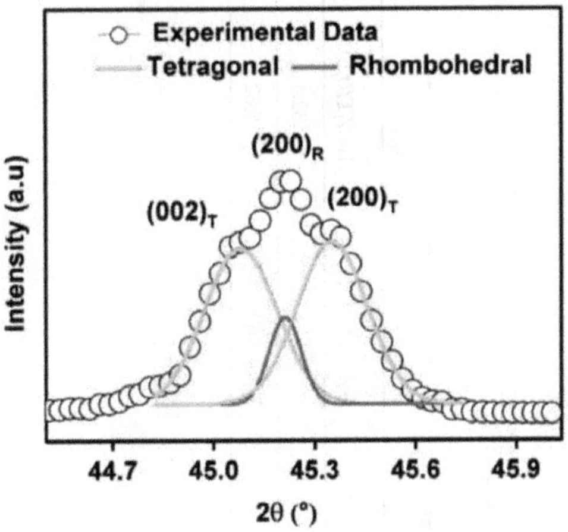

Figure 14.
Expanded XRD in the region of (44–46)° for BCZT/0.00 composition [26].

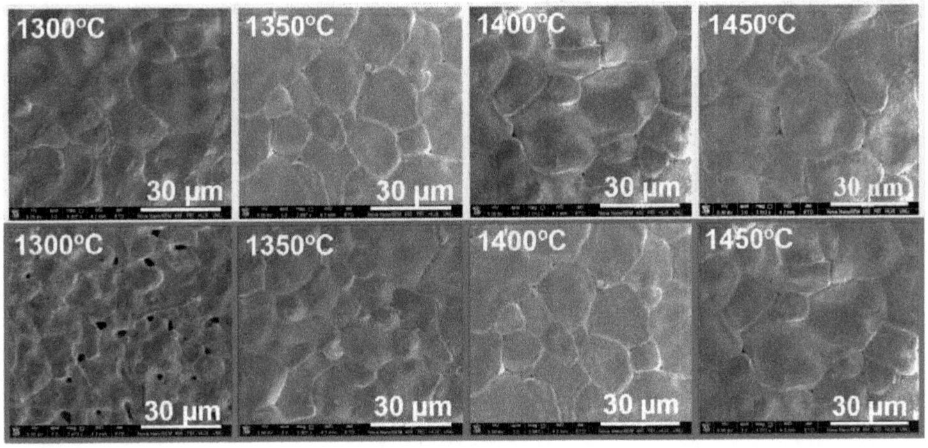

Figure 15.
SEM images of BCZT/0.15 (yellow border) and BCZT/0.00 (red border) ceramics sintered at various temperatures.

fraction was 71.7% [26]. It could be concluded that ZnO nanoparticle content of 0.15 wt.% did not change crystalline symmetry but only varied the fraction of phases in the samples. This could affect the piezoelectric properties of BCZT/0.00 and BCZT/0.15 samples sintered at 1450°C.

Figure 15 illustrates surface morphologies of BCZT/0.00 and BCZT/0.15 materials sintered at various temperatures. Grain size and density were calculated and listed in **Table 2**.

Porous microstructure with small grains was received for BCZT/0.00 ceramic sintered at 1300°C (**Figure 15**). However, when 0.15 wt.% of ZnO nanoparticle content was added, a denser microstructure with larger particles is viewed at the same sintering temperature. Moreover, both grain size and density of BCZT/0.00 and BCZT/0.15 samples were raised as sintering temperature increased (**Figure 16**). Especially, grain size and density of BCZT/0.15 ceramics are larger than that of BCZT/0.00 ceramics for each sintering temperature. It could be concluded that a small amount of ZnO nanoparticles can be improved sintering behavior.

Figure 17 presents the comparison of the piezoelectric parameters for BCZT/0.00 and BCZT/0.15 materials. As the sintering temperature rises, the

T (°C)	1300		1350		1400		1450	
Material	**BCZT/0.15**	**BCZT/0.00**	**BCZT/0.15**	**BCZT/0.00**	**BCZT/0.15**	**BCZT/0.00**	**BCZT/0.15**	**BCZT/0.00**
G (μm)	18.2	9.7	21.6	12.1	31.4	28.5	34.6	32.4
ρ (g/cm³)	5.52	5.01	5.60	5.31	5.64	5.57	5.69	5.62

Table 2.
Grain size, G, and density, ρ, for BCZT and BCZT/0.15 ceramics sintered at different temperatures.

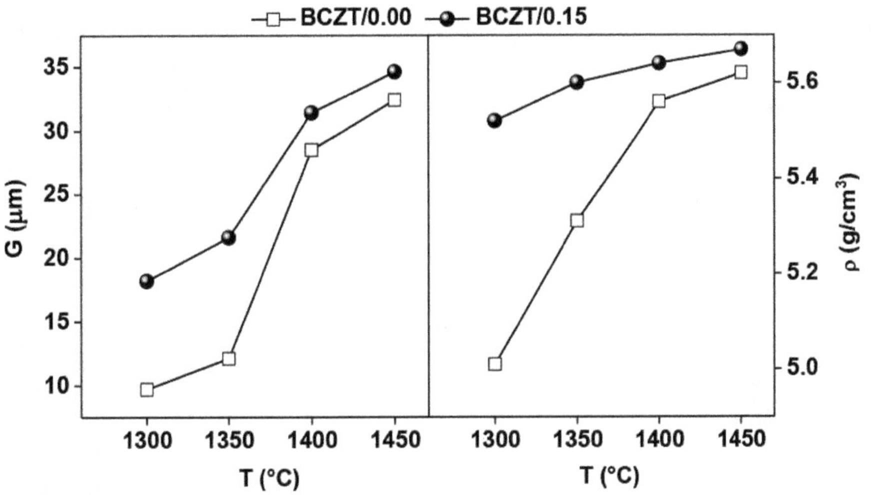

Figure 16.
Grain size and density of BCZT/0.00 and BCZT/0.15 materials as a function sintering temperature.

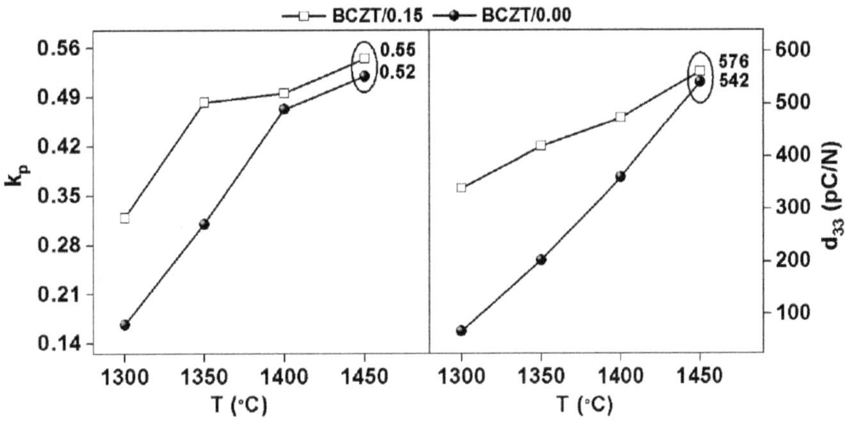

Figure 17.
Electromechanical factor, k_p, and piezoelectric coefficient, d_{33}, of BCZT/0.00 and BCZT/0.15 materials as a function sintering temperature.

piezoelectric parameters of both samples increase. Moreover, the piezoelectric parameters of BCZT/0.15 samples are higher than the ones of BCZT materials. These results may be due to the improvement in microstructure of the materials.

For the materials sintered at 1450°C, d_{33} = 576 pC/N and k_p = 0.55 were obtained for BCZT/0.15 system, whereas these parameters for BCZT/0.00 system were 542 pC/N and 0.52, respectively [9]. As known, in the MPB region, there are 14 possible polar directions with the low potential energy barrier that includes 6 in tetragonal phase and 8 in rhombohedral phase [27]. These polar directions are the optimum orientation during polarization process leading to the excellent piezoelectric properties. It could be supposed that there are only tetragonal and rhombohedral phases in two mentioning compositions and then their volume fractions are quantified as in **Table 3**. It can be seen that the nanostructured ZnO addition has heightened concentration of rhombohedral phase leading to enhance directional ability for polarization vectors. It is believed that it was the reason for the higher piezoelectric properties of BCZT/0.15 material than the BCZT/0.00 one. In other words, a competition between

Material	Tetragonal volume fraction	Rhombohedral volume fraction
BCZT/0.15	63.7%	36.3%
BCZT/0.00	71.7%	29.3%

Table 3.
Volume fraction of tetragonal and rhombohedral phases for BCZT and BCZT/0.15 materials sintered at 1450°C calculated by Eqs. (1) and (2).

structure phases with addition of ZnO nanoparticles could induce the difference in piezoelectric response for BCZT/0.00 and BCZT/0.15 samples.

4. Conclusion

The addition of ZnO nanoparticles with grain size of 59 nm has aided to successfully synthesize the BCZT/x ceramics at a relatively low sintering temperature of 1350°C. The added ZnO particles in nanoscale influenced the relaxor ferroelectric phase change of the materials. As a result, BCZT/0.15 composition possessed the highest diffuseness characteristic. Remanent polarization was improved and reached to a maximum value of 6.19 μC/cm^2 at x = 0.15, whereas the coercive field went up continuously under increasing doping concentration. The ZnO addition has also improved the quality of the piezoelectric material, and best quality was observed for BCZT/0.15 composition, given that the values of d_{33}, d_{31}, k_p, k_t, and k_{33} are 420, −174, 0.483, 0.423, and 0.571 pC/N, respectively. The obtained results suggested that the lead-free BCZT/x material could be an expected lead-free piezoelectric ceramic for applications.

Besides, the influence of sintering temperature on structure, microstructure, and some piezoelectric parameters of BCZT/0.15 sample was examined. As the sintering temperature increased, improved sintering behavior and very high piezoelectric properties of d_{33} = 576 pC/N and k_p = 0.55 were obtained for the sample sintered at 1450°C. As a comparison, corresponded properties of BCZT without ZnO nanoparticles or BCZT/0.00 specimen were investigated. The received results show that sintering behavior and some piezo-parameters of BCZT/0.15 samples are better than that of BCZT/0.00 samples at each sintering temperature. Especially, the difference in properties for samples sintered at 1450°C is attributed to competition between structure phases occurred in materials.

Acknowledgements

This work was carried out in the framework of the National Project in Physics Program until 2020 under no. ĐTĐLCN.10/18.

Author details

Dang Anh Tuan[1], Vo Thanh Tung[2*], Le Tran Uyen Tu[2] and Truong Van Chuong[2]

1 Ha Nam Provincial Department of Science and Technology, Vietnam

2 University of Sciences, Hue University, Vietnam

*Address all correspondence to: vttung@hueuni.edu.vn

References

[1] Panda PK. Review: Environmental friendly lead-free piezoelectric materials. Journal of Materials Science. 2009;**44**:5049-5062. DOI: 10.1007/s10853-009-3643-0

[2] Rodel J, Wook J, Seifert KTP, Eva-Maria A, Granzow T, Damjanovic D. Perspective on the development of lead free piezoceramics. Journal of the American Ceramic Society. 2009;**92**(6):1153-1177. DOI: 10.1111/j.1551-2916.2009.03061.x

[3] Wenfeng L, Xiaobing R. Large piezoelectric effect in Pb-free ceramics. Physical Review Letters. 2009;**103**:257602-257605. DOI: 10.1103/PhysRevLett.103.257602

[4] Xue D, Zhou Y, Bao H, Zhou C, Gao J, et al. Elastic, piezoelectric, and dielectric properties of Ba $(Zr_{0.2}Ti_{0.8})$ O_3-50 $(Ba_{0.7}Ca_{0.3})TiO_3$ Pb- free ceramic at the morphotropic phase boundary. Journal of Applied Physics. 2011;**109**:054110. DOI: 10.1063/1.3549173

[5] Damjanovic D, Biancoli A, Batooli L, Vahabzadeh A, Trodahl J. Elastic, dielectric, and piezoelectric anomalies and Raman spectroscopy of $0.5Ba(Ti_{0.8}Zr_{0.2})$ O_3-$0.5(Ba_{0.7}Ca_{0.3})TiO_3$. Applied Physics Letters. 2012;**100**:192907. DOI:10.1063/1.4714703

[6] Wu J, Xiao D, Wu W, Chen Q, Zhu J, Yang Z, et al. Composition and poling condition-induced electrical behavior of $(Ba_{0.85}Ca_{0.15})(Ti_{1-x}Zr_x)$ O_3 lead-free piezoelectric ceramics. Journal of the European Ceramic Society. 2012;**32**:891-898. DOI: 10.1016/j. jeurceramsoc.2011.11.003

[7] Wang P, Li Y, Lu Y. Enhanced piezoelectric properties of $(Ba_{0.85}Ca_{0.15})$ $(Ti_{0.9}Zr_{0.1})O_3$ lead-free ceramics by optimizing calcinations and sintering temperature. Journal of the European Ceramic Society. 2011;**31**:2005-2012. DOI: 10.1016/j.jeurceramsoc.2011.04.023

[8] Su S, Zuo R, Lu S, Xu Z, Wang X, Li L. Poling dependence and stability of piezoelectric properties of $Ba(Zr_{0.2}Ti_{0.8})$ O_3-$(Ba_{0.7}Ca_{0.3})TiO_3$ ceramics with huge piezoelectric coefficients. Current Applied Physics. 2011;**11**:S120-S123. DOI: 10.1016/j.cap.2011.01.034

[9] Tuan DA, Tinh NT, Tung VT, Chuong TV. Ferroelectric and piezoelectric properties of lead-free BCT- xBZT solid solutions. Materials Transactions. 2015;**56**(9):1370-1373. DOI: 10.2320/matertrans.MA201511

[10] Dung QTL, Chuong VT, Anh PD. The effect of TiO_2 nanotubes on the sintering behavior and properties of PZT ceramics. Advances in Natural Sciences: Nanoscience and Nanotechnology. 2011;**2**:025013. DOI: 10.1088/2043-6262/2/2/025013

[11] Hayati R, Barzegar A. Microstructure and electrical properties of lead free potassium sodium niobate piezoceramics with nano additive. Materials Science and Engineering B. 2010;**172**:121-126. DOI: 10.1016/j. mseb.2010.04.033

[12] Ying-Chieh L, Tai-Kuang L, Jhen-Hau J. Piezoelectric properties and microstructures of ZnO-doped $Bi_{0.5}Na_{0.5}TiO_3$ ceramics. Journal of the European Ceramic Society. 2011;**31**:3145-3152. DOI: 10.1016/j.jeurceramsoc.2011.05.010

[13] Ramajo LA, Taub J, Castro MS. Effect of ZnO addition on the structure, microstructure and dielectric and piezoelectric properties of $K_{0.5}Na_{0.5}NbO_3$ ceramics. Materials Research. 2014;**17**(3):728-733. DOI: 10.1590/S1516-14392014005000048

[14] Saeri MR, Barzegar A, Moghadama HA. Investigation of nano particle additives on lithium doped KNN lead free piezoelectric ceramics. Ceramics International. 2011;**37**(8):3083-3087. DOI: 10.1016/j. ceramint.2011.05.044

[15] Bari AR et al. Effect of solvent on the particle morphology of nanostructured ZnO. Indian Journal of Pure and Applied Physics. 2009;**47**:24-27

[16] Yingying L, Leshu Y, Heyong H, Yuying F, Dongzhen C, Xin X. Application of the soluble salt-assisted route to scalable synthesis of ZnO nanopowder with repeated photocatalytic activity. Nanotechnology. 2012;**23**(6):065402-065409. DOI:10.1088 /0957-4484/23/6/065402

[17] Quintana-Nedelcos A, Fundora A, Amorín H, Siqueiros JM. Effects of Mg addition on phase transition and dielectric properties of $Ba(Zr_{0.05}Ti_{0.95})O_3$ system. The Open Condensed Matter Physics Journal. 2009;**2**:1-8. DOI: 10.2174/1874186X00902010001

[18] Kraus W, Nolze G. POWDER CELL - a program for the representation and manipulation of crystal structures and calculation of the resulting X-ray powder patterns. Journal of Applied Crystallography. 1996;**29**:301-303. DOI: 10.1107/S0021889895014920

[19] Tuan DA, Tung VT, Phuong LV. Analyzing 2D structure images of piezoelectric ceramics using ImageJ. International Journal of Materials and Chemistry. 2014;**4**(4): 88-91. DOI: 10.5923/j.ijmc.20140404.02

[20] The Institute of Electrical and Electronics Engineers, Inc. IEEE Standard on Piezoelectricity, ANSI/ IEEE Std 176-1987

[21] Gao J, Xue D, Wang Y, Wang D, Zhang L, Wu H, et al. Microstructure basis for strong piezoelectricity in Pb-free $Ba(Zr_{0.2}Ti_{0.8})O_3$-$(Ba_{0.7}Ca_{0.3})TiO_3$ ceramics. Applied Physics Letters. 2011;**99**:092901. DOI: 10.1063/1.3629784

[22] Hao J, Bai W, Li W, Zhai J. Correlation between the microstructure and electrical properties in high-performance $(Ba_{0.85}Ca_{0.15})(Zr_{0.1}Ti_{0.9})O_3$ lead-free piezoelectric ceramics. Journal of the American Ceramic Society. 2012;**95**(6):1998. DOI: 10.1111/ j.1551-2916.2012.05146.x

[23] Dobal PS, Dixit A, Katiyar RS. Effect of lanthanum substitution on the Raman spectra of barium titanate thin films. Journal of Raman Spectroscopy. 2007;**38**:142-146. DOI: 10.1002/jrs.1600

[24] Mudinepalli VR, Feng L, Wen-Chin L, Murty BS. Effect of grain size on dielectric and ferroelectric properties of nanostructured $Ba_{0.8}Sr_{0.2}TiO_3$ ceramics. Journal of Advanced Ceramics. 2015;**4**(1):46-53. DOI: 10.1007/s40145-015-0130-8

[25] Wu J, Xiao D, Wu W, Chen Q, Zhu J, Yang Z, et al. Role of room-temperature phase transition in the electrical properties of (Ba, Ca)(Ti, Zr)O_3 ceramics. Scripta Materialia. 2011;**65**:771-774. DOI: 10.1016/j. scriptamat.2011.07.028

[26] Tuan AD, Tung TV, Chuong VT, Tinh TN, Huong TMN. Structure, microstructure and dielectric properties of lead-free BCT-xBZT ceramics near the morphotropic phase boundary. Indian Journal of Pure and Applied Physics. 2015;**53**:409-415

[27] Shrout TR, Zhang SJ. Lead-free piezoelectric ceramics: Alternatives for PZT? Journal of Electroceramics. 2007;**19**:111-124. DOI: 10.1007/ s10832-007-9047-0

5

Perovskite Nanoparticles

Burak Gultekin, Ali Kemal Havare, Shirin Siyahjani, Halil Ibrahim Ciftci and Mustafa Can

Abstract

2D perovskite nanoparticles have a great potential for using in optoelectronic devices such as Solar Cells and Light Emitting Diodes within their tuneable optic and structural properties. In this chapter, it is aimed to express "relation between chemical structures and photo-physical behaviours of perovskite nanoparticles and milestones for their electronic applications". Initially, general synthesis methods of perovskite nanoparticles have been explained. Furthermore, advantages and disadvantages of the methods have been discussed. After the synthesis, formation of 2D perovskite crystal and effects on shape factor, particle size and uniformity of perovskite have been explained in detail. Beside these, optic properties of luminescent perovskite nanoparticles have been summarized a long with spectral band tuning via size and composition changes. In addition, since their different optical properties and relatively more stable chemical structure under ambient conditions, a comprehensive compilation of opto-electronic applications of 2D perovskite nanoparticles have been prepared.

Keywords: perovskite, nanoparticle, nanocrystal, opto-electronics, solar cells, OLEDs, fluorescence

1. Introduction

Inorganic and/or hybrid perovskites have become prominent in solution-processed optoelectronics. Thousands of reports have been proposed about photovoltaics [1–3], light-emitting diodes (LEDs) [4–6], lasers [6, 7] and photodetectors [8–12] containing perovskite semiconductors per annum over the past decade. Halide perovskite nanostructures exhibit tremendous optic and electrical properties such as strong absorption and/or emission, higher photoluminescence quantum yields (PLQYs) [13–15], higher exciton binding energies and tuneable bandgaps than those of bulk perovskites and other nanomaterials. With these outstanding properties, they may offer many scopes for optoelectronics. Particularly, hybrid lead halide perovskite nanostructures (PNSs) offer unique opportunities for light-emitting diode (LED) applications. PNSs represent larger exciton binding energies and longer carrier decay times than those of bulk crystals. In addition to this, their narrow emission band makes them good candidates for LED [16] and laser [17]. Since Schmidt et al. reported the synthesis of first colloidal PNS by a simple procedure under ambient conditions in 2014 [18], many research groups have started working on these materials. In the synthesis of PNSs, it is possible to use organic ligands for capping to stop crystals` growth in the nanometer scale. Furthermore, capping ligands can reduce surface defects in the same way of the traditional NC preparation. Thus, size and shape of PNSs can be tune finely from a single perovskite layer or below the exciton Bohr radius for using the effect to quantum confinement to multilayers which exhibits bulk-like properties. With this method, it is possible to prepare nanostructures like quantum dots (QDs), nanoplatelets (NPLs), nanosheets (NSs), and nanowires (NWs) [19, 20].

In a simple PNS growth process, the methylammonium cations are embedded in the voids of the corner-sharing PbX6 octahedra, the long alkyl chain cations only at the periphery of the octahedra with their chains hanging it. Therefore, long alkyl-ammonium ions can be used as the capping agents to limit the growth of the PNSs in three-dimension (3D). There are five common methods proposed in the literature which exhibits good prospects for obtaining various uniform and defect free PNSs. These are named as solvent-induced precipitation, hot injection, template assisted, ligand assisted reprecipitation and emulsion methods [13, 14].

In order to determine the structural and optic properties of PNSs, some characterization techniques such as small angle x-ray scattering (SAXS), scanning electron microscope (SEM), absorption and emission spectroscopy have been used extensively. It is crucial to know the shape and optical response of the nanocrystals for the further opto-electronic application of them.

As it is mentioned above, well-defined PNSs were used in many applications such as optical lasing and LEDs. Some of those applications have been investigated intensively by many groups all around the world. Some milestone studies have been presented comparatively.

2. Synthesis of nano-crystalline perovskites

There are many proposed methods for synthesis of PNSs by many groups in the literatures. Here we would like to express five of most common and essential methods (**Figure 1**) [21–23].

2.1 Solvent-induced precipitation

The physical properties of perovskite cluster such as size of NPs can be arranged by using long alkyl chain amine derivatives while oleic acid ensures the colloid stability via preventing the aggregation. In order to initiate the solvent-induced precipitation, and obtain colloidal $MAPbBr_3$ NPs, lead bromide ($PbBr_2$) and methyl ammonium bromide (CH_3NH_3Br) were mixed withOctylammoniumbromide (OABr) in acetone with oleic acid (OAc) and octylamine and the solution were kept at 80°C. The PLQY of obtained NPSs was about 20% as well as stable over three months. After this first attempt, PLQY of NPSs was increased up to 83% by optimization of the molar ratios of starting materials [22, 24].

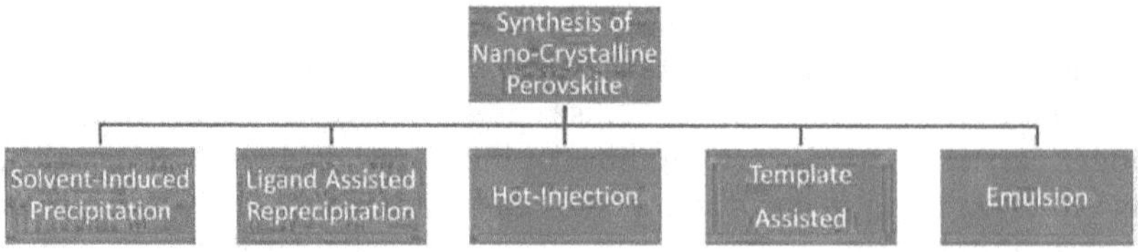

Figure 1.
Common synthesis methods for perovskite nano-crystals.

2.2 Ligand assisted reprecipitation (LARP) technique

In this method, a polar solvent such as dimethylformamide (DMF) which dissolve all starting materials and capping ligands have been used to prepare the precursor solution [25, 26]. This solution is added dropwise into vigorously stirred toluene which is not a good solvent for starting materials and perovskite crystal. Zhang et al. has demonstrated the synthesis of colour- tuneable PNSs (average particle size of 3.3 ± 0.7 nm) with a PLQYs of 50–70% (**Figure 2**) [25]. It was claimed that a slight descending of PLQY was observed by the ascending of the size of the perovskite crystal (2–8 μm, PLQY < 0.1%). In the nanometer scales, surface defects of the crystals can easily be passivated by ligands. Thus, most of the photo-generated charges can recombine before there are trapped by defects on the surface. However, due to the less ligand passivation, the number of defects on the bulk perovskite structure`s surface, which increased the number of trapped charges, are significantly high, resulting very low PLQY. In another study, well-defined cubic and thermally stable $FAPbX_3$ nanocrystals (about 10 m) has been prepared by LARP method. The reported PLQY for the NPS was 75% [27]. A new procedure, which was used to obtain a core-shell shape by using the LARP approach has been recently demonstrated. With this proposed method, a solution-processed, stable core–shell-type Methyl ammonium (MA^+) + Octyl ammonium (OA^+) lead bromide perovskite NPs (≈5–12 nm) with good PLQY was prepared. In addition to this, Core–shell-type NPs was accomplished by systematically changing the molar ratio of capping ligands, OABr, and MABr without altering total amount of alkylammonium bromide and synthesis conditions. The color tunability of NPs in the blue to green spectral region (438–521 nm), high PLQY, and reasonable stability under ambient condition are credited to the quantum confinement imparted by the crystal engineering associated with coreshell NP formation [28–30].

2.3 Hot injection method

The stability of inorganic perovskites is significantly higher than that of hybrid perovskite crystals. By changing the organic cation with an inorganic one (e.g., Cs),

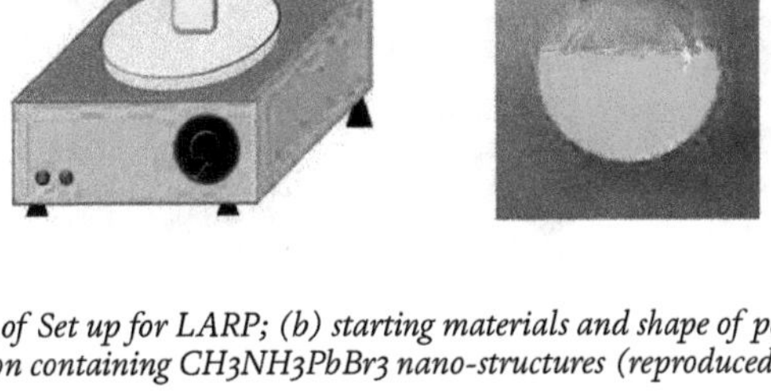

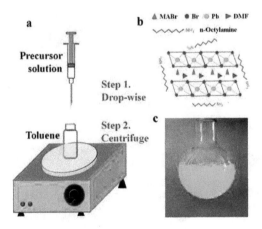

Figure 2.
(a) Schematic illustration of Set up for LARP; (b) starting materials and shape of perovskite nano-crystals; (c) image of typical solution containing CH3NH3PbBr3 nano-structures (reproduced with permission of Ref. [26]).

chemical and thermal stability of Perovskite material can be increased greatly (**Figure 3**) [31, 32]. It is possible to obtain various crystal phases by changing the temperature resulting a shifting in the optical responses. The hot injection method is a widespread method for synthesizing inorganic PNSs. In a typical synthesis procedure, a solution of PbX_2 (X = I/Br/Cl), in octadecene (ODE) along with oleic acid and oleyl amine is prepared. During stirring, Cs-oleate is injected into that solution quickly under dry condition at 140–200°C [32]. To quench the reaction (after 5–10 sec), the reactor is cooled with an ice bath. With this method, it is easy to obtain uniform nanocubes with the size of 4–15 nm edge length representing high PLQY, up to 50–90% and a very narrow emission band (12–42 nm) in the visible region (410–700 nm). As it was mentioned before, the optical properties of the NPSs are related to shape, size,

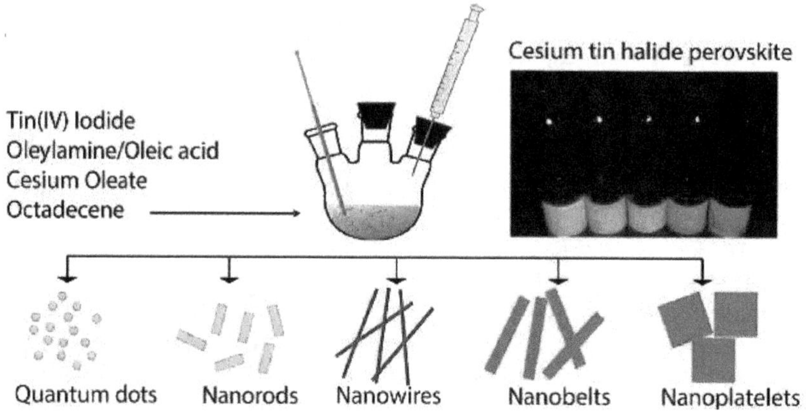

Figure 3.
Schematic illustration of Hot-injection method (left-hand), image of Cs2SnI6 samples under UV-light (right-hand) and possible crystal shapes obtained by hot-injection method (reproduced with permission of Ref. [33]).

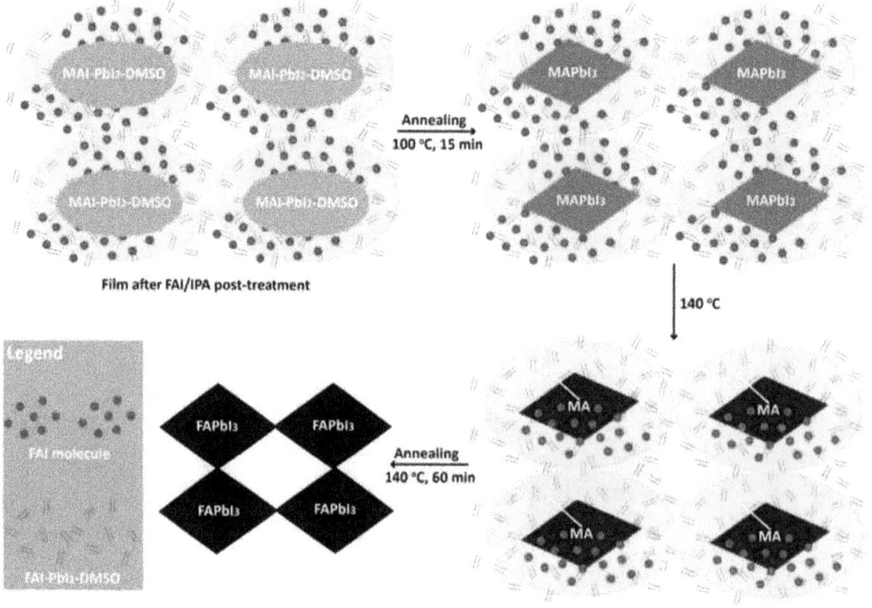

Figure 4.
Schematic illustration of perovskite nano-crystal preparation by the template-assisted method (reproduced with permission of Ref. [36]).

and surface chemistry of them due to the significant changes in the band structures. Therefore, by the variation of the surfactants, ligands, reaction temperature and time, the perovskite precursor composition can be adjust [33] to realize the formation of nanowires, nanoplatelets, spherical dots, and nanorods (**Figure 3**) [33–35]. A critical point for the hot injection is the reaction temperature effecting the size of the NPSs. To overcome this disadvantage, new methods have been developed for hybrid perovskites.

2.4 Template-assisted method

In this synthesis approach, NP formation is induced by a specific substrate such as mesoporous silica and aluminum oxide film as a template [36–38]. NPSs is obtained on the films which exhibits intensive light. This technique is suitable for formation of mono disperse NPSs with various narrow emissions (full wind of half maximum (FWHM) is less than 40 nm) bands from green (FWHM: 22 nm) to near infrared (FWHM: 36 nm) by template optimization [37–39]. In a recent study, perovskite nanocrystals have been prepared in a nano-porous structure. By this kind of strategies, it is possible to confine PNSs (<10 nm) without any capping agents (**Figure 4**). In other words, the emission wavelength of perovskite nanoparticles can be arranged precisely for sophisticated photonic applications such as lasers [36, 37].

2.5 Emulsion method

In order to control the crystallization of perovskite and obtain uniform NPs, emulsion synthesis method was modified [40]. By using this method, it is possible to tune the size of PNSs under 10 nm with an PLQY up to 92% [41]. In this method, an emulsion is prepared with two immiscible solvents. After that, a demulsifier is added into this for initiating solvent mixing and start crystallization (**Figure 5**). For this procedure, DMF and n-hexane are very good candidates as immiscible solvents while tertbutanol or acetone is used as demulsifier solvent [41]. This method is suitable to obtain PNSs in solid state which can be used in another solvent matrix for an application later. Furthermore, long alkyl ammonium halides are used as capping agent.

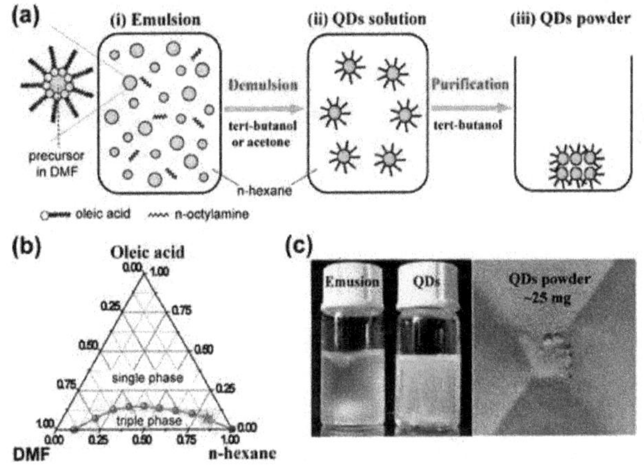

Figure 5.
(a) Schematic illustration of preparation of nano-crystals by emulsion method; (b) ternary phase diagram (DMF/OA/n-hexane); (c) image of a typical CH3NH3PbBr3 emulsion, colloidal NP solution and solid-state powder of CH3NH3PbBr3 NPs (reproduced with permission of Ref. [41]).

3. Form factor of perovskite nano-crystals

3.1 Crystal shape of perovskite nano structures

As it is well known, semiconductor perovskite crystalline structure has a general formula of $ABX_3$where A is an organic or inorganic cation, organic methylammonium (MA^+), formamidinium (FA^+) or inorganic cesium (Cs^+); B is a metal cation typically lead (Pb^{2+}) or tin cations (Sn^{2+}); and X is a halide anion. Cation A is slightly larger than centred cation B having 6 co-ordination number. There is a significant relationship, which is called "Goldschimidt Tolerance Factor" between the size of the ions and the formation and the shape of the crystal structure:

$$t' = \frac{r_A + r_X}{\sqrt{2}\left(r_B + r_X\right)} \tag{1}$$

Where, r_A and r_B are cationic diameters of A and B respectively, and r_X is the anionic diameter of halide in Å. When cation A is too big or cation B is too small (>1), hexagonal perovskite crystal is formed. If cation A and has the ideal size ($0.9 < t < 1$), cubic perovskite crystal is formed. Finally, if cation is too small to fit into the space between metal cations (B) ($0.7 < t < 0.9$), orthorhombic perovskite structure is formed [42].

3.2 Geometry of perovskite nano-crystals

A typical "bulk" perovskite structure is called three-dimensional (3D) which is usually comply to the ABX_3 formula completely [43–45]. In this situation, "3D" refers to the growth in every dimension without any confinement. X anions are combined through corner sharing to form a 3D network. Beside this, the cation A occupies the site in the middle of eight octahedra, and each element needs to owe the proper valence state to keep a whole charge balance [43]. To obtain nano crystal-line perovskites, crystal growth must be restricted with at least one dimension by a capping agent or a matrix. Thus, It is possible to prepare various dimensioned PNSs such as zero, one or two dimension (0D, 1D or 2D respectively) (**Figure 6**) [1, 46].

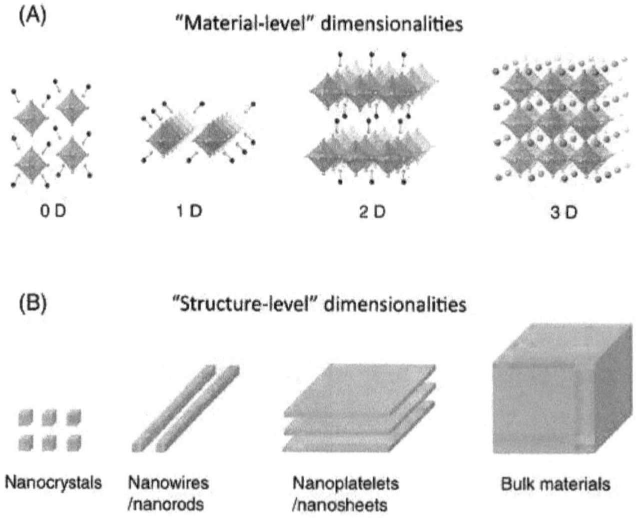

Figure 6.
Schematic illustration of low-dimensional perovskites and 3D perovskite crystal structures (reproduced with permission of Ref. [47]).

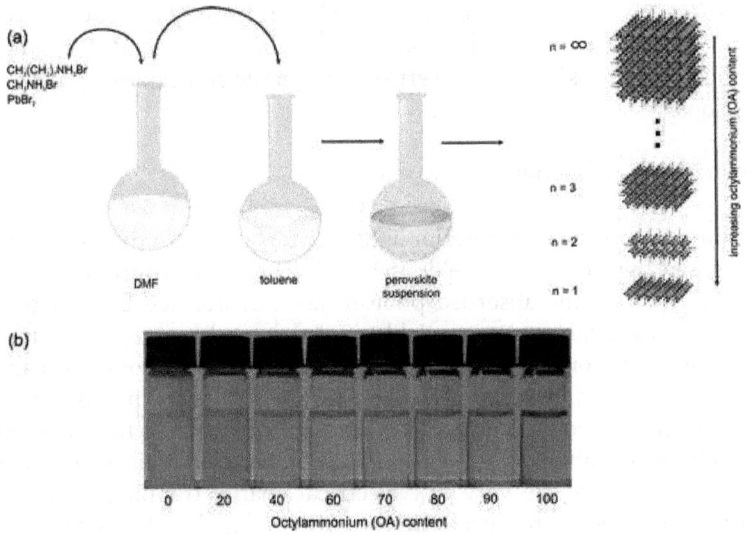

Figure 7.
(a) Schematic illustration of the perovskite nanopelets preparation. (b) Image of OA/MA perovskite suspensions in toluene under ambient light (reproduced with permission of Ref. [49]).

0D PNSs have been synthesized, by LARP [25], hot-injection [32] and tem-plate-assisted method [36] by different groups. Crystal growth is prevented in all dimensions by a ligand or a metal-oxide matrix to obtained quantum-dot-like NPs representing good Photoluminescence (PL). Prepared nanostructures have been used in many opto-electronic applications such as solar cells, LEDs and laser.

Perovskite nanowires or nanorods are basically *1D PNSs* which represent outstanding anisotropic optical and electrochemical properties with very high PLQFs [41]. 1D PNSs have been proposed by Horvath et al. for the first time in the literature. Nanowires with 50–400 nm wide and 10 μm length has been prepared in DMF solution phase. Obtained NPSs have been used in Solar Cells [48], LED [4] and photodetector [8] applications.

Consequently, *2D PNSs* are nano pellet or nano sheet shaped materials which consist of several unit cells leading larger binding energy for the exciton and more intense PL. These types of NPs have similar networks with corresponded 3D perovskite, however the general formula of ABX_3 is changed when the NP is very thin. Number of sheets in a nano pellet NPS can be determine by using absorption and emission spectroscopy. Differences of Absorption and emission maximum of nanosheets (n = 1, 2, 3 and 4) are very significant. Maximum absorption peak is red shifted with the increasing of number of unit cell (nano- pellet) (**Figure 7**).

4. Applications of perovskite nanoparticles

4.1 Optical properties of perovskite nanoparticles

In recent years, perovskite is a very important milestone in solar cell research, thanks to its perfect exciton and charge carrier properties. This excellent performance has allowed perovskite to be used as outstanding light emitters in Light Emitting Diodes (LEDs) and other optoelectronic applications [4, 7, 25, 50–56]. One of the most attractive features of perovskites is their emissions, which can be easily adjusted in the visible range compared to traditional III–V and II–VI groups. All inorganicperovskite ABX_3 emissions, including quantum dots and nanoplatelets, can cover the entire visible area, even close to the infrared or ultraviolet region, by substituting halide elements from chloride to iodine [26, 34, 51, 57, 58]. Another way to adjust the emission is to insert other organic molecules into it or replace anions/cations.

4.2 High quantum efficiency

Perovskites are considered superior light emitters owing to their large absorption coefficients and high quantum yields [59, 60]. The high quantum yield generally indicates that most of the absorbed photons are transformed by radiative recombination processes. High quantum yields (90%) have been reported in inorganic ABX_3 and organic-inorganic methyl-ammonium halide perovskite nanocrystals without further surface treatment [26, 34]. However, the main reason for reduced quantum efficiency in conventional III –V and II –VI groups is that these nanocrystalline structures are often affected by surface defects or donorreceptor levels. In perovskite, where there are very few electrical charge trapping conditions, high quantum efficiency is the result of the formation of a clear band gap that greatly supports exciton radial recombination efficiency [61].

4.3 Quantum confinement effect

Optical absorption and emission characteristics of a semiconductor can be adjusted by changing the size of the semiconductor. If the size of such materials is in the nanocrystal range, changes in band gaps can be observed. The reduction of the crystal size causes the quantum capture effect to be observed and the bandwidth to shift to blue. If the semiconductor size is too small to compare with the Bohr radius of excitons, quantum trapping can be seen in the optical properties of the semiconductors. For example, the quantum capture effect is quite evident in completely inorganic $CsPbBr_3$ perovskite nanocrystals and in organic-inorganic methyl ammonium lead halide perovskite nanocrystals. This can usually be observed when the nanocrystalline size is comparable to the exciton Bohr radius. **Figure 8a** and **b** demonstrate that the emission of CsPbBr3 perovskite nanocrystals can actually be adjusted from 2.7 eV to 2.4 eV, with a size ranging from 4 nm to 12 nm, which is compatible with the theoretical calculation [34].

4.4 Linear absorption and emission

Many groups have focused on improving the band spacing, excitonic characteristic and optical properties of photoluminescent quantum yields of halide perovskite nanocrystals for optoelectronic applications in recent years [18, 62]. The most interesting of these features is that bandwidth is adjustable. It is possible to adjust the bandwidth by changing the individual components of the metal halide perovskites (MHP). Optical properties of bulk perovskite thin films could be changed across the entire visible spectrum. Thus, it has been shown that the optical properties of $MAPbBr_3$ nanocrystals, which have an emission of approximately 529 nm, can also be altered throughout the entire visible spectrum [63, 64]. For $CsPb(X = Cl, Br \text{ or } I)_3$ nanocrystals, using halide components, the emission wavelength is from 410 nm (X = Cl), (X = Br) to 512 nm, (X = I) It has been shown that it can be shifted to 685 nm (**Figure 9**).

Adjustable optical features of perovskite nanocrystals are based on the electronic structure of these materials (**Figure 10**). The conduction band consists of external p orbitals of halid and antibonding orbitals of hybridization of Pb 6p orbitals. The valence band consists of antibonding of the hybridization of Pb 6s

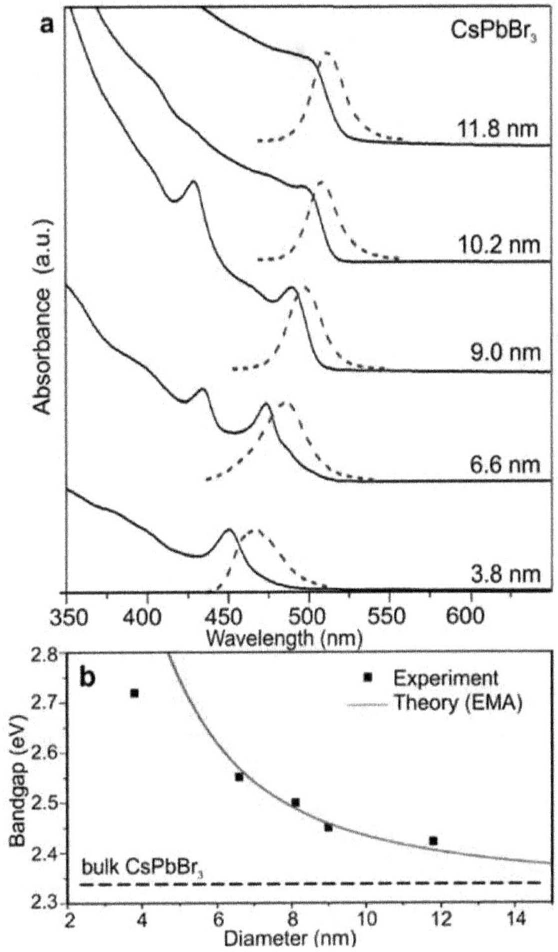

Figure 8.
(a) The emission spectra of CsPbBr3 NCs Quantum-size effects in the absorption and (b) experimental versus effective mass approximation size (theoretical technique) with respect to the band gap energy range (reproduced with permission of Ref. [34]).

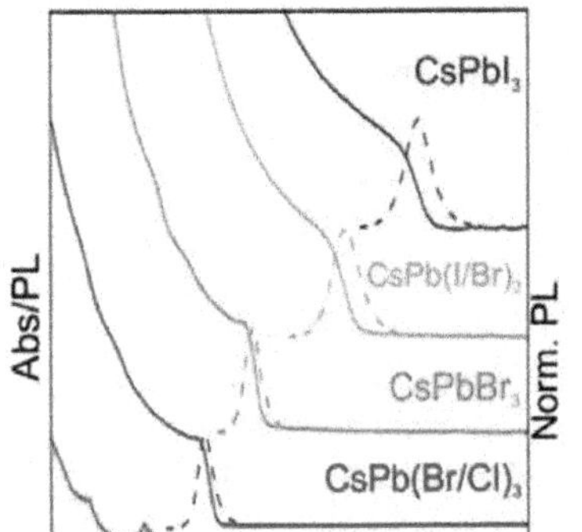

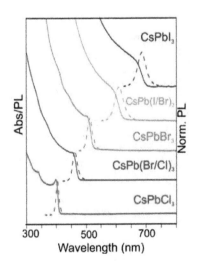

Figure 9.
UV–vis and photoluminescence spectra shows that the band gap could be tuned by controlling of CsPbX3 NCs as a function of halide (reproduced with permission of Ref. [34]).

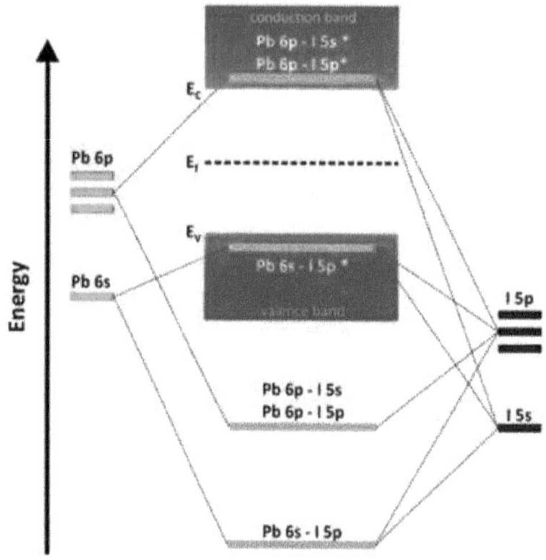

Figure 10.
The energy bands forms in a lead iodide perovskite by the crossing of lead and iodide orbitals (reproduced with permission of Ref. [23]).

and the same halide p-orbitals. The conduction band is generally p-like owing to the high density energy bands lead density contribution [65]. However, as the opposite of this situation, the band gap of gallium arsenide occurs between bond-ing and antibonding orbitals. As the halide component changes, the valence energy band shifts at the limit value, while only small changes occur in the energy limit value of the transmission band [66]. The cation A does not contribute considerably to the conduction and valence orbitals, but has an important effect on the band gap of perovskite [67]. As a result, emission energies of MA-based perovskite nanocrystals have been demonstrated to range from halide to 407 to 734 nm. It is understood that the FA emissions of nanocrystalline FAs shifted to 408 nm with Cl, to 535 nm with Br and to I and 784 nm, that is, to red [68–70]. In addition, the B cation has a significant mission in changing the optical properties of metal halide perovskites nanocrystals. As is known, the lead is a harmful element for the nature, instead of using the band [71], the band gap and PL emissions of the metal halide perovskites nanocrystals shifted from Cl to 443 nm and from I to 953 nm. The reason is probably a result of the higher electronegativity of Sn_2 than Pb_2 [72]. But, the stability of Sn^{2+} and similarly Ge^{2+} based on perovskite compounds is too weak owing to the reduction of non-interacting electron pair effects corresponding to a decrease in the stability of the divalent oxidation state [73]. As a result, PL emissions or energy band gaps of nanocrystalline structures obtained by various methods depend only on stoichiometry [74]. Stokes shift is an important parameter at the absorption and emission spectra that LHP nanocrystals show typically small, ranging from 20 to 85 meV [75–77]. Stokes shift increases as nanocrystals decrease in size. This is clarified by the creation of a compatible hole state that can be delocalized across the whole nanocrystal [78]. PL line widths of metal halide perovskites are another important point, particularly for LED applications. In fact, the line widths are commonly in the range of 70-110 meV and have been found to vary significantly with respect to the halide content. In many articles, the halide component greatly varies the PL spectra in terms of wavelength, for instance for Cl-perovskites reaches to 10–12 nm and for I-perovskites 40 nm. In terms of photoluminescent quantum yields, LHP nanocrystals display high values with the more epitaxial shell range the more chalcogenide QDs without electronic passivation [69]. Some article abot MA based halide perovskites have been reached to 80 % to 95 % for photoluminescent

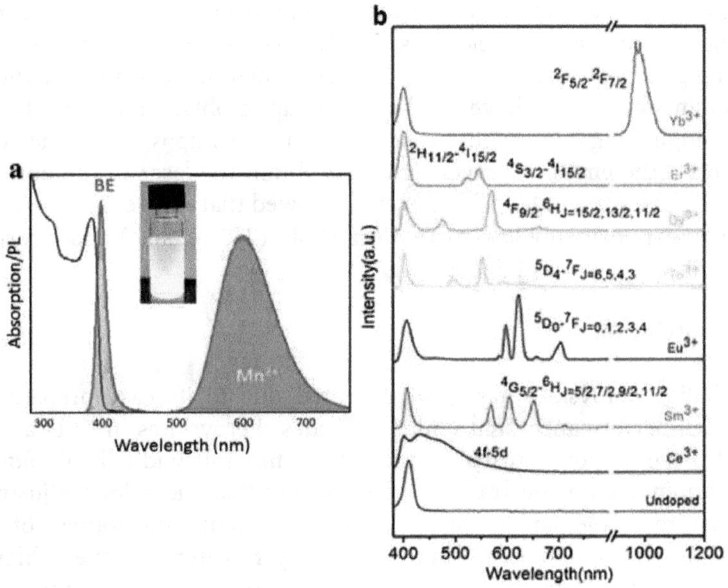

Figure 11.
(a) Photoluminescence emission & absorption of Mn-doped CsPbCl3 NCs, (b) photoluminescence curves of CsPbCl3 NCs doped with different lanthanide ions (reproduced with permission of Refs. [23, 87]).

quantum yields, bromide and iodides, respectively [79–82]. The defect tolerance of halide perovskite gives new properties due to orbital structure and the bandgap that forms between two antibonding orbitals. FA-based nanocrystals also reach to 70–90% photoluminescent quantum yield. There is alternative method that doping with additional ions to handle the optical emission of LHP nanocrystals. The addition of Mn^{2+} ions leads to a strong Stokes shift emission by the band gap given by the perovskite matrix and the emission from the atomic states of the Mn^{2+} ions (**Figure 11a**) [83]. Mn^{2+}doping can result in a pair of controllable emissions from both localized Mn^{2+}states and band gap recombination [84, 85]. In contrary, $CsPbBr_3$ nanocrystals have been shown to cause blue shift of doping, band boundary and PL emission with other divalent cations such as Sn^{2+}, Cd^{2+} and Zn^{2+}. In these cases, an important portion (0.2–0.7%) of the original Pb ions have been exchanged by new metal cations that produce alloy nanocrystals. CsPbBr3 alloy nanocrystals with 0.2% content of Al^{3+}ions have a blue shear PL emission with a centre of 456 nm and a relatively high 42% photoluminescence quantum yield [86].

In all these cases, perovskite nanocrystals act as an absorbent host that stimulates dopants through energy transfer. In addition, by selecting specific dopant atoms, the emission wavelengths of the nanocrystals obtained can be easily adjusted. When lanthanide ions are doped, the emission of $CsPbCl_3$ nanocubes ranges from 400 nm to 1000 nm and their quantum yields are 15% and 35%, respectively (**Figure 11b**) [87].

5. Optoelectronics applications of perovskite

5.1 Optical lasing

High absorption coefficient and strong photoluminescence is the most powerful side of metal halide perovskite. It is possible to obtain a laser with a high quantum efficiency material and a suitable optical band spacing. With the understanding that amplified spontaneous emission can be obtained from perovskite material, research has focused on this subject. Especially, laser obtaining studies were carried out by using perovskite crystal. It has also been shown that perovskite nanowires in high crystalline feature can have a FabryPerot gap to obtain a laser in the material [88]. By manipulating the content of the halide in the composition of the perovskite lead to control the emission wavelength and obtain the laser in the entire visible spectrum (**Figure 12a** and **b**) [57]. Zhu et al. showed that laser exposure in CH_3NH_3 PbX_3 with an exponentially low current threshold (220 nJ cm^{-2}) and a corresponding carrier density as low as 1.5×10^{16} cm^{-3}.

5.2 Light emitting diodes

Metal halide perovskite has great potential due to its easy preparation, low cost and high-performance light emitting diodes. Perovskites in LED applications show well a high colour purity, usually 15–25 nm full width half-colour purity for electroluminescence spectra. The point is here that the colour adjustment can cover the entire visible part of the spectrum by changing the content of different halides within the compounds. Therefore, many researchers have achieved high performance in perovskite quantum dot LED because of their quantum dots, strong luminescence, and high external quantum yields (**Figure 13a**) [25]. The stability problem of organic-inorganic hybrid perovskite can be eliminated by synthesizing the inorganic perovskites (ie $CsPbX_3$) as quantum dots for LED applications (**Figure-13b**) [89]. The performance of perovskite nanoplatelets in LED applications was lower than the quantum dots (**Figure 13c**). As a result, perovskite has a great potential for lighting and display applications as a new generation LED material.

5.3 Alternative applications

FETs, photodetectors, and single photon emitters are the other potential optoelectronic application devices. Liu et al. [90] used perovskite nanoplatelets to produce a FET on a Si/SiO_2. For this device, the current-voltage curve showed ohmic contact between the perovskite nanoplatelets and electrodes, and a linear dependency was noted. In that study, they revealed a strong light-material interaction and broadband light harvesting capability of perovskite. Many photodetectors were fabricated based on a horizontal $CH_3NH_3PbI_3$ nanowire array [49]. The

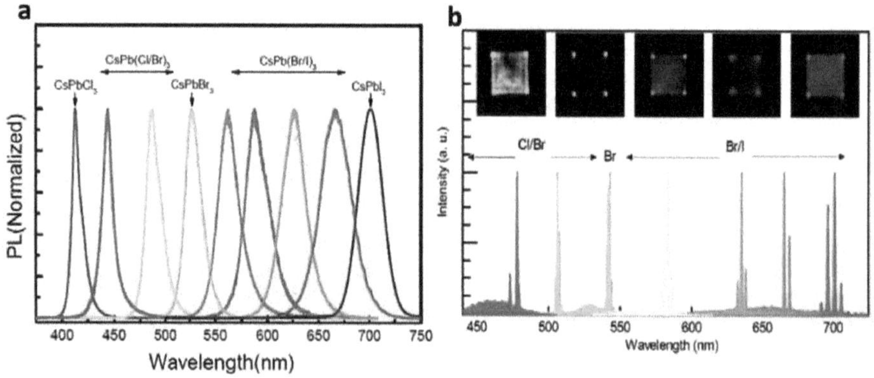

Figure 12.
(a) Photoluminescence spectra of CsPb (X = Cl, Br, I)3 nanoplatelets and (b) wavelength adjustment of perovskite lasing by controlling the content of halide in CsPbX3 perovskite [6, 7, 57]. There is really an extraordinary point about nanowires is that they have very little carrier capture area and the laser quantum efficiency reaches to 100% (reproduced with permission of Ref. [6]).

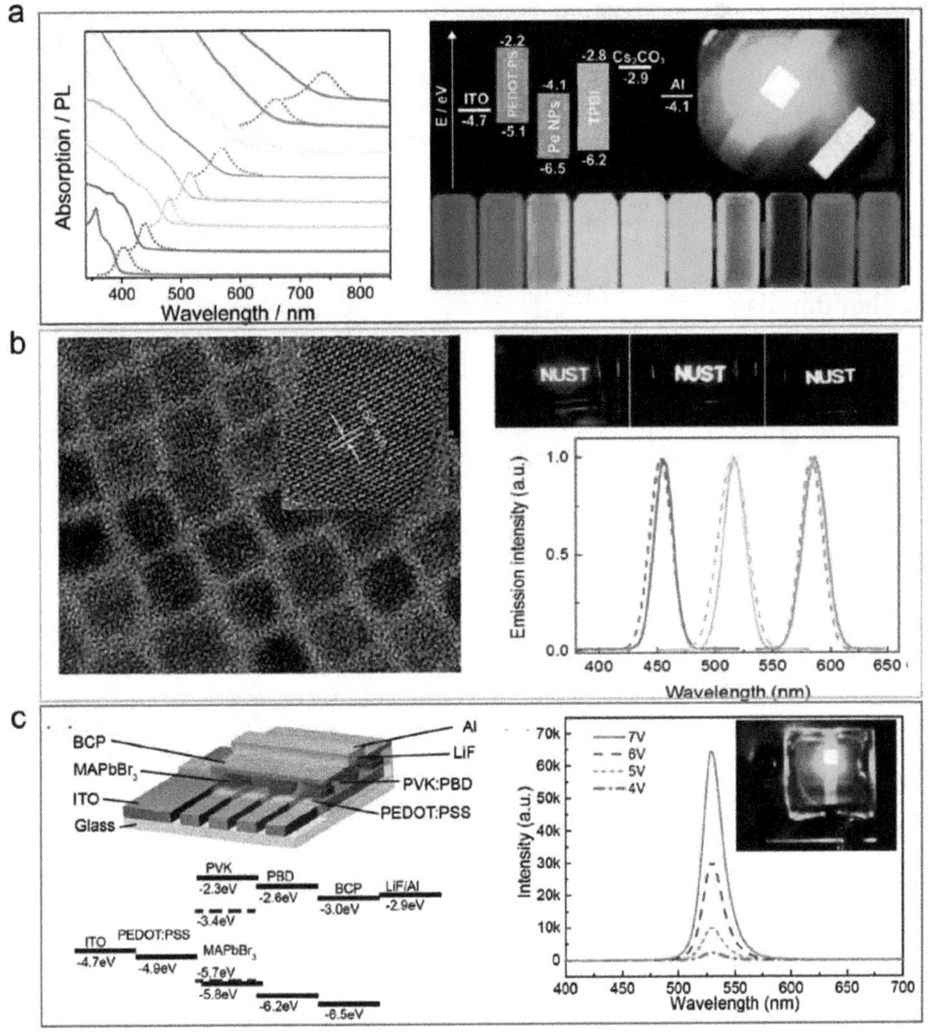

Figure 13.
In LED applications are generally used a low dimensional perovskite. (a) CH3NH3PbX3 quantum dots are applied to fabricate LED, (b) CsPbX3 quantum dots are used to form LED, (c) CH3NH3PbX3 nanoplatelets are performed to obtain LED (reproduced with permission of Refs. [25, 89, 91]).

response time compared to the obtained photodetectors bulk perovskite and other inorganic nanowire photodetectors is higher in terms of response time of 0.3 ms, 1.3 A W^{-1} response and a detectivity of 2.5 × 10^{12} Jones. Park et al. show a remarkable perovskite nanomaterials application that Quantum dots were used as a single photon emitter at standard conditions. $CsPbX_3$ perovskite was used as the leading material to synthesize cubic shapes and quantum dots with an average size of 10 nm. Perovskite quantum dots displayed an excellent photon beam of emitted light and photoluminescence (PL) intensity fluctuations associated with PL life. It is defined that phenomenon as "A-type flashing" that is popular in the quantum dot system.

6. Conclusion

In this chapter, preparation methods and applications of 2D perovskite nanoparticles were reviewed. The most crucial points for synthesis method are uniformity and form factor of synthesised nanoparticles. Furthermore, structural, optic, and electrochemical properties of 2D perovskites have been introduced in detail. Due to the low ionic interaction in the crystal structures, organo-halide perovskites exhibit low stability under ambient conditions. However, 2D perovskite nanoparticles still offer a great potential due to the structure-dependent optic and electronic properties.

Author details

Burak Gultekin[1*], Ali Kemal Havare[2], Shirin Siyahjani[1], Halil Ibrahim Ciftci[3] and Mustafa Can[4]

1 Ege University, Solar Energy Institute, Izmir, Turkey

2 Faculty of Engineering, Electric and Electronics Engineering, Photoelectronic Lab. (PEL), Toros University, Mersin, Turkey

3 Faculty of Life Sciences, Medicinal and Biological Chemistry Science Farm Joint Research Laboratory, Kumamoto University, Kumamoto, Japan

4 Faculty of Engineering and Architecture, Izmir Katip Celebi University, Izmir, Turkey

*Address all correspondence to: burakgultekin@gmail.com

References

[1] Tsai H et al. *High-efficiency two-dimensional Ruddlesden–Popper perovskite solar cells*. Nature.2016;**536**(7616):312-316

[2] Liu M, Johnston MB, Snaith HJ. *Efficient planar heterojunction perovskite solar cells by vapour deposition*. Nature. 2013;**501**(7467):395-398

[3] Burschka J et al. *Sequential deposition as a route to high-performanceperovskite-sensitized solar cells*. Nature. 2013;**499**(7458):316-319

[4] Yuan M et al. *Perovskite energy funnels for efficient light-emitting diodes*. Nature nanotechnology. 2016;**11**(10):872-877

[5] Tan Z-K et al. *Bright light-emitting diodes based on organometal halide perovskite*. Nature nanotechnology. 2014;**9**(9):687-692

[6] Zhu H et al. *Lead halide perovskite nanowire lasers with low lasing thresholds and high quality factors*. Nature materials. 2015;**14**(6):636-642

[7] Zhang Q et al. *Room-temperature near-infrared high-Q perovskite whispering-gallery planar nanolasers*. Nano letters. 2014;**14**(10):5995-6001

[8] Miao J et al. *Highly Sensitive Organic Photodetectors with Tunable Spectral Response under Bi-Directional Bias*. Advanced Optical Materials. 2016;**4**(11):1711-1717

[9] Miao J, Zhang F. *Recent progress on highly sensitive perovskite photodetectors*. Journal of Materials Chemistry C. 2019;**7**(7):1741-1791

[10] Miao J et al. *Photomultiplication type organic photodetectors with broadband and narrowband response ability*. Advanced Optical Materials. 2018;**6**(8):1800001

[11] Li L et al. *Achieving EQE of 16,700% in P3HT: PC 71 BM based photodetectors by trap-assisted photomultiplication*. Scientific reports. 2015;**5**(1):1-7

[12] Li L et al. *Revealing the working mechanism of polymer photodetectors with ultra-high external quantum efficiency*. Physical Chemistry Chemical Physics. 2015;**17**(45):30712-30720

[13] Jung S et al. *Enhancement of Photoluminescence Quantum Yield and Stability in CsPbBr3 Perovskite Quantum Dots by Trivalent Doping*. Nanomaterials. 2020;**10**(4):710

[14] Zhang D et al. *Increasing Photoluminescence Quantum Yield by Nanophotonic Design of Quantum-Confined Halide Perovskite Nanowire Arrays*. Nano letters. 2019;**19**(5):2850-2857

[15] Khan Y et al. *Waterproof perovskites: high fluorescence quantum yield and stability from a methylammonium lead bromide/formate mixture in water*. Journal of Materials Chemistry C. 2020

[16] Dong, Q., et al., *Operational stability of perovskite light emitting diodes*. Journal of Physics: Materials, 2020.**3**(1): p. 012002.

[17] Stylianakis MM et al. *Inorganic and hybrid perovskite based laser devices: a review*. Materials. 2019;**12**(6):859

[18] Schmidt LC et al. *Nontemplate synthesis of CH3NH3PbBr3 perovskite nanoparticles*. Journal of the American Chemical Society. 2014;**136**(3):850-853

[19] Chen D, Chen X. *Luminescent perovskite quantum dots: synthesis, microstructures, optical properties and applications*. Journal of Materials Chemistry C. 2019;**7**(6):1413-1446

[20] Zhang C, Kuang DB, Wu WQ. *A Review of Diverse Halide Perovskite*

Morphologies for Efficient Optoelectronic Applications. Small Methods. 2020;**4**(2):1900662

[21] Ha S-T et al. *Metal halide perovskite nanomaterials: synthesis and applications*. Chemical science. 2017;**8**(4):2522-2536

[22] Gopinathan N et al. *Solvents driven structural, morphological, optical and dielectric properties of lead free perovskite CH 3 NH 3 SnCl 3 for optoelectronic applications: Experimental and DFT study*. Materials Research Express. 2020

[23] Shamsi J et al. *Metal halide perovskite nanocrystals: synthesis, post-synthesis modifications, and their optical properties*. Chemical reviews. 2019;**119**(5):3296-3348

[24] Shen D et al. *Understanding the solvent-assisted crystallization mechanism inherent in efficient organic–inorganic halide perovskite solar cells*. Journal of Materials Chemistry A. 2014;**2**(48):20454-20461

[25] Xing J et al. *High-efficiency light-emitting diodes of organometal halide perovskite amorphous nanoparticles*. ACS nano. 2016;**10**(7):6623-6630

[26] Zhang F et al. *Brightly luminescent and color-tunable colloidal CH3NH3PbX3 (X= Br, I, Cl) quantum dots: potential alternatives for display technology*. ACS nano. 2015;**9**(4):4533-4542

[27] Pathak S et al. *Perovskite crystals for tunable white light emission*. Chemistry of Materials. 2015;**27**(23):8066-8075

[28] Tyagi P, Arveson SM, Tisdale WA. *Colloidal organohalide perovskite nanoplatelets exhibiting quantum confinement*. The journal of physical chemistry letters. 2015;**6**(10):1911-1916

[29] Hassan Y et al. *Structure-Tuned Lead Halide Perovskite Nanocrystals*. Advanced materials. 2016;**28**(3):566-573

[30] Luo B et al. *Synthesis, optical properties, and exciton dynamics of organolead bromide perovskite nanocrystals*. The Journal of Physical Chemistry C. 2015;**119**(47):26672-26682

[31] Wang Y et al. *Thermodynamically stabilized β-CsPbI3–based perovskite solar cells with efficiencies> 18%*. Science. 2019;**365**(6453):591-595

[32] Shi J et al. *Efficient and stable CsPbI 3 perovskite quantum dots enabled by in situ ytterbium doping for photovoltaic applications*. Journal of Materials Chemistry A. 2019;**7**(36):20936-20944

[33] Wang A et al. *Controlled synthesis of lead-free and stable perovskite derivative Cs2SnI6 nanocrystals via a facile hot-injection process*. Chemistry of Materials. 2016;**28**(22):8132-8140

[34] Protesescu L et al. *Nanocrystals of cesium lead halide perovskites (CsPbX3, X= Cl, Br, and I): novel optoelectronic materials showing bright emission with wide color gamut*. Nano letters. 2015;**15**(6):3692-3696

[35] Lim S-C et al. *Binary halide, ternary perovskite-like, and perovskite-derivative nanostructures: hot injection synthesis and optical and photocatalytic properties*. Nanoscale. 2017;**9**(11):3747-3751

[36] Shi, P., et al., *Template-Assisted Formation of High-Quality α-Phase HC (NH2) 2PbI3 Perovskite Solar Cells*. Advanced Science, 2019. **6**(21): p.1901591.

[37] Ashley MJ et al. *Templated synthesis of uniform perovskite nanowire arrays*. Journal of the American Chemical Society. 2016;**138**(32):10096-10099

[38] Malgras V et al. *Hybrid methylammonium lead halide perovskite nanocrystals confined in gyroidal silica templates*. Chemical Communications. 2017;**53**(15):2359-2362

[39] Ananthakumar S, Babu SM. *Progress on synthesis and applications of hybrid perovskite semiconductor nanomaterials—A review*. Synthetic Metals. 2018;**246**:64-95

[40] Pileni M-P. *The role of soft colloidal templates in controlling the size and shape of inorganic nanocrystals*. Nature materials. 2003;**2**(3):145-150

[41] Huang H et al. *Emulsion synthesis of size-tunable CH3NH3PbBr3 quantum dots: an alternative route toward efficient light-emitting diodes*. ACS applied materials & interfaces. 2015;**7**(51):28128-28133

[42] Wali Q et al. *Advances in stability of perovskite solar cells*. Organic Electronics. 2020;**78**:105590

[43] Wang, R., et al., *A review of perovskites solar cell stability*. Advanced Functional Materials, 2019. **29** (47): p. 1808843.

[44] Heo JH et al. *Efficient inorganic–organic hybrid heterojunction solar cells containing perovskite compound and polymeric hole conductors*. Nature photonics. 2013;**7**(6):486

[45] Etgar L. *The merit of perovskite's dimensionality; can this replace the 3D halide perovskite?* Energy & Environmental Science. 2018;**11**(2): 234-242

[46] Ma C et al. *High performance low-dimensional perovskite solar cells based on a one dimensional lead iodide perovskite*. Journal of Materials Chemistry A. 2019;**7**(15):8811-8817

[47] Zhu P, Zhu J. *Low-dimensional metal halide perovskites and related optoelectronic applications*. InfoMat. 2020;**2**(2):341-378

[48] Kulkarni SA et al. *Perovskite Nanoparticles: Synthesis, Properties, and Novel Applications in Photovoltaics and LEDs*. Small Methods. 2019;**3**(1):1800231

[49] Sichert JA et al. *Quantum size effect in organometal halide perovskite nanoplatelets*. Nano letters. 2015;**15**(10):6521-6527

[50] Wang J et al. *Interfacial control toward efficient and low-voltage perovskite light-emitting diodes*. Advanced Materials. 2015;**27**(14):2311-2316

[51] Hong WL et al. *Efficient low-temperature solution-processed lead-free perovskite infrared light-emitting diodes*. Advanced Materials 2016;**28**(36):8029-8036

[52] Bade SGR et al. *Fully printed halide perovskite light-emitting diodes with silver nanowire electrodes*. ACS nano. 2016;**10**(2):1795-1801

[53] Li G et al. *Highly efficient perovskite nanocrystal light-emitting diodes enabled by a universal crosslinking method*. Advanced Materials. 2016;**28**(18):3528-3534

[54] Veldhuis SA et al. *Perovskite materials for light-emitting diodes and lasers*. Advanced Materials. 2016;**28**(32):6804-6834

[55] Liang D et al. *Color-pure violet-light-emitting diodes based on layered lead halide perovskite nanoplates*. ACS nano. 2016;**10**(7):6897-6904

[56] Yantara N et al. *Inorganic halide perovskites for efficient light-emitting diodes*. The journal of physical chemistry letters. 2015;**6**(21):4360-4364

[57] Zhang Q et al. *High-quality whispering-gallery-mode lasing from cesium lead halide perovskite nanoplatelets*. Advanced Functional Materials. 2016;**26**(34):6238-6245

[58] Fu Y et al. *Nanowire lasers of formamidinium lead halide perovskites and their stabilized alloys with improved stability*. Nano letters. 2016;**16**(2):1000-1008

[59] Nedelcu G et al. *Fast anion-exchange in highly luminescent nanocrystals of cesium lead halide perovskites (CsPbX3, X= Cl, Br, I)*. Nano letters. 2015;**15**(8):5635-5640

[60] Akkerman QA et al. *Tuning the optical properties of cesium lead halide perovskite nanocrystals by anion exchange reactions*. Journal of the American Chemical Society. 2015;**137**(32):10276-10281

[61] Wu K et al. *Ultrafast interfacial electron and hole transfer from CsPbBr3 perovskite quantum dots*. Journal of the American Chemical Society. 2015;**137**(40):12792-12795

[62] Huang H et al. *Colloidal lead halide perovskite nanocrystals: synthesis, optical properties and applications*. NPG Asia Materials. 2016;**8**(11):e328-e328

[63] Noh JH et al. *Chemical management for colorful, efficient, and stable inorganic–organic hybrid nanostructured solar cells*. Nano letters. 2013;**13**(4):1764-1769

[64] Tanaka K et al. *Comparative study on the excitons in lead-halide-based perovskite-type crystals CH3NH3PbBr3 CH3NH3PbI3*. Solid state communications. 2003;**127**(9-10):619-623

[65] Umebayashi T et al. *Electronic structures of lead iodide based low-dimensional crystals*. Physical Review B. 2003;**67**(15):155405

[66] Schulz P et al. *Interface energetics in organo-metal halide perovskite-based photovoltaic cells*. Energy & Environmental Science.2014;7(4):1377-1381

[67] Eperon GE et al. *Formamidinium lead trihalide: a broadly tunable perovskite for efficient planar heterojunction solar cells*. Energy & Environmental Science. 2014;7(3):982-988

[68] Protesescu L et al. *Monodisperse formamidinium lead bromide nanocrystals with bright and stable green photoluminescence*. Journal of the American Chemical Society. 2016;**138**(43):14202-14205

[69] Lignos I et al. *Unveiling the shape evolution and halide-ion-segregation in blue-emitting formamidinium lead halide perovskite nanocrystals using an automated microfluidic platform*. Nano letters. 2018;**18**(2):1246-1252

[70] Minh DN et al. *Room-temperature synthesis of widely tunable formamidinium lead halide perovskite nanocrystals*. Chemistry of Materials. 2017;**29**(13):5713-5719

[71] Hao F et al. *Lead-free solid-state organic–inorganic halide perovskite solar cells*. Nature photonics. 2014;**8**(6):489

[72] Jellicoe TC et al. *Synthesis and optical properties of lead-free cesium tin halide perovskite nanocrystals*. Journal of the American Chemical Society. 2016;**138**(9):2941-2944

[73] Chen Q et al. *Under the spotlight: The organic–inorganic hybrid halide perovskite for optoelectronic applications*. Nano Today. 2015;**10**(3):355-396

[74] Zhang D et al. *Synthesis of composition tunable and highly luminescent cesium lead halide nanowires through anion-exchange reactions*. Journal of the American Chemical Society. 2016;**138**(23):7236-7239

[75] Tong Y et al. *From precursor powders to CsPbX3 perovskite nanowires: one-pot synthesis, growth mechanism, and oriented self-assembly* . Angewandte

Chemie International Edition. 2017;**56**(44):13887-13892

[76] Brennan MC et al. *Origin of the size-dependent stokes shift in CsPbBr3 perovskite nanocrystals*. Journal of the American Chemical Society. 2017;**139**(35):12201-12208

[77] Brennan MC, Zinna J, Kuno M. *Existence of a size-dependent Stokes shift in CsPbBr3 perovskite nanocrystals*. ACS Energy Letters. 2017;**2**(7):1487-1488

[78] Huang G et al. *Postsynthetic Doping of MnCl2 Molecules into Preformed CsPbBr3 Perovskite Nanocrystals via a Halide Exchange-Driven Cation Exchange*. Advanced materials. 2017;**29**(29):1700095

[79] Gonzalez-Carrero S et al. *The luminescence of CH3NH3PbBr3 perovskite nanoparticles crests the summit and their photostability under wet conditions is enhanced*. Small. 2016;**12**(38):5245-5250

[80] Liu F et al. *Highly luminescent phase-stable CsPbI3 perovskite quantum dots achieving near 100% absolute photoluminescence quantum yield*. ACS nano. 2017;**11**(10):10373-10383

[81] Tong Y et al. *Highly luminescent cesium lead halide perovskite nanocrystals with tunable composition and thickness by ultrasonication*. Angewandte Chemie International Edition. 2016;**55**(44):13887-13892

[82] Koscher BA et al. *Essentially trap-free CsPbBr3 colloidal nanocrystals by postsynthetic thiocyanate surface treatment*. Journal of the American Chemical Society. 2017;**139**(19):6566-6569

[83] Meinardi F et al. *Doped halide perovskite nanocrystals for reabsorption-free luminescent solar concentrators*. ACS Energy Letters. 2017;**2**(10):2368-2377

[84] Giansante C, Infante I. *Surface traps in colloidal quantum dots: a combined experimental and theoretical perspective*. The journal of physical chemistry letters. 2017;**8**(20):5209-5215

[85] He M et al. *Mn-Doped cesium lead halide perovskite nanocrystals with dual-color emission for WLED*. Dyes and Pigments. 2018;**152**:146-154

[86] Liu M et al. *Aluminum-Doped Cesium Lead Bromide Perovskite Nanocrystals with Stable Blue Photoluminescence Used for Display Backlight*. Advanced Science. 2017;**4**(11):1700335

[87] Pan G et al. *Doping lanthanide into perovskite nanocrystals: highly improved and expanded optical properties*. Nano letters. 2017;**17**(12):8005-8011

[88] Xing J et al. *Vapor phase synthesis of organometal halide perovskite nanowires for tunable room-temperature nanolasers*. Nano letters. 2015;**15**(7):4571-4577

[89] Song J et al. *Quantum dot light-emitting diodes based on inorganic perovskite cesium lead halides (CsPbX3)*. Advanced materials. 2015;**27**(44):7162-7167

[90] Liu J et al. *Two-dimensional CH3NH3PbI3 perovskite: Synthesis and optoelectronic application*. ACS nano. 2016;**10**(3):3536-3542

[91] Ling Y et al. *Bright light-emitting diodes based on organometal halide perovskite nanoplatelets*. Advanced materials. 2016;**28**(2):305-311

6

Optimal Temperature Sensor Based on a Sensitive Material

Asma Bakkali, José Pelegri-Sebastia, Youssef Laghmich and Abdelouahid Lyhyaoui

Abstract

The context of this chapter is the development of passive sensors for temperature sensing applications. The purpose is to successfully reduce the energy consumption in wireless sensor networks. The sensor is based on the electromagnetic transduction principle, and its originality is based on the integration of a high temperature-sensitive material into passive structure. Variation in temperature makes the dielectric constant of this material changing, and such modification induces variation in the resonant frequencies of high-Q whispering-gallery modes in the millimeter-wave frequency range. In this way, the proposed device shows a linear response to the increasing temperature, and these variations can be remotely detected from a radar interrogation of an antenna loaded by the whispering-gallery mode resonator. Proposed device is a powerful tool for many interesting applications since it offers very low power consumption and provides environmentally friendly temperature measurement. The sensor is simulated in order to outline its performance and to show the benefit of the batteryless sensing device.

Keywords: energy consumption, wireless sensor networks, lifetime, energy conservation, whispering gallery mode, PLZT material, temperature, dielectric resonator

1. Introduction

From sensing motion to identifying a gas and measuring temperature, sensors are a key element in our daily lives for analytical, monitoring, and diagnostic applications [1–3]. Following the progress of technology and current concerns for the protection of the environment and people, the development of these devices is expanding significantly, to transform chemical, mechanical, and thermal phenomena into a measurable quantity: electrical signal. Nowadays, we are facing an explosion in the sensor market, and the number of applications is expanding in parallel with advances in electronics and wireless communication technologies.

Temperature detection is currently one of the most expected needs, as it is generally not well controlled at a low cost. Temperature sensors have been one of the first fields of application of micro-systems, and they now represent a very important part of this market due to the increasing demand in the consumer and domestic application sectors but also in production, aeronautics, and health. The main characteristics currently required of these components are most often to be miniature, efficient, and economical and can be integrated into complex electronic systems.

Several research projects focus on optimizing the energy consumption of sensors by using innovative conservation techniques to improve the network's performance, including maximizing its lifespan. The sensor proposed in this chapter presents an interesting technological solution for temperature detection, thus allowing an extremely low consumption compared to conventional techniques. This new, highly integrated device requires no onboard power supply and uses electromagnetic transduction for temperature measurement.

2. Passive and wireless temperature sensor design

For decades, dielectric resonators (RDs) have been very important in the microwave field for many applications, such as oscillators and filtering devices [4]. The significant progress in the development of dielectric materials, both in terms of reliability and in improving the loss tangent at microwave frequencies, makes it possible to use them from microwave frequencies to millimetric frequencies. As the dimensions of these resonators can nowadays be small, they can be integrated into many telecommunication systems, in particular filtering systems where dielectric resonators have made it possible to maintain very good characteristics while reducing their size. These fields have led to a mature technology that allows the realization of reliable devices.

We have included the device proposed by Guillon et al. [5], composed of a silicon platform with coplanar lines on membrane. The second part is a dielectric resonator mounted on a support between the two coplanar lines. The entire device was coupled to an Monolithic Microwave Integrated Circuit (MMIC) amplifier, which subsequently made it possible to design a millimetric oscillator [6]. We can see that the device designed by Guillon was intended for the realization of an oscillator, which is not the case for us. The objective was to explore the interest that this device can provide in measuring small fluctuations in the dielectric permittivity of a material sensitive to temperature change. Therefore, the proposed structure is designed with more powerful simulation tools than those that existed in the 1990s, since they allow us to simulate the entire structure: coplanar lines, sensitive material, and dielectric resonator. In addition, we have resized the device to have the best possible transmission signal.

The temperature sensor shown in **Figure 1** consists of two parts: the micromachined coplanar lines and the dielectric resonator covered with the sensitive material maintained by a support between and above these two lines. In the rest of this chapter, we will detail the design of this device before presenting the simulation results of the optimized structure.

2.1 Presentation of the gallery modes

Dielectric resonators operating in conventional electric (TE) or magnetic (TM) transverse modes radiate a significant portion of energy at millimeter wavelength frequencies [7]. In order to avoid dimensional problems and radiation losses at millimeter frequencies, we have chosen to use an excited dielectric resonator on gallery modes (whispering gallery modes).

From an electromagnetic point of view, one of the essential characteristics of WGMs is the distribution of energy in the resonator. The energy of the gallery modes has the particularity of being confined in a region close to the air—dielectric interface. Moreover, one of the main advantages offered by this type of mode is the possibility of exciting a dielectric resonator with an oversized geometry while remaining at the millimetric frequencies [1, 2]. The dimensions of a resonator

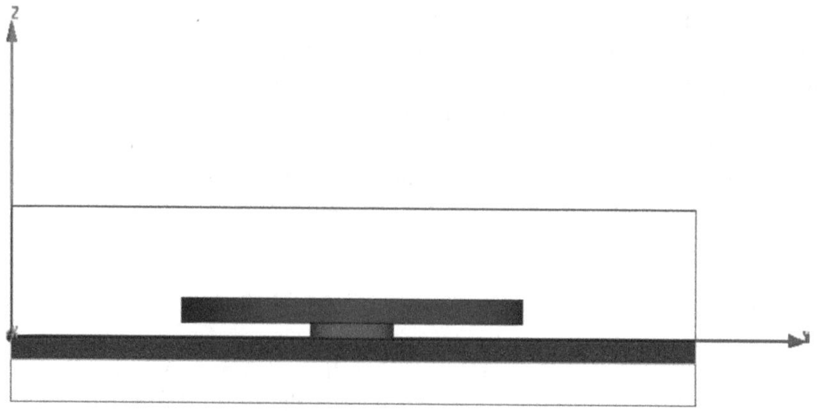

Figure 1.
Design of the temperature sensor: cross-section view.

excited in a gallery mode are much larger than those of the resonators used in conventional TE or TM modes. This makes it easier to use them at high frequencies since, given the dimensions of the resonator, it can be handled more easily.

The gallery modes of dielectric resonators are classified into two families: $WGH_{n,m,l}$ (magnetic field gallery modes) and $WGE_{n,m,l}$ (electric field gallery modes). This nomenclature makes it possible to identify each mode by taking into account the state of polarization and the importance of the transverse components of the electromagnetic field [3]. Thus, we are able to distinguish, on the one hand, WGE modes where the axial component of the field is essentially magnetic, while the transverse components are mainly electric. On the other hand, we distinguish the WGH modes, which correspond to the dual modes of the WGE. The three integers n, m, and l indicate the spatial configuration of the electromagnetic field inside the resonator (number of variations of the field in the three directions of the cylindrical reference mark):

- n: number of variations along the azimuthal direction
- m: number of variations according to radial direction
- l: number of variations according to the axial direction

It is important to mention that the azimuth number has an influence on the caustic radius. Indeed, a high azimuth number results in a higher caustic radius and therefore in a confinement of electromagnetic energy closer to the lateral surface of the resonator.

2.2 Advantages of gallery modes for temperature detection

Among the main advantages of gallery modes, we will note here the most important for our study. First of all, the dimensions of the resonator excited on a gallery mode are much larger than in the case of conventional TE or TM modes. This oversizing makes it possible to consider the use of this resonator at millimeter frequencies by facilitating temperature detection. On the other hand, thanks to the high-energy confinement in the dielectric, the vacuum quality factors are practically limited only to the loss tangent of the material used [3]. The latter will thus be very sensitive to the presence of a variation in the dielectric properties of the sensitive material, which will improve its detection thanks to an offset in the resonance frequency of a gallery mode.

In addition, it is important to note that the gallery modes have the particularity of not having any energy at the center of the resonator. As mentioned earlier, the larger the azimuth number, the larger the area. The central part can therefore be used to fix the resonator; we can add a shim to the circuit in order to maintain it or adjust its position in relation to the coupling circuit (lines) without disturbing its operation. Finally, these modes are no longer stationary but progressive when the resonator is excited by a progressive wave source, whose propagation constant is close to that of the gallery mode. The mode then propagates in the azimuthal direction. This type of excitation for gallery modes results in a directional coupling with the line. Therefore, this coupling will make it possible to consider the design of a directional filter with a narrow bandwidth, which in turn will allow temperature measurement thanks to the shift in its resonance frequency.

3. Sensor based on millimeter band resonant gallery modes

In order to define an innovative temperature sensor with high RF performance, we will take advantage of the dielectric properties of the sensitive material, the machining of coplanar membrane lines, and the characteristics of gallery modes. This new detection device is based on the modification of the resonance frequency of the dielectric resonator at the presence of a temperature variation. This is done by means of a perovskite material deposited above this resonator whose dielectric constant varies with a change in temperature.

The general concept chosen is to modify the electromagnetic coupling that exists in the device by temperature (see **Figure 2**). This concept allows for a very high sensitivity. Indeed, variable coupling can be achieved by modifying (globally or locally) the environment around the electromagnetic field lines by directly modifying the electromagnetic properties of the sensitive material.

In other words, a change in temperature leads to a change in the permittivity of the sensitive material and therefore a change in the electromagnetic field. This change affects the electrical parameters of the resonator, namely, the resonance frequency and the quality factor. Thus, a resonator designed to operate at a center frequency f_0 for a particular gallery mode sees this shift as a function of the variation in the dielectric constant of the sensitive material. The electromagnetic field is therefore used to measure the temperature variation. The detection principle is then based on the shift in the resonance frequency of the gallery modes in the dielectric resonator. The latter and the sensitive material are used here to measure the influence of the permittivity of this material and to deduce the temperature variation.

In addition, this type of detection technique based on electromagnetic transduction has shown interesting results for the realization of an oscillator in the first place and subsequently for specific gas and pressure detection applications. These passive microwave sensors have been designed using the relaxation phenomenon present in sensitive materials. In particular, the installation of a functional TiO_2-based

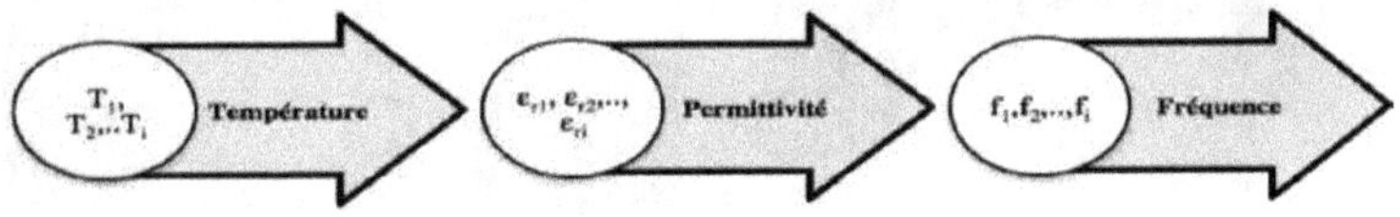

Figure 2.
Principle of electromagnetic transduction.

microwave sensor in the Ka-band has been proposed for gas concentration detection. This detector is based on the direct variation of the dielectric properties of the resonator at the presence of a gas.

4. Study of the geometric parameters of the sensor

There are parameters that can modify the dielectric characteristics of a material such as temperature, for example. In our case, the aim is for this modification to generate a frequency variation in the microwave domain. The basic element is the sensing material whose dielectric constant varies with temperature. The choice of the sensitive material is not obvious because of the requirements imposed, which complicate the integration of dielectric materials into microelectronic circuits or their use in the manufacture of the sensor. To achieve this, we have carried out an in-depth biblio-graphical study in order to find the appropriate material that meets our specifications. Among the materials proposed in the literature, we chose lead-lanthanum-zirconatetitanate (Abbreviated in PLZT). As a result, and as part of our collaboration with the Faculty of Computer Science and Materials Sciences, Silesian University of Poland (see **Figure 3**) [8, 9], we have several samples taken in his laboratory.

Also referred to as lanthanum-doped lead zirconate titanate, this material meets our needs in terms of temperature range and operating frequency. In particular, it has a dielectric property that depends on the change in temperature (see **Figure 4**) [10, 11]. Indeed, the variation in temperature has an impact on the properties of this material since it causes a relatively large change in its dielectric permittivity [12]. This PLZT will then detect the temperature change through its integration into the microwave circuit described in **Figure 1**, with a radius of RPLZT = 3.25 mm and a thickness of HPLZT = 10 μm.

The system is based relatively on the use of a dielectric resonator and two coplanar membrane lines. These lines serve as an excitation support for the RD gallery modes. The study of the mechanism of operation of these two components was widely discussed in the literature [13–15]. Thus, the RD has a radius of R_{RD} = 3.25 mm, a thickness of HRD 360μm, and a relative dielectric permittivity of 80. This dielectric resonator is held on the line plane by an Alumina (Al_2O_3) wedge with $R_{Support}$ = 0.8 mm radius and $H_{Support}$ = 230 μm height.

The RD's gallery mode excitation mechanism and sensitive material were selected, and a radar interrogation was carried out to transmit temperature information.

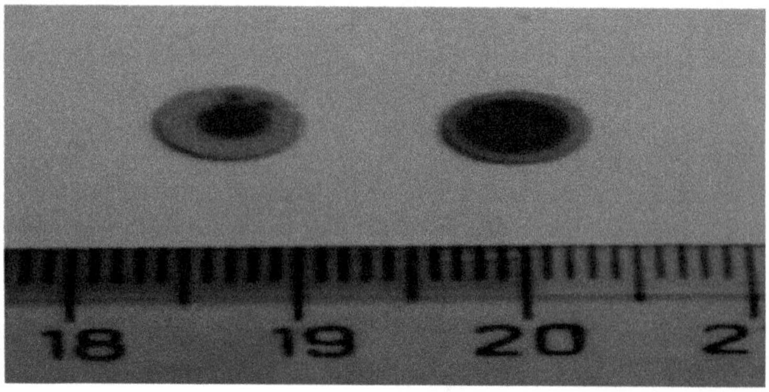

Figure 3.
Samples of PLZT material taken by Wawrzała and Korzekwa at the Silesian University of Poland.

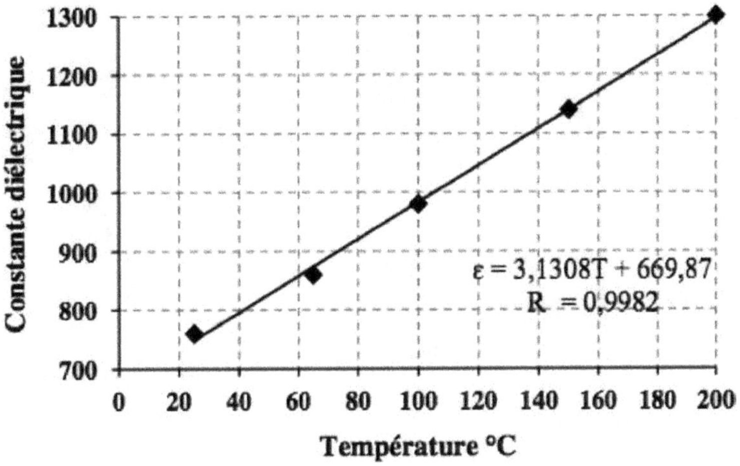

Figure 4.
Dielectric constant of PLZT as a function of temperature.

5. Sensor interrogation method

The transduction mode, size, and frequency of operation of this sensor are important characteristics that represent a technological break with the existing systems of passive wireless temperature sensors RFID and SAW.

To be remotely accessible, the sensor requires a reader that is compatible with its operating characteristics. Technical criteria for the use of a reader must be defined to satisfy the detection but also to ensure that the interrogation range is as long as possible. The existing readers for passive sensor interrogation, present in RFID and SAW technologies, do not meet the problems imposed on our study in terms of high frequencies of use and a range greater than 10 m.

As a result, the characteristics of our sensor (wide range of detection, analysis, and processing of high frequency signals) guide us to consider a radar technology reader. Its operating principle, as with any radar, is to send a flow of electromagnetic waves to the sensor, which will return an echo whose power amplitude and will depend on the measured temperature. Indeed, radar is used in many applications such as level measurement, obstacle detection for automobiles, meteorology, or the military [16]. Its use for passive sensor network interrogation with RF transduction presents an innovative solution.

The proposed temperature sensor uses a millimeter radiofrequency transduction. The resonant frequencies of the sensor are included in the Ka band and shift from a bandwidth of a few hundred MHz to a few GHz. An antenna to communicate remotely with the reader will connect the sensor. To interrogate this sensor, we turned to a radar technology reader developed during Chebila's thesis [17], according to precise technical criteria in terms of operating frequency satisfying wireless communication over a range greater than 20 m. This distance remains a key point because many applications in the aeronautics, construction, and nuclear sectors refer to it for the installation of sensor networks. The modulation technique of this radar and its architecture based around a voltage-controlled oscillator (VCO) facilitated its realization and adjustment (see **Figure 5**).

The radar developed in 2011 is a frequency-modulated continuous radar (FMCW), used in the Ka band around 30 GHz (see **Figure 6**) [17]. This HF radar will be used to remotely detect the temperature sensor measurements. The signals received by the reader must therefore inform us about the distance between the radar and the sensor but also about the temperature value coming from the

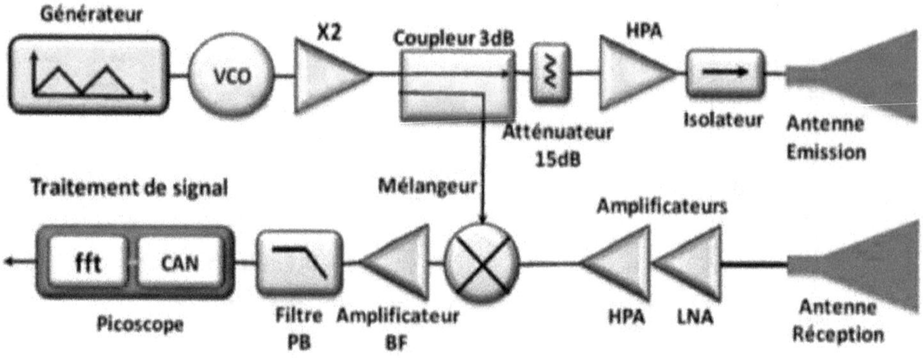

Figure 5.
Synoptic diagram of the 30 GHz radar.

Figure 6.
Picture of the radar.

questioned measuring cell. In conclusion, the radar in question satisfies three important parameters for remote reading: its range is greater than 20 m, works at a frequency compatible with the proposed sensor, and contains a system for identifying cells within a network.

The potential advantages of this type of transducer are:

- A significant reduction in signal losses thanks to the direct modulation of the microwave signal by the quantity to be measured

- A high sensitivity of the electromagnetic propagation to the environment used to perform the sensor function

- High spatial and temporal resolution due to the high operating frequency

- A more flexible choice of operating frequency that can be adapted to the different operating constraints of the sensor

- Easy integration into a measurement chain (radar and antennas)

The following section is devoted to the results of microwave measurements made using a high-performance simulator, allowing the frequency offset to be monitored and a direct relationship to be established between the temperature variation and the observed frequency offset. In this way, a temperature measurement is carried out via an electromagnetic transduction.

6. Simulation results of the complete sensor

Series of free oscillation simulations (eigenmode) using the HFSSTM software, applied to the dielectric resonators, determine the diameter, thickness, permittivity of the dielectric resonator, as well as the distribution of the electromagnetic field necessary to define the caustic and the optimal coupling with the coplanar lines (see **Figure 7**). The determination of caustic makes it possible to establish the distribution of the electromagnetic field in the RD and leads to the definition of the position of the coplanar lines. In addition, the diameter of the resonator and the confinement of the electromagnetic field in it impose the distance between these two micro-machined lines.

In this study, we studied a dielectric resonator with a relative permittivity of 80 and no losses. The thickness and diameter of the dielectric resonator are used to determine the resonance modes and frequencies associated with them. Subsequently, we are interested in the modification of the physical properties of the PLZT material in the presence of a change in temperature, more precisely the variation of its dielectric permittivity. As shown in **Figure 8**, we observe the distribution of the electric field of the $WGE_{8.0.0}$ mode in the dielectric resonator; it is thus isolated around 30 GHz.

We should also mention that the overall circuit of our detection system represents a directional filter consisting of two parts:

- Coplanar lines (CPWs) used for RD excitation and field propagation electromagnetic

- Dielectric resonator used for coupling and excited in WGM as well as the material PLZT as an element of recognition

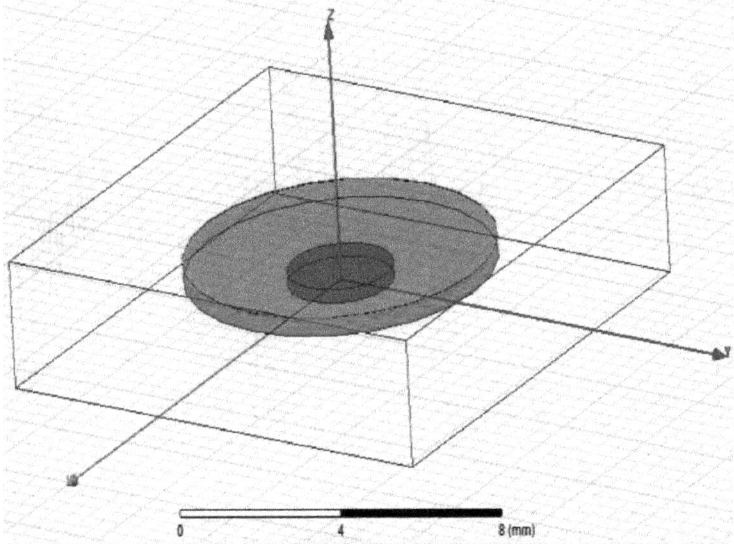

Figure 7.
The sensitive material and the resonator in the cavity with a holder.

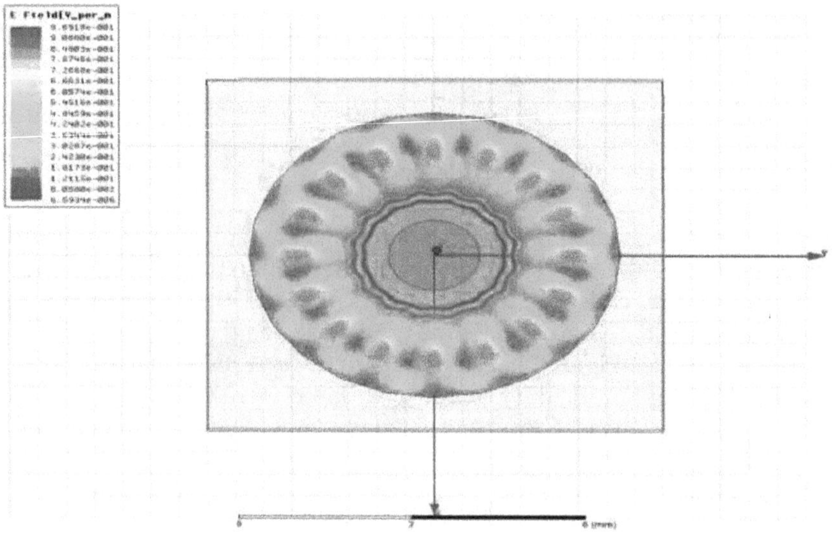

Figure 8.
The field distribution of the $WGE_{8.0.0}$ *to 30 GHz mode for eigenmode calculation (HFSS™).*

From the coupling coefficient S_{12} between access 1 and 2 (simulated) given as a function of frequency, the gallery modes WGE and WGH were identified over a frequency range of 25–40 GHz. **Figure 9** shows the look of the transmission parameter between access 1 and 2 in the Ka-band.

Figure 10(a) and **(b)** shows examples of simulation results corresponding to amplitudes of the magnetic field of the gallery modes at their resonant frequencies.

Based on the results obtained as well as those in the literature, PLZT appears to be the right candidate for frequency temperature transduction. Indeed, it has been previously demonstrated that the dielectric permittivity of this material can be modified in the presence of a temperature variation. We therefore aim to analyze the impact of such a modification on the resonance frequency of a gallery mode. For this purpose, the dielectric constant $\varepsilon_r$was varied between 700 and 900 with a

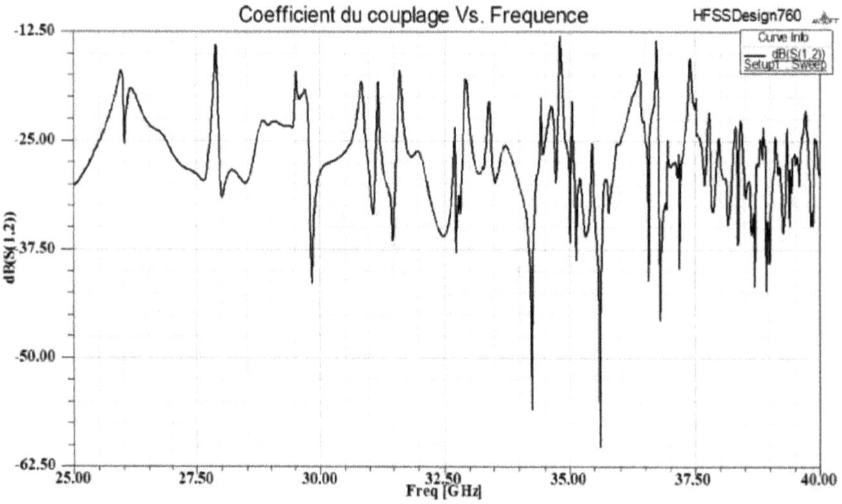

Figure 9.
Coupling coefficient S_{12} *as a function of frequency for* $\varepsilon_{PLZT} = 760$.

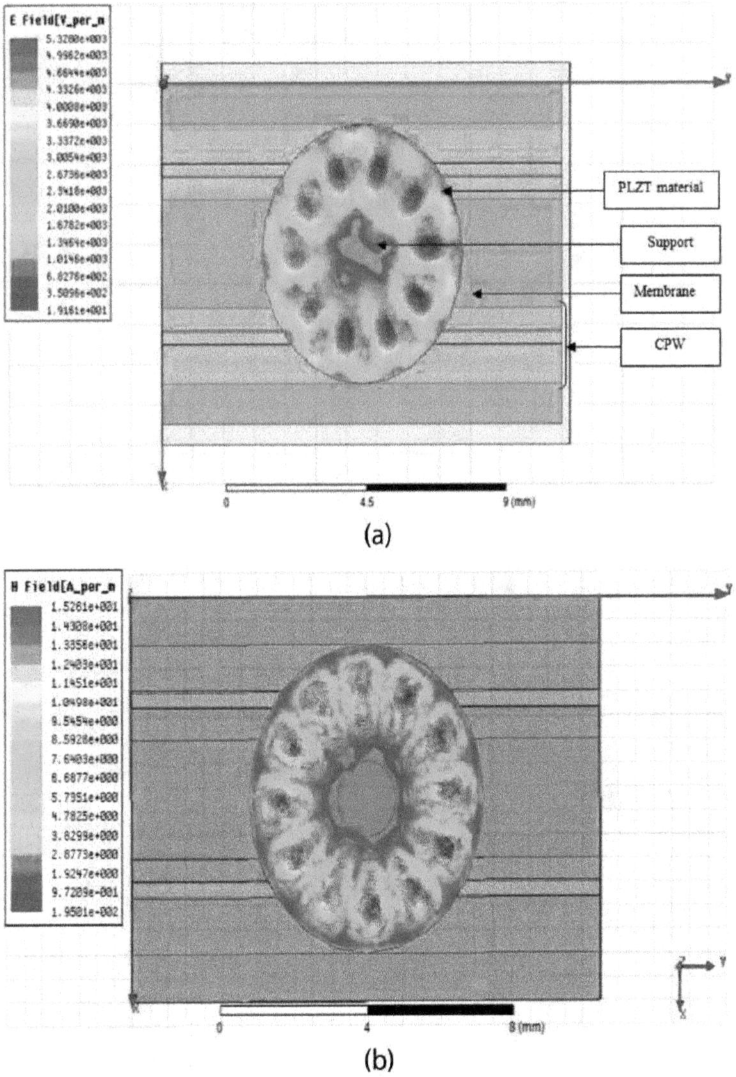

Figure 10.
Amplitude of the magnetic field of the gallery modes: (a) $WGH_{6,0,0}$ *and (b)* $WGE_{5,0,0}$.

14% variation. As shown in **Figure 11**, the variation in PLZT permittivity produces measurable changes in resonance frequency, reflected in a shift to low frequencies of about 1 GHz, for example, in the case of $WGE_{4.0.0}$ mode.

These modifications on the resonance frequency for variations in the permittivity of the PLZT material highlight the high sensitivity of this type of device. This sensitivity represents that of the electromagnetic transducer, which transforms a variation in permittivity into a variation in the resonance frequency of a WGM. In order to evaluate this sensitivity, we have shown in **Figure 12(a)** and **(b)** the resonance frequencies of the $WGH_{7.0.0}$ and $WGE_{9.0.0}$ modes, respectively, as a function of the permittivity of the PLZT when there is a temperature variation. The results obtained with a cylindrical dielectric resonator have shown that the resonance frequencies of the gallery modes are very sensitive to the change in the permittivity of the sensitive material. The relationship between permittivity and resonance frequencies is approximately linear.

The sensor sensitivity is the combination of the transducer sensitivity with the variation in PLZT permittivity as a function of temperature.

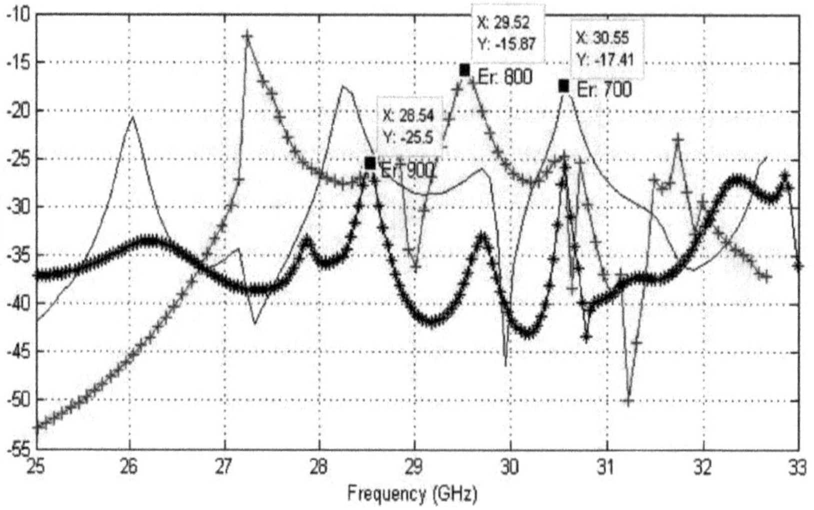

Figure 11.
Transmission coefficient (S_{12}) as a function of frequency for variations in PLZT permittivity (ε_r = 700, 800, and 900).

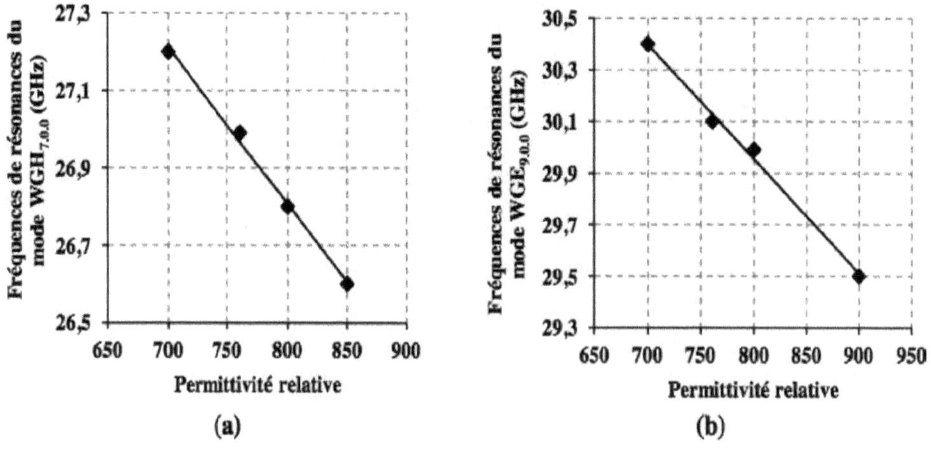

Figure 12.
Resonance frequencies as a function of PLZT permittivity: (a) $WGH_{7,0,0}$ mode and (b) $WGE_{9,0,0}$ modes.

The gallery modes in the dielectric resonator and the PLZT material are evidently used here to measure the permittivity change and deduce the temperature change. From rigorous electromagnetic numerical simulations (HFSS™), we show here that a small change in the permittivity of the sensitive material induces a large variation in the resonance frequency of the dielectric resonator gallery modes.

Based on the characteristics of the sensitive material used and the previous results, it is therefore possible to deduce the relationship between the resonance frequency of the detected gallery modes and the temperature. As a result, a linear dependence between frequency and temperature is clearly observed for all modes detected.

In other words, the relatively linear dependence between temperature and dielectric constant of the sensitive material leads to a change in the coupling coefficient between the two coplanar waveguides, and this is subsequently reflected in a shift in the resonance frequency of the system. As a result, a temperature variation usually results in a shift in the resonance frequency of the excited mode in the dielectric resonator.

7. Application of the sensor for marine fire detection

In terms of application areas, sensor networks have a potential that is revolving around many sectors of our economy and our daily lives; from environmental monitoring and preservation to industrial manufacturing, automation in the transport and health sectors, and the modernization of medicine, agriculture, telematics, and logistics. This new technology promises to revolutionize the way we live, work, and interact with the physical environment around us. Wirelessly communicating sensors with computing capabilities facilitate a series of applications that were impossible or too expensive a few years ago. Today, these tiny and inexpensive devices can be literally scattered over roads, structures, walls, or machines, capable of detecting a variety of physical phenomena. Many fields of application are then considered, such as disaster detection and monitoring, environmental monitoring and biodiversity mapping, intelligent buildings, precision agriculture, machine monitoring and preventive maintenance, medicine and health, logistics, and intelligent transport.

Today, the use of these sensors is increasingly required for supervision and safety. Industrial companies then propose wireless sensors that can inform the user about the evolution of different physical quantities, so they constitute a very fertile research axis. In addition, the development of temperature sensors has several advantages, the most important of which is safety. The current trend, given the new applications that are emerging, is to oversize sensors and make them compatible with signal processing systems in order to obtain fully integrated systems. Environmental objectives and firefighting are the most targeted applications today. The lists of applications (safety, control, analysis, comfort, etc.) and fields of application (environment, safety, medical, automotive, home automation, etc.) are very long, reflecting the great interest in the development of temperature sensors.

The development of these systems generally includes a miniature, low-cost, and high-performance sensor. This is what drives our current research. Indeed, miniaturization is important to be able to easily embed autonomous systems that are increasingly distributed in networks. The cost price is of course an important factor and will be decisive for the marketing development of these sensors. The quest for performance is to make the information obtained by these sensors more reliable and affordable. This is practically interesting for the intended application: improving firefighting in marine applications.

Obviously, of all the disasters that can happen to a ship, fire is certainly the most horrible. Ashore, occupants of burning buildings can rely on fire pumps and ladders that can be on site within minutes of the first alarm signal. A ship at sea, on the other hand, must rely solely on itself in fighting the fire as in everything else, and from the first fire signal to retirement or hard-won victory, there may be no chance of a canoe rescue in bad weather.

The consequences of a fire on board a ship are always costly and sometimes tragic. It is therefore essential to have effective firefighting systems, but it is now known that conventional temperature sensors, which were widely used in the past, have significant disadvantages. The present system aims to solve this problem by using simple, autonomous, and inexpensive means. As a result, this design greatly reduces the overall energy consumption of the temperature sensor network on board ships.

To effectively control ship temperature, the use of this passive temperature sensor makes it possible to keep an eye on the temperature at all times, for long periods of time, and to alert staff in the event of a problem. If the temperature in a monitoring zone suddenly exceeds the threshold value, the sensor detects it in real time and transmits the information to the supervisor for intervention. This avoids the risk that the fire will remain ignored for a long time and therefore take on such a magnitude that any action to fight it will be too late and therefore futile.

This marine fire protection system therefore makes it possible to detect any fire risk in time, before or quickly after it is triggered, and to manage alarms in real time throughout the journey. Clearly, the deployment of such a network can provide an alarm system to detect intrusions, and it has a great advantage for long-term use on board ships without the need to charge or change the battery.

8. Conclusion

The results presented in this chapter are very encouraging: the simulation of a passive temperature sensor based on an electromagnetic transduction gave very good performances. These simulation results obtained, using previous work, allow us to consider the realization of a new high-performance temperature sensor. The optimization of this type of device was done using global electromagnetic simulations on HFSSTM including all system elements. The combination of PLZT material, dielectric resonator, and coplanar lines makes it possible to produce a narrowband filter that we excite with an electromagnetic wave at microwaves in gallery modes.

The key point of our application is based on the performance of the perovskite material and the properties of the RD gallery modes. Indeed, the dielectric resonator, covered with the sensitive material, is excited in a gallery mode that allows its oversizing with millimeter waves and its association with transmission lines in order to have a band-pass directional filter. Thus, we designed the entire 3D device on HFSS™. This software has the advantage of being rigorous and allows, a priori, taking into account all the physical and geometric characteristics of the device.

In the second part of this chapter, we have presented some of the results obtained. Several gallery modes have been observed over a wide frequency band; the Ka-band. We have therefore shown through these simulations that the measurement of the resonance frequency of a gallery mode in the dielectric resonator translates, in principle, a temperature variation with a remarkable sensitivity.

The implementation of such a device, which offers passive temperature detection, makes it possible to consider the design of a temperature sensor with high sensitivity electromagnetic transduction. The characterization of such a sensor makes it possible to evaluate the frequency in terms of geometric dispersions such as the influence of material thicknesses, and diameters will be presented in the next chapter.

Finally, our contribution provides innovative technology to meet the need to improve fire protection in maritime transport. Temperature control and traceability still require manual procedures dependent on onboard energy. The proposed system simplifies and automates relatively all these manual interventions and particularly does not require any onboard power supply. Thereafter, this firefighting device has a sensitivity of about 10 MHz/°C and allows the temperature to be controlled at any time and to react quickly in the event of a problem.

Author details

Asma Bakkali[1,2], José Pelegri-Sebastia[2*], Youssef Laghmich[3]
and Abdelouahid Lyhyaoui[1]

1 Laboratory of Innovative Technologies, Abdelmalek Essaadi University, Tangier, Morocco

2 Research Institute for Integrated Management of Coastal Areas, Universitat Politècnica de Valencia, Valencia, Spain

3 Polydisplinary Faculty, Hassan 1st University, Khouribga, Morocco

*Address all correspondence to: jpelegri@eln.upv.es

References

[1] Peng H. Study of whispering gallery modes in double disk sapphire resonators. IEEE Transactions on Microwave Theory and Techniques. 1996;**44**:848-853

[2] Boriskina VS, Nosich IA. Radiation and absorption losses of the whispering-gallery-mode dielectric resonators excited by a dielectric waveguide. IEEE Transactions on Microwave Theory and Techniques. 1999;**47**:223-231

[3] Mouneyrac D. Utilisation des modes de galerie pour la caractérisation des semi-conducteurs et la réalisation d'un oscillateur ultra faible bruit de phase. Thèse de doctorat de l'Université de Limoges; 2011

[4] Guillon B. Conception et réalisation de circuits millimétriques micro-usines sur silicium: Application à la réalisation d'un oscillateur à résonateur diélectrique en bande Ka. Thèse de doctorat de l 'Université Paul Sabatier de Toulouse; 1999

[5] Guillon B, Cros D, Pons P, Cazaux JL, Lalaurie JC, Plana R, et al. Ka band micromachined dielectric resonator oscillator. IEEE Electronics Letters. 1999;**35**(11):909-910

[6] Guillon B, Grenier K, Pons P, Cazaux JL, Lalaurie JC, Cros D, et al. Silicon micromachining for millimeter-wave applications. Journal of Vacuum Science & Technology, A: Vacuum, Surfaces, and Films. 2000;**18**(2):743-745

[7] Krupka J, Derzakowski K, Abramowicz A, Tobar ME, Geyer RG. Use of whispering-gallery modes for complex permittivity determinations of ultra-low-loss dielectric materials. IEEE Transactions on Microwave Theory and Techniques. 1999;**47**(6):752-759

[8] Wawrzała P, Korzekwa J. PLZT type ceramics as a material for applications in power pulse capacitors. IEEE Proceedings. 2012;**10**(1109):1-4

[9] Kuscer D, Korzekwa J, Kosec M, Skulski R. A- and B-compensated PLZT x/90/10: Sintering and microstructural analysis. Journal of the European Ceramic Society. 2007;**27**:4499-4507

[10] Lee FY, Jo HR, Lynch CS, Pilon L. Pyroelectric energy conversion using PLZT ceramics and the ferroelectric-ergodic relaxor phase transition. Smart Materials and Structures. 2013;**22**:1-16

[11] Kozielski L, Plonska M, Bucko MM. Electrical and mechanical examination of PLZT/PZT graded structure for photovoltaic driven piezoelectric transformers. Materials Science Forum. 2010;**636-637**:396-373

[12] Trivijitkasem S, Koyvanich K. Characterization of lead lanthanum zirconate titanate (PLZT) ceramics sintered at various temperatures. Kasetsart Journal (Natural Science). 2007;**41**:192-197

[13] Jatlaoui MM. Capteurs passifs a transduction électromagnétique pour la mesure sans fil de la pression. Thèse de doctorat de l'INPT de Toulouse; 2009

[14] Saadaoui M. Optimisation des circuits passifs micro-ondes suspendus sur membrane diélectrique. Thèse de doctorat de l'Université Paul Sabatier de Toulouse; 2005

[15] Hallil H. Conception et réalisation d'un nouveau capteur de gaz passif communicant à transduction RF. Thèse de doctorat de l'Université Paul Sabatier de Toulouse; 2010

[16] Lacomme P, Hardange JP, Marchais JC, Normant E. Radar applications and roles. Air and Spaceborne Radar Systems. 2001;PM**108**:347-369

[17] Chebila F. Lecteur Radar pour Capteurs Passifs à Transduction Radio Fréquence. Thèse de doctorat de l'INPT de Toulouse; 2011

7

Organic Inorganic Perovskites: A Low-Cost-Efficient Photovoltaic Material

Madeeha Aslam, Tahira Mahmood and Abdul Naeem

Abstract

Organic-inorganic perovskite materials, due to the simultaneous possession of various properties like optical, electronic and magnetic beside with their structural tunability and good processability, has concerned the attention of researchers from the field of science and technology since long back. Recently, the emergence of efficient solar cells based on organic-inorganic perovskite absorbers promises to alter the fields of thin film, dye-sensitized and organic solar cells. Solution processed photovoltaics based on organic-inorganic perovskite absorbers $CH_3NH_3PbI_3$ have attained efficiencies of over 25%. The increase in popularity and considerable enhancement in the efficiency of perovskites since their discovery in 2009 is determined by over 6000 publications in 2018. However, although there are broad development prospects for perovskite solar cells (PSCs), but the use of $CH_3NH_3PbI_3$ results in lead toxicity and instability which limit their application. Therefore, the development of environmental friendly, stable and efficient perovskite materials for future photovoltaic applications has long-term practical significance, which can eventually be commercialized.

Keywords: organic-inorganic, Perovskite, photovoltaic material, optical properties, solar cells

1. Introduction

Organic-inorganic perovskites materials have emerged as a promising material for high-efficiency nanostructured devices such as light-emitting diodes (LEDs), detectors, field-effect transistors, and photovoltaic devices etc. [1, 2]. Organic-inorganic perovskites have attracted extensive attention due to their promising optical and electronic properties, excellent crystallinity, adjustable bandgap, long charge diffusion length, electroluminescence, and conductivity [3, 4]. As the most fascinating new-generation photovoltaic materials, organic–inorganic perovskite due to their facile synthesis, low temperature deposition, and capability to make flexible devices has been considered as a vigorous component of the efficient, low-cost, lightweight and flexible Perovskite solar cells [3, 4]. Perovskite solar cells (PSCs) have rapidly become the leading edge of third generation 3G photovoltaic technologies [5, 6]. PSCs based on the organic inorganic perovskite materials have fascinated great consideration, with their power conversion efficiencies (PCEs) reaching 25.2% certified [7, 8]. Over the past several months, it has been observed a surprising revolution and rapid progress in the field of emerging photovoltaic, with the understanding of highly efficient solar cells based on organic inorganic perovskite materials. This perovskite technology is now well-matched with the 1G and 2G technologies and is thus probably be embraced by the conventional photovoltaic community and industry [9, 10].

1.1 Organic inorganic perovskite materials in solar cells

The advent of organic inorganic perovskite based solar cells has resulted in rapid growth in photovoltaic history. Organic inorganic perovskite materials have recently, fascinated greater attention due to its outstanding light-harvesting features [7].

Organic-inorganic perovskite absorbers have appeared in the field of DSSCs since 2009. The first perovskite-sensitized DSSCs were developed by Kojima et al. [11] which obtained PCE of 3.13% using liquid electrolytes. However, continuous irradiation produced a photocurrent decay in an open cell when exposed to air. Later, the electron transporting layer (Titania) surface and perovskite process-ing were optimized, and in 2011, Im et al. [12] developed first stable PSC, using $CH_3NH_3PbI_3$-based iodide liquid electrolyte offered a PCE of 6.5%. However, the perovskite nanocrystals dissolved in the liquid iodide electrolyte solution, and the cell degraded within 10 minutes. To avoid the problem of perovskite dissolution in an electrolytic solution, the liquid electrolyte was replaced by a solid in 2012, and a PCE of 9% was achieved showing good stability up to 500 h without significant losses [13, 14]. Afterward, Al_2O_3 an insulating network was used to substitute conducting nano porous TiO_2. By using mixed $MAPbI_{3-X}Cl_X$ as the sensitizer, an enhanced open-circuit voltage (V_{OC}) and PCE (10.9%) was achieved [15]. In 2013, a successive deposition method for the perovskite layer within the porous metal oxide film was developed. The fabrication technique for solid-state mesoscopic solar cells greatly improved the reproducibility of cell performance and produced a high PCE of 15%. Many PSC devices are now attaining PCE > 20% since 2015 and 25% in 2019 [4, 16]. National Renewable Energy Laboratory (NREL), on 3rd August 2019 declared a new world record PCE of 25.2% for PSCs. This PCE value is improved up to ~28% for perovskite-silicon tandem structures [4].

1.2 Device architecture of perovskite solar cells

PSC consists of a perovskite absorbing material sandwiched between electron transporting layer (ETL) and hole-transporting layer (HTL) along with the transparent conducting oxide substrate (FTO) and a top electrode such as gold, silver [17]. In PSCs, the effective charge separation and the light harvesting efficiency are significantly affected by the properties like particle size, porosity, surface area, surface morphology, band gap, thickness of semiconductor materials, and the nature of organometal halide perovskites [18].

The primary function of an **ETL** is to extract a photo generated electron from perovskite and then transfer to electrodes. The basic criterion for an ideal ETL is high optical transmittance, excellent electron mobility, high conductivity, and an appropriate work function. ETL also performs as hole-blocking layer (HBL) [19]. The configuration and the choice of ETL are essential factors to understand the electronic mechanisms in PSCs, which control processes such as carrier separation, extraction, transport, and the recombination. Hence, the configuration of device structure is critical to alter different materials for ETL, electrode contacts, and the barrier layer of insight these processes and mechanisms [18]. TiO_2, which has a wide band gap, has been extensively studied as an efficient electron transport material

(ETM). Moreover, ZnO and other *n*-type semiconductors such as SnO_2, Nb_2O_5 and $BaSnO_3$ are frequently used as ETMs and are used in flexible perovskite solar cells [20, 21].

The HTL lies in the heart i.e. in between the metal electrode and perovskite of device. It plays a central-role in the PSC and extracts holes from the perovskite and transfer them to top-electrode. It avoids the direct contact of perovskite and top electrode [22]. For efficient hole transport, the highest occupied molecular orbit (HOMO) must match the valence band (VB) of perovskite materials. According to the chemical composition, HTMs in PSCs can be divided into two types: organic and inorganic HTMs. Spiro-OMeTAD is the most used organic HTM, which displays good penetration in perovskite and is an appropriate match with the VB energy of perovskite, though its hole mobility is not as superior as that of other organic HTMs. [21, 23]. Hence, in order to improve the hole mobility, polymers are doped with *p*-type (i.e., cobalt or Lithium salts) or some additives (i.e., bis(trifluoromethane) sulfonimide lithium, LiTFSI, and 4-tert-butyl pyridine, TBP) [21, 24]. Other organic materials reported as HTMs in PSCs are PTAA, PEDOT:PSS, P_3HT etc. [25]. Inorganic *p*-type semiconductor materials, due to their advantages such as high hole mobility, wide band gap, and easy solvent treatment process as compared with organic HTMs exhibit the possibility to replace organic HTMs. The reported inorganic HTMs for PSCs are NiO, CuSCN, CuI, $CsSnI_3$ etc. [21, 25].

A PSC includes an organic-inorganic perovskite material as the light-harvesting active layer. Amongst the component's PSCs, perovskite materials perform a key role. Perovskite is comprised of earth abundant and inexpensive materials. It is processed at lower temperature rather via the printing techniques [26]. The organic–inorganic perovskites can exhibit appropriately good ambipolar charge transport and the primary functions of photovoltaic operation comprising light absorption, generation of charges, and transport of both electrons and holes. They perform both as efficient light absorbers and charge carriers [21]. The commonly used perovskites are Methylammonium lead triiodide ($CH_3NH_3PbI_3$) and formamidinium lead triiodide ($CH_3(NH_2)_2PbI_3$) [27]. Moreover, The PSC architecture is represented in **Figure 1**.

The two main device architectures of PSC are

1. mesoscopic

2. planar

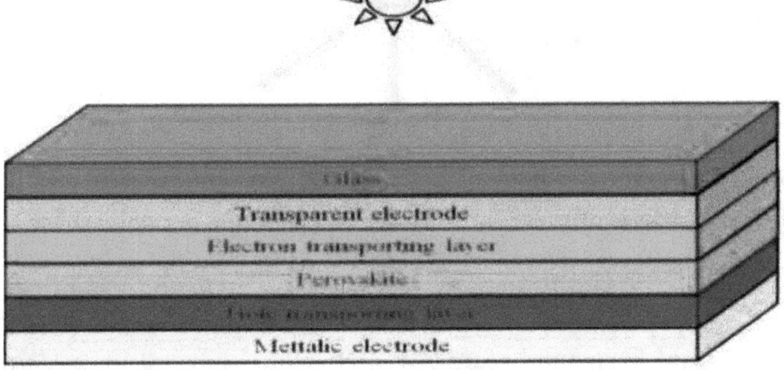

Figure 1.
Representative architecture of PSC.

The conventional PSC consists of mesoscopic n-i-p structure and is the novel architecture of PSC devices which consists of an FTO, an electron transport layer (ETL), a mesoporous oxide layer such as TiO_2, or SnO_2, perovskite (light absorb-ing) layer, a hole transport layer (HTL), and an electrode layer. The mesoporous TiO_2 layer played a significant role in the electron transfer process and as a scaffold providing mechanical support of the perovskite crystal. The use of mesoporous materials in PSC permit the perovskite material to adhere to the mesoporous metal oxide framework to increase the light-receiving area of the photosensitive material and results in improving the efficiency of the device. The mesoporous layer was usually less than 300 nm. The presently mesoporous structure of PSCs is one of the most common structures with a power conversion efficiency (PCE) greater than 20% [28]. The mesoscopic structure due to the fabrication ease and outstanding best efficiencies is the most extensively adopted in research labs. However, high temperature (˃450°C) sintering is required for mesoporous layer-based devices, which prevents the use of plastic substrates [29–31]. To overcome this problem, the planar perovskite solar cell was developed that showed comparable performance for mesoporous perovskite solar cell. Planar heterojunction PSCs have been reported by several researchers in which only compact layers of ETM and HTM is used without a mesoporous layer at a temperature lower than 200°C [19, 21, 32]. Hence the planar structure turns out to be very attractive for basic research purposes. The mesoscopic and planar structures of PSC are represented in **Figure 2**.

1.3 Structure of perovskite materials

Perovskites materials are designated by the formula ABX_3, where A and B are cations of different sizes (A being larger than B) and X is an anion [7]. The crystal structure of perovskites is depicted in **Figure 3** and it has a cubic crystal structure with three-dimensional (3D) framework sharing BX6 octahedron with the A ion placed at the octahedral interstices [33, 34]. In organic-inorganic materials, the A is organic cations generally methylammonium, ethylammonium and formamidinium and B is usually metal ions of group IV such as Pb^{2+}, Sn^{2+} and Ge^{2+} whereas the X are VII group anions I^-, Cl^- and Br^- [2, 7, 34].

The crystallographic stability and probable structure of perovskite can be inferred by studying a "tolerance factor" t and an "octahedral factor" μ. A "tolerance factor" is defined as the "ratio of the A-X distance to the B-X distance in an idealized solid-sphere model" and is represented by the formula:

$$t = \frac{(R_A + R_X)}{\left[\sqrt{2}(R_B + R_X)\right]} \quad (1)$$

where R_A, R_B and R_X are the ionic radii of the corresponding ions.

An "octahedral factor" is defined as "the ratio $\frac{R_B}{R_X}$".

For halide (X = F, Cl, Br, I) perovskites, generally $0.81 < t < 1.11$ and $0.44 < \mu < 0.90$ [35]. If t value lies in the narrow range 0.89–1.0, the structure is cubic, but, if it is lower, symmetric tetragonal or orthorhombic structures is expected [2]. Regardless of these limitations, conversions between these structures are common on heating, at the high-temperature cubic phase is generally obtained.

For the organic–inorganic perovskites, organic cation A usually methylammonium ($CH_3NH_3^+$) with R_A = 0.18 nm, ethylammonium ($CH_3CH_2NH_3^+$) (R_A = 0.23 nm) and formamidinium ($NH_2CHNH_2^+$) (R_A = 0.19–0.22 nm) are used. The cation B is commonly Pb (R_B = 0.119 nm); however, Sn (R_B = 0.110 nm) forms similar compounds with more ideal bandgap but exhibits lower stability

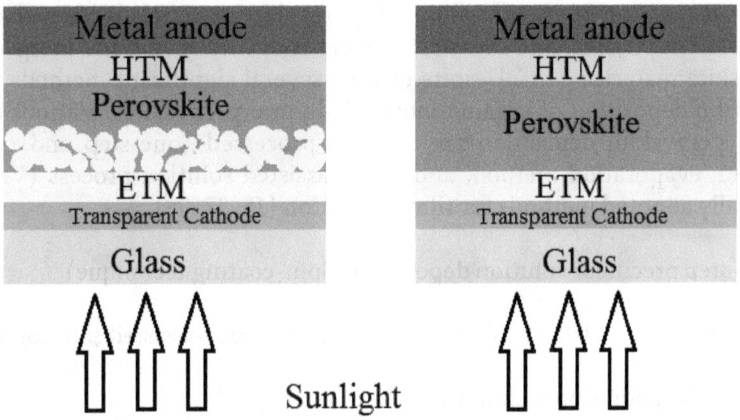

Figure 2.
Representative scheme of a mesoporous (right) and planar PSC (left).

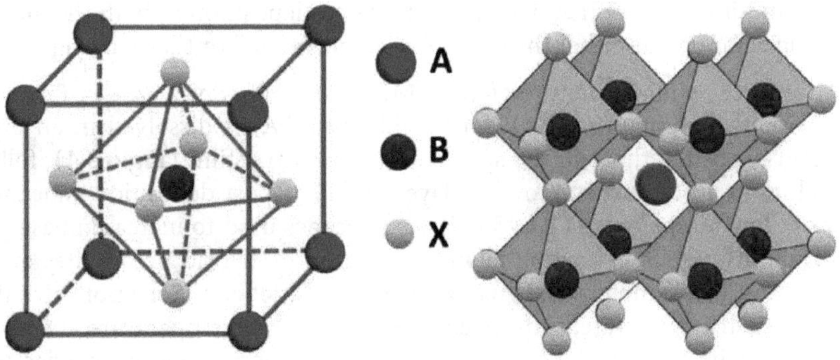

Figure 3.
Structure of perovskite.

(ascribed to the oxidation of Sn to SnI_4in the iodide perovskite). The anion X is a halogen, generally iodine (R_X = 0.220 nm) is used, however Br and Cl are also used (R_X = 0.196 nm and 0.181 nm) [35, 36]. The commonly used organic inorganic perovskite material is methylammonium lead triiodide ($CH_3NH_3PbI_3$).

$MAPbX_3$ perovskite show multiple phases as a function of composition and temperature. These different phases have markedly different optical and electrical properties as well as stability. MAPbI displayed α-phase, δ-phase, and γ-phases with transition temperatures of 400 K, 333 K, and 180 K, respectively. Generally, the δ-phase $MAPbI_3$is used as absorber in solar cell due to its thermodynamically stable nature at room temperature and its increased conductivity and absorption coefficient (>26 mm^{-1}) in contrast to the α-phase. Though, a phase transition from δ-phase to α-phase may occur under continuous 1 sun illumination [15].

2. Synthesis of inorganic–organic solar cells materials

The deposition technique of organic-inorganic perovskites films is quite an important issue for perovskite studies, because the possible use of perovskite materials depends on the availability of simple and perfect thin film deposition method. As concerns the preparation methods of organometallic halide perovskite $CH_3NH_3PbX_3$ thin films, solution-based procedures have been proposed to manufacture thin films.

However, deposition of organic-inorganic perovskite materials is often challenging due to different physical and chemical properties of the organic and inorganic parts of perovskite materials [15]. Despite of this, several significant methods are used for thin film deposition of organic-inorganic hybrid perovskites. Various methods used for perovskite deposition are solution-processed (one-step and two-step) deposition, evaporation method, and vapor assisted solution process (VASP) are the typically adopted methods for film deposition [15, 37, 38].

1. one-step precursor solution deposition (Spin-coating technique)

2. two-step sequential deposition (including the vapor-assisted solution process)

3. thermal evaporation technique

2.1 One-step precursor solution deposition (spin-coating technique)

One-step processing (spin-coating) is a suitable technique extensively applied for uniform thin film deposition and is based on the co-deposition of both the inorganic and organic components either through solution processing or thermal evaporation.

In solution processing, a mixture of both MX2 (M ¼ Pb, Sn; X ¼ Cl-, Br-, I-) and AX (A ¼ methylammonium MA); formamidinium, FA) is dissolved in an organic solvent and deposited through the spin coating to form a film (**Figure 4**), followed by annealing to produce the perovskite layer [15]. The post deposition annealing of the films at low temperature (T < 250°C) is sometimes used to increase phase purity and crystallinity [6]. Spin-coating allows deposition of hybrid perovskites on various substrates, containing glass, quartz, plastic, and silicon. Selection of suitable parameters such as substrate, spin speed and the substrate temperature are essential for this technique and can be selected accordingly. The wetting properties of the solution on the chosen substrate can be improved by pre-treating the substrate with a suitable adhesion agent. The spin-coating technique does not involve cumbersome equipment and it gives high-quality films in quite short time at room temperature.
It is considered as a distinct case of solution crystal growth, which results in the formation of highly oriented perovskites layer on a substrate. In order to obtain a layer with the desired thickness, optimization of various parameters such as concentration of perovskites solution, and spin-coating parameters (spin speed, acceleration and spin duration) can be carried out. Generally, 2D homogeneous perovskites films with a thickness ranging 10 nm to 100 nm can be obtained by carefully choosing the parameters. The selection of solvent is also important by considering the solubility for both the organic ammonium and the inorganic lead halide. The usually used solvents for spin coating technique are Dimethylformamide (DMF) or Dimethyl sulfoxide (DMSO) [39]. These spin-coated perovskites films are very reproducible, and this technique is suitable for all PSC structures (mesoporous vs. planar) [39].

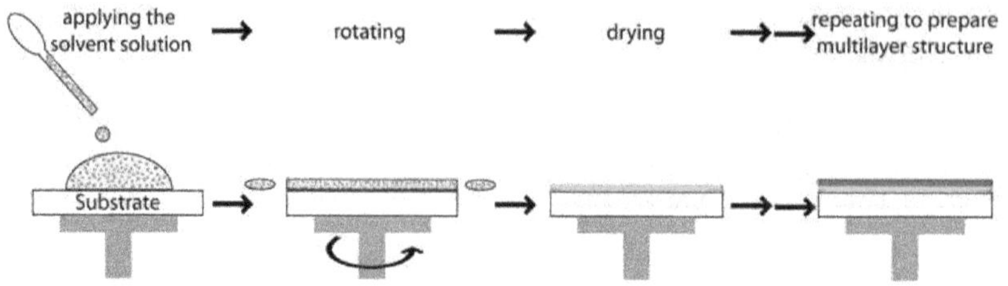

Figure 4.
Schematic of the spin-coating process [4].

2.2 Two-step dip-coating

Mitzi [40] first time reported the two-step dipping technique in 1998, and later by Burschka et al. [41] in 2013. In a two-step dip-coating deposition process, a metal halide PbI_2 layer is first deposited by vacuum evaporation or spin-coated on a substrate. Then this coated film is altered into the perovskite by dipping into an organic MAI solution as it is shown in **Figure 4**. This method offered PCE of 15% and certified 14.14% [39]. Suitable selection of solvent is important for the dipping process. The solvent is selected such that can dissolve organic salt but cannot metal halide and the final organic-inorganic perovskite, toluene/2-propanol mixture is an appropriate solvent for the organic salt. The organic cations in solution intercalate into and react with metal halide on the substrate and form a crystalline film [6]. The dipping times are quite short: several seconds to minutes, depending on the system. This method is a suitable method for a variety of inorganic and organics, even if they have an incompatibility in solubility. This process effectively reduces the chemical reaction between the perovskite and the underlying ETL. The development of successive deposition methods has offered a variety of ETL options, though allowing for perovskite films to be prepared successfully at room temperature [15]. In addition, Chen et al. [42] developed a vapor assisted solution processing (VASP) method that used the reaction between MAI vapor and pre-deposited PbI_2 to form the completed perovskite film. The resulting $MAPbI_3$ exhibits excellent film quality.

2.3 Thermal evaporation technique

M. Era et al. [43] first used thermal evaporation method. They used the dual source vapor deposition by using ammonium iodide RNH_3I and lead iodide PbI_2, organic and inorganic source were co-evaporated and deposited on quartz. The pressure of evaporation chamber was about 10^{-6} Torr. By using this method, it is possible to precisely control the smoothness and thickness of the films. However, it is often hard to balance the organic and inorganic rates, which is important in attaining the correct composition of the resultant perovskite films. Furthermore, Mitzi et al. [40] developed another method, by using a single evaporation source to deposit perovskites thin films called single source thermal ablation (SSTA) technique. This consists of a vacuum chamber, with an electrical feed-through to a thin tantalum sheet heater. A suspension of insoluble powders in a drying solvent is placed on the heater. Under a suitable vacuum, the temperature goes to approximately 1000°C in 1–2 second, the whole starting charge ablates from the heater. After ablation, the organic and inorganic parts reassemble on the substrates to yield films of the chosen product. Liu et al. [44] in 2013, improved this technique as a dual-source vapor deposition method for pinhole-free $MAPbI^{1-}{}_{x}Cl_x$ perovskite films with a thickness of hundreds of nanometers for planar PSCs.

Later on, the chemical vapor deposition (CVD) method was reported by Leyden et al. [45], which precisely control the crystallization process. Vapor deposition methods are appropriate for multi-layered thin-film and a variety of substrates, though needs high vacuum [39]. However, this method has drawbacks of yield and therefore is not very effectively employed at industrial scale [46]. Though great achievements have been attained, researchers still meet some challenges, involving reproducibility and grain boundaries of perovskite films which are considered as a defect region initiates carrier recombination and accelerates device degradation. Hence, efforts to increasing grain size and reducing grain boundary of films are critical for stable and highly efficient PSCs.

3. Advances in perovskite solar cells

The possibility of merging the properties of inorganic with those of organic solids has inspired intensive research into the versatile properties. Organic-inorganic perovskite materials have been widely used in PSCs using different ETLs and HTLs. The optimization of materials and structures is one of the solutions to improve the PCE. **Table 1** shows some representative devices and their architectures and performance.

Methylammonium Lead halide perovskites ($MAPbX_3$) are mostly regarded as promising light absorbers owing their many advantages comprising high absorption coefficients, optimal bandgaps, and long-range exciton diffusion lengths. These perovskites have led to solar cells with PCEs upto15% in combination with meso-structure metal oxides and deposition methods (such as sequential and vapor deposition) [34]. There were few attempts to synthesize new perovskites by changing halide anions (X) in the $MAPbX_3$ structure, but these materials did not result too much improvement in device efficiency. Optical and electronic properties of organo lead halide perovskites have been considered by replacing MA cation with other organic cations such as ethyl ammonium and formamidinium [55].

Dkhissi et al. [32] fabricated an efficiently $CH_3NH_3PbI_3$-based planar perovskite solar cells on polymer substrates at 150°C or below. The hole blocking layer employed is a TiO_2 layer. The devices showed an average efficiency of 10.6 ± 1.2%, and a maximum efficiency of 12.3% for flexible perovskite solar cells, presenting great potential for further enhancement of the low-cost, low-temperature processing solar technology.

In 2014, Choi et al. [55] modified perovskite material with Cesium (Cs) by doping methyl ammonium lead iodide perovskites by Cesium to improve the performance of inverted-type perovskite/fullerene planar heterojunction hybrid solar cells. $Cs_xMA_{1-x}PbI_3$ perovskite devices achieved improvement in device efficiency from 5.51–6.8% with an optimized 10% Cs doping concentration. The devices exhibited an outstanding increase in efficiency due to increases in short-circuit current density and open-circuit voltage.

ETL	HTL	Perovskite	PCE	References
Gr/ZnO-QDs	Spiro-OMeTAD	$CH_3NH_3PbI_3$	9.73	[18]
TiO_2-Al_2O_3	Spiro-OMeTAD -Li-TFSI	—	10	[24]
ZnO-NPs	Spiro-OMeTAD	$CH_3NH_3PbI_3$	10.2	[34]
TiO_2	Spiro-OMeTAD	$CH_3NH_3PbI_{3-x}Cl_x$	11.7	[47]
TiO_2	Spiro-OMeTAD	$CH_3NH_3PbI_3$	15.4	[48]
TiO_2	—	$CH_3NH_3PbBr_3$−n	8.54	[49]
SnO_2 QD	Spiro-OMeTAD-Li-TFSI	$Cs_{0.05}(MA_{0.17}FA_{0.83})0.95Pb(I_{0.83}Br_{0.17})_3$	20.79	[50]
SnO_2 QD	Spiro-OMeTAD-Li-TFSI	$CH_3NH_3PbI_3$	19.73	[50]
SnO_2	Spiro-OMeTAD	Cs/MA/FA perovskite	20.7	[51]
SnO_2 QD	Spiro-OMeTAD	$CH_3NH_3PbI_3$	19.12	[52]
TiO_2	Spiro-OMeTAD	$(FAI)_{0.81}(PbI_2)_{0.85}(MAPbBr_3)_{0.15}$	21.02	[53, 54]

Table 1.
Comparison of different organic inorganic perovskite materials with different hole and electron TLs in PSCs.

$CH_3NH_3PbBr_3$ and $CH_3NH_3PbI_3$, were used as sensitizers for TiO_2 in a liquid junction solar cell, with open-circuit voltages of 0.61 and 0.71 V were achieved. $CH_3NH_3PbI_3$ on mesoporous TiO_2 showed good charge transport properties, where the perovskite is both the absorber and the hole conductor. Further $CH_3NH_3PbX_3$ (X = Br, I), mixed perovskite lead halides i.e., $CH_3NH_3PbI_2Cl$, $CH_3NH_3PbBr_3 - xClx$, and $CH_3NH_3PbI_3 - xCl_x$ were studied [27].

Giacomo et al. [47] fabricated PSCs using $CH_3NH_3PbI_3$-xCl_x with different hole-transporting materials. The mostly used Spiro-OMeTAD has been compared to the P3HT. By changing the energy level of P_3HT and optimizing the device fabrication, PCE reached to 9.3%. They showed that P_3HT can be used a suitable low-cost hole transport material for efficient perovskite based solar cells.

NiO has been tried as a substitute for organic molecular or polymeric HTMs (spiro-MeOTAD), displaying encouraging results in the $TiO_2/CH_3NH_3PbI_3$ configuration, a PCE of 9.5% was attained with nanocrystalline NiO layer. As the valence band edge (5.4 eV) for NiO is near to that of iodide perovskite (5.3 eV), so post-treatment of NiO film by means of UV light or oxygen plasma is vital to progress hole injection efficiency due to an increase in the work function of NiO by such post treatments. UV-ozone post-treated NiO usually has a greater photovoltaic performance than untreated NiO, due to change in work function and an enhancement in wettability indicating a better chemical interaction between perovskite and NiO [36].

The use of perovskites with mixed cations and halides has become significant for PV applications which are mainly MAPbX ,3 $FAPbX_3$ and $CsPbX_3$ (X = Br or I). on the introduction of MA into FA brings the crystallization of FA perovskite (because MA is slightly smaller than FA) which allows a large fraction of the yellow phase to continue. MA/FA compounds show notable PCEs and therefore the development of these compounds is an opportunity in the advancement of PSCs. Saliba et al. [56] introduces an innovative approach using a triple Cs/MA/FA cation mixture where Cs is used to progress MA/FA perovskite compounds. A small amount of Cs is enough to efficiently suppress yellow phase impurities allowing the preparation of pure, defect-free perovskite films.

Song et al. [57] reported that the combination of FA decreases the release of organic species but does not stop the formation of I/HI. Though, the addition of Cs successfully overcomes the release of all volatile gases. The best photostability is found with FA/Cs mixed perovskites, presenting the complete removal of MA from mixed-cation perovskite is favored for more photostable perovskites.

As $CH_3NH_3PbI_3$ has ambipolar properties and is slightly more p-type than n-type and is satisfactory to develop p-n junction-like devices without an HTM, known as HTM-free photovoltaic cells. $CH_3NH_3PbI_3$ could act both as light absorber and hole transporter in a $CH_3NH_3PbI_3$/mesoporous TiO_2 heterojunction device with a PCE of 5.5%. It was observed that HTM-free perovskite solar cells had a poor FF and a low Voc as compared to those with an HTM, which is related to the larger shunt current along with a lower IPCE for these devices [2].

Different proportions of inorganic (Pb, Sn) cations, organic cations and halide anions (I, Br, Cl) can be combined in mixed perovskites, permitting their proper-ties to be fine-tuned [35]. Tuning of bandgap of $MAPbX_3$ has been attained through the substitution of I with Cl/Br ions, which occurs from a dependence of electronic energies on the effective exciton mass. The optical absorption can be tuned by bandgap engineering to comprise the whole visible spectrum. In the meantime, the combination of Cl/Br into iodide-based structure has markedly advanced the charge transport and the separation kinetics within the perovskite layer. Hence, by tuning the composition of perovskite resulted in improved efficiency and the stability of PSCs. It was observed that an increase in the size of perovskite cation materials resulted in a reduction in the bandgap. A tunable bandgap can be obtained (between 1.48 and 2.23 eV) by replacing the methylammonium with a slightly larger formamidinium cation. Significantly, the reduced bandgap led to a PCE of up to 14.2% and high short circuit currents (>23 mA cm^{-2}) [31].

$CH_3NH_3SnI_3$ is demanded to be a low-carrier-density p-type metal. Theoretical calculations on perovskite recommended that their electronic properties intensely depend on the structure of the inorganic cage and formation of the perovskite octahedral network. By changing the inorganic and organic components and their stoichiometric ratio, it is probable to control the system dimensionality and electronic and optical properties. Furthermore, the presence of weak bonds in the perovskite structures ensures malleability and flexibility that could permit the deposition of thin films on flexible substrates [26].

4. Toxicity and stability issue

4.1 Lead: the toxicity problem

Regardless of the excellent properties and high efficiencies, the poor stability of organic–inorganic perovskite materials are yet a serious challenge, inhibiting PSCs from being commercialized. To be marketable for commercial purposes, PSCs need to be capable of work constantly for over 20 years under outdoor conditions. Thus, large consideration has recently been centered to overcome barriers associated with stability and environmental compatibility of perovskite materials [10].

Presently there is a debate on the use of lead (main component) in PSCs, which causes toxicity problems during device manufacture, placement, and disposal. Hence, the toxicity of lead-based perovskites is an obvious problem due to leaching of lead into the environment [9]. Lead toxicity has been pointed out as one of the most challenging barriers towards the commercialization of solar cells, as compared to stability issues and cost-effective production ways. The environmental impact benefits of lead-free (or lead-reduced) solar cells have been analyzed by Life Cycle Assessment (LCA) [15].

4.1.1 Lead-free perovskite solar cells

Up to now, several research groups have ambiguously proven their solution to this challenge. Thus, it is critical to test alternatives to attain similar optical and photovoltaic performances for the commercialization of PSCs. Several research groups have tried to replace lead with other elements (Sn, Ge) and organic cations with inorganic cations to form new appropriate non-toxic and stable perovskite materials, which may be a long journey before the final commercialization of PSCs [9].

It is worth studying alternatives using lead-free PSCs, but Lead-free PSCs reached a PCE of only 6% at a time when lead-based PSCs produced efficiency of 17%. Moreover, the Sn-based solar cells display poorer stability than Pb-based solar cells. [14]. Bivalent Sn is the most favorable choice for replacing Pb as they both are in the same group and possess analogous lone-pair s orbitals [10]. Both Sn- and Pb-based materials have a tetragonal structure under ambient conditions; however, Sn-based perovskite have a higher symmetrized α phase as compared to the Pb-based materials lower symmetrized ß phase [47]. Chung et al. [54] first demonstrated $CsSnI_3$ as a solid electrolyte in DSSCs. Afterward Chen et al. [58] fabricated a photovoltaic device, ITO/CsSnI /Au/$Ti_{,3}$ attaining very low PCE of 0.88%. However, Sn^{2+} based perovskite undergoes oxidation from Sn^{2+} to Sn^{4+}, which is destructive for the charge transport properties, and PCE. Recently, Lv 2019 [59] reported the replacement of spiro-OMeTAD by a Zn-derivative porphyrin in a lead-free solar cell has resulted in stability up to 60 h for water and 100 h for thermal stability.

There is another approach of mixed Pb/Sn perovskite Solar cell have also been reported. Lead and tin were revealed to be arbitrarily spread in the [MX6] octahedra in the perovskite and percentage of tin could be altered from 0 to 1 [15]. These devices presented the best photocurrent at a 50% mixing ratio. SnO used as ETL has also resulted with good PCE of (13%) and stability (>700 h storage) [15, 60]. LCA showed the replacement of lead did not decrease the environmental impacts, meanwhile the loss of PCE and stability generates an environmental burden. However, those studies are also interesting because they draw consideration to other toxicity problems occurring from the solvent use during processing of charge transport layers (ETLs) [15].

For lead-free inorganic perovskites Tetravalent cations have also been thought to replace Pb. A new chemical formula of A_2BX_6 structure is designed by eliminating half of the B-site ions in the ABX_3 perovskite for adjusting the heterovalent cation substitution as shown in **Figure 1**. Due to the lack of connectivity in the [BX6] octahedral structure, the A_2BX_6 can be considered as a 0D non-perovskite which results in different optical and optoelectronic properties of the A2BX6 from those of the ABX_3. [3, 9, 10]. Amongst the A_2BX_6 perovskites, Cs_2SnI_6, Cs_2TiBr_6 and Cs_2PdBr_6 have been employed in photovoltaic devices [9, 10]. Chung et al. [54] first utilized this material as a solid hole transport material in DSSCs.

Furthermore, a special concern for toxicity must be upraised during experimental work in the laboratory, since hazards arise primarily by the absorption of the toxic lead when used in solution, which is significantly higher, particularly through the dermal and respiratory routes; some of the lead derivatives are soluble both in water and fat, posing a high risk. Solvents such as dimethylformamide (DMF) and dimethylsulfoxide (DMSO) are not only toxic, but also raise the risk of bio incorporation as they are miscible in all ratios with water. Thus, these solvents have also been considered as a major contributor to environmental impact [61].

4.2 Stability

Perovskite solar cells (PSCs) have been established with promising PCEs. Regardless of the great potential as PV material in terms PCE, the instability of the PSCs is one of the core barriers for larger scale applications [8, 9, 39]. At present, PSCs can only perform for several months under active conditions, whereas traditional silicon cells can operate for more than 25 years. Therefore, stability issues must be reasonably dealt with before its actual use and commercialization [62]. Poor stability of PSCs is due to several affiliated factors resulting from exposure to moisture, oxygen, light, and heat [63, 64].

Nevertheless, the importance of stability has been highlighted and recognized as the foremost problem, in the past five years to solve for the perovskite solar cells (PSCs) to be able to challenge in the market arena. So, how to increase the stability of perovskite solar cells is the most significant issue in this field [15].

In this section, the effect of environmental factors will be discussed on PSCs along with approaches developed to improve stability of perovskite solar cells.

4.2.1 Moisture

The first environmental factor observed to degrade perovskites was Moisture/water. The instability of perovskite at high humidity is the serious issue that needs to be focused. Solar cells when exposed to moisture (water), due to the hygroscopic nature of the organic components of perovskite materials are spontaneously affected by moisture access and then degrade [62]. It has basically been supposed that moisture-induced degradation is the leading issue, imitating MNH_3PbI_3 stability under ambient conditions.

Prolonged exposure of perovskite material to water vapor activates an irreversible degradation which eventually leads to transformation of the perovskite back to the initial precursors (such as PbI_2). In detail, perovskite forms hydrate complexes with water such as $(CH_3NH_3)_4PbI_6{\cdot}2H_2O$ and leaves out PbI_2, which tend to crystallize, forcing the forward reaction. Moreover, MA^+ is slightly acidic and reacts with water to form volatile methylamine (CH_3NH_3) and hydroiodic acid (HI), according to the following reaction (1): [63].

$$(CH_3NH_3)_4 PbI_6.2H_2O \rightarrow PbI_2 + 4CH_3NH_3 + 4HI + 2H_2O \qquad (2)$$

Some researchers have reported that the compositions, microstructures (such as grain size) also affected the moisture stability of perovskite devices and concluded that larger grains resulted in a smaller area density of grain boundaries, which can be correlated with the improved stability [65].

In demand to progress the chemical stability of MAPbI -based$_3$ PSCs against moisture, scientists have proposed replacing the organic cation MA^+ with alternative components at the A position. For example, $FAPbI_3$ has been presented to be further thermally stable than $MAPbI_3$ because of its larger tolerance factor.
Though, $FAPbI_3$ suffers a phase transition from the a-$FAPbI_3$ (black triangular) phase to the d-$FAPbI_3$ (yellow hexagonal) phase due to the presence of moisture. Furthermore, degradation of $FA_{0.9}Cs_{0.1}PbI$ is prevented by adding a small amount of cesium (Cs) into orbital lead-iodine to form $FA_{0.9}Cs_{0.1}PbI$ in high humidity environment [33].

Smith et al. [66] discussed that Low-dimensional 2-D perovskites exhibited better moisture stability than 3D perovskites due to the hydrophobic nature of organic cations. Though, the insulating aspect of the organic cations with poorer charge transport resulted in lower PCE as compared to 3D perovskites. Therefore, various efforts have been made to form a quasi-2D (or 2D–3D mixture) and 2D on top of 3D (2D@3D) to use the benefits of both 2D and 3D perovskites. The use of 2D perovskite is mostly to improve the moisture stability, a thin 2D layer was deposited on top of 3D $MAPbI_3$ perovskite to cover it fully and shield the 3D perovskite from moisture. The highest PCE for 2D@3D perovskite solar cell was observed to be of 18.0%, with an enhanced device stability under both inert (90% of initial PCE for 32 d) and ambient conditions (72% of initial PCE for 20 d) without encapsulation.

Polymers, such as poly(4-vinylpyridine) (PVP), poly (methyl methacrylate) (PMMA) covering p-type and n-type semiconductors, or insulators, were also reported to improve stability. These long chain polymer acts as defect passivator and a moisture blocker by forming a network along perovskite grains and resulted in improved device efficiency and stability [58, 60, 64].

4.2.2 Light

Light-induced perovskite solar cell degradation and environmental stability are the most frequently cited villains. Early on, stability of PSC was a big issue. But just as there were quick improvements in efficiency of PSCs, there has also been similar quick progresses in stability. Ultraviolet light (UV) can also cause the degradation of MAPbI3 perovskite. For e.g. the commonly used TiO_2 electron transport layer

(ETL) for these PSCs is responsible for UV-induced degradation. According to the international standards for climate chamber tests (IEC 61646), solar cells need to tolerate long-term stability at 85°C.

Bryant et al. [67] demonstrated that contact of $MeNH_3PbI_3$ films to both light and molecular oxygen can initiate quick degradation. Particularly, this reaction is started by the deprotonation of the methylammonium cation of the perovskite by a photogenerated reactive oxygen species (superoxide, O_2^-). The stability of $MeNH_3PbI_3$ based devices was checked under different operating (e.g. light and dark) and environmental conditions and infer that oxygen induced degradation, is relatively dominant than moisture induced degradation and limits the working stability of $MeNH_3PbI_3$ containing devices under ambient conditions. Moreover, they pointed out that this fall in device performance can be reduced by the addition of electron acceptor layers within device architecture. Such layers are exposed to augment electron extraction from the absorber (perovskite material) before they react with oxygen, hence decreasing the amount of superoxide O_2^- and increasing the device stability.

It was noticed that by replacing MA with Cs and FA resulted in improved photostability of the PSCs. By systematically monitoring the development of PL intensity of perovskites, light-induced formation and annihilation of defects were reported to induce photo-instability [68]. Photostability can be improved through defect control by passivating which acted as a defect reservoir on the surface and grain boundaries. To stabilize surface defects, polyethylene oxide was applied and thus improved photostability was achieved. By substituting MA with FA, the degradation became slow with small pores forming on the surface after exposure to light. Moreover, Addition of Cs into the MAFA (forming CsMAFA) further lessen the degradation. XPS, XRD, Fourier transform infrared (FT-IR) spectrometry, and ultravioletvisible absorption spectrometry were used to investigate the variation of MAPbI3 films under illumination. The result showed that light induced degradation is the main cause of degradation. Using polymer such as PTAA (Poly(triarylamine)) as the HTM, it was observed that pure $MAPbI_3$ devices retained nearly 100% of their initial efficiency after 1000 h aging under constant illumination at room temperature. PTAA which act as a protection layer, inhibited the discharge of gaseous degradation products enhanced stability. However, for devices using Spiro as the HTM, their stability under illumination was lesser than that using PTAA [69, 70].

4.2.3 Heat

Heat is also another factor that influences stability due to the inherent mat-ter with low formation energies, and perovskites thus have a great response to a small increase in external temperature [71]. Organic-inorganic perovskites tend to decompose due to the instability of organic A^+ cations under thermal atmosphere. Commercial solar cells should be able to work efficiently above 85°C, to have any influence in the market.

$MAPbI_3$ is basically unstable upon thermal stress which produces a discharge of I_2 and the presence of metallic Pb at 40°C in the dark [63]. This is produced by the decomposition reactions (2) and (3):

$$CH_3NH_3PbI_3 \rightarrow CH_3NH_3I + PbI_2 \tag{3}$$

$$PbI_2 \rightarrow Pb0 + I_2 \tag{4}$$

Although reaction (3) is reversible at just 80–85°C, methylammonium iodide decomposes into more volatile compounds as represented by reactions (4) and (5):

$$CH_3NH_3I \rightarrow CH_3I + NH_3 \quad (5)$$

$$CH_3NH_3I \rightarrow CH_3NH_2 + HI \quad (6)$$

It was found that HI(g) and CH $NH_{3\,2(g)}$ were dominant products during the decomposition of $MAPbI_3$ and only trace amounts of CH_3I and NH_3 were found. Though, the ratio of CH I_3 and NH_3 increased at higher temperature and lesser than $HI_{(g)}$ and $CH_3NH_{2(g)}$. In short, $HI_{(g)}$ and $CH_3NH_2(g)$ were the dominant decomposition products at ambient temperature under vacuum while CH I_3 and NH_3 gases were obtained at high temperature. Both processes occurred simultaneously near ambient temperature in vacuum and the later was favored at high temperature.

To find out the decomposition temperature of perovskites, Thermogravimetric analysis (TGA) was used. From the mass loss of TGA curve for $MAPbI_3$, the decomposition onset temperature was found to be 234°C [62]. This indicates that as the practical application temperature usually is less than 100°C, so this high decomposition temperature made the stability of $MAPbI_3$ not a big issue. The as prepared film did not show any changes in XRD patterns when stayed inside the vacuum for up to three days. This might be owing to the purer perovskite films without any exposure to the ambient atmosphere. Though, the commonly degradation of the perovskite solar cell was apparent even with encapsulation. This could be inadequate to estimate the long-term stability of a photovoltaic material, which is essential to work for a long time at temperatures lower than the decomposition temperature [72].

The fact that inert condition and encapsulation cannot completely avoid $MAPbI_3$ perovskite degradation. At low temperature, the degradation of $MAPbBr_3$ was found by only releasing HBr and CH_3NH_2 gases [69]. The encapsulation of devices is essential not only to prevent exposure to oxygen and moisture, but also to avoid leakage of volatile decomposition products. Photostability can also be increased by replacing MA cation with more stable Cs/FA combination.

Substituting organic cations with inorganic Cs^+ or Rb^+ cations is also valuable to stabilize perovskite solar cells [73, 74]. Grancini et al. [74] stated an ultra-stable 2D/3D $(HOOC(CH_2)_4NH_3)_2PbI_4/CH_3NH_3PbI_3$ perovskite, presenting a PCE of 12.9% with carbon electrodes and 14.6% with the normal mesoporous structure and stability of one-year.

By introducing n-butylammonium iodide (BAI) to $MAPbI_3$ perovskite, a mixed 2D $(BA)_2PbI_4$ structure is formed, which probably provide an improved protection for the 3D perovskite against heat stress [75]. Octylammonium (OA) cation has also been reported to enhance the thermal stability of perovskites and keep 80% of their initial efficiency for 760 h aged at 85°C in ambient atmosphere without encapsulation [76]. Other additives, such as π-conjugated polymer, nonvolatile ionic liquids, bifunctional hydroxylamine hydrochloride guanidinium isothiocyanate, have also been reported to improve the thermal stability of various perovskites [77–79].

5. Conclusion and perspective

The discovery and development of organic inorganic perovskite materials have become a hot research topic in the field of photovoltaics. This chapter deals with a comprehensive discussion on the properties and applications of organic inorganic perovskites materials in PSCs. The extraordinarily outstanding performances of organic inorganic perovskites result of their excellent properties. Solar cells based on organic inorganic perovskite materials have achieved much advancement, both in PCE and stability, in a short very time. To PSCs, though great progress has been attained, there are still various obstacles in terms of stability and toxicity until its practical usage in the PV market. However, for large-scale performance, are required to be overcome. So far, great research has been made to overcome these issues by changing the composition of organic inorganic perovskite material either by replacing Pb with Sn or Ge or organic methyl with other organic or inorganic cations. However, commercialization of an organic inorganic perovskite solar cell needs further development in both efficiency and long-term stability, with low-cost photovoltaic materials and ease of printability. To increase stability, various methods such as the use of buffer layers, varying the composition of organic inorganic perovskite materials, and better techniques of encapsulation. In inference, the research which has been enduring for the past five years has attained significant results. Future research needs to endeavor for longer stability with high efficiency.

Author details

Madeeha Aslam, Tahira Mahmood* and Abdul Naeem
National Centre of Excellence in Physical Chemistry, University of Peshawar, Peshawar, Pakistan

*Address all correspondence to: tahiramahmood@uop.edu.pk

References

[1] Shi PP, Lu SQ, Song XJ, Chen XG, Liao WQ, Li PF, Tang YY, Xiong RG. Two-Dimensional Organic-Inorganic Perovskite Ferroelectric Semiconductors with Fluorinated Aromatic Spacers. J Am Chem Soc. 2019; 141: 18334-18340. DOI: 10.1021/jacs.9b10048.

[2] Yang Z, Zhang WH. Organolead halide perovskite: A rising player in high-efficiency solar cells. Chin J Catalysis. 2014; 35: 983-988. DOI: 10.1016/S1872-2067(14)60162-5.

[3] Zhang C, Arumugam GM, Liu C, Hu J, Yang Y, Schropp REI, Mai Y. Inorganic halide perovskite materials and solar cells. APL Mater 2019;7. DOI: 10.1063/1.5117306.

[4] Mali SS, Patil JV, Hong CK. Making air-stable all-inorganic perovskite solar cells through dynamic hot-air. Nano Today. 2020; 33. DOI: 10.1016/j.nantod.2020.100880.

[5] Ji C, Liang C, Zhang H, Sun M, Song Q, Sun F, Feng X, Liu N, Gong H, Li D, You F, He Z. Secondary grain growth in organic-inorganic perovskite films with ethylamine hydrochloride additives for highly efficient solar cells. ACS Appl Mater Interfaces. 2020;1-30. DOI:10.1021/acsami.9b23468.

[6] Kandjani SA, Mirershad S, Nikniaz A. Inorganic–Organic Perovskite Solar Cells. In: editor. Solar Cells - New Approaches and Reviews. 10.5772/58970.

[7] Wang L, Fan B, Zheng B, Yang Z, Yin P, Huo L. Organic functional materials: recent advances in all-inorganic perovskite solar cells. Sustainable Energy Fuels. 2020. DOI: 10.1039/D0SE00214C.

[8] Hadadian M, Smått J-H, Correa-Baena J-P. The role of carbon-based materials in enhancing the stability of perovskite solar cells. Energy Environ Sci. 2020; 13: 1377-1407.

[9] Yi Z, Ladi NH, Shai X, Li H, Shen Y, Wang M. Will organic–inorganic hybrid halide lead perovskites be eliminated from optoelectronic applications? Nanoscale Adv. 2019; 1:1276-1289. DOI: 10.1039/C8NA00416A.

[10] Bo Li, Fu L, Li S, Li H, Pan L, Wang L, Chang B, Yin L. Pathways toward high-performance inorganic perovskite solar cells: challenges and strategies. J Mater Chem A. 2019. DOI: 10.1039/C9TA04114A.

[11] Kojima A, Teshima K, Shirai Y, Miyasaka T. Organometal Halide Perovskites as Visible-Light Sensitizers for Photovoltaic Cells. J Am Chem Soc 2009; 131: 6050-6051. DOI:10.1021/ja809598r.

[12] Im JH, Lee CR, Lee JW, Park SW, Park NG. 6.5% efficient perovskite quantum-dot-sensitized solar cell. Nanoscale. 2011; 3: 4088-4093. DOI:10.1039/c1nr10867k.

[13] Kim H-S, Lee C-R, Im J-H, Lee K-B, Moehl T, Marchioro A, Moon S-J, Humphry-Baker R, Yum J-H, Moser JE, Gratzel M, Park N-G. Lead Iodide Perovskite Sensitized All-Solid-State Submicron Thin Film Mesoscopic Solar Cell with Efficiency Exceeding 9%. Sci Rep. 2012; DOI: 2. 10.1038/srep00591.

[14] Venugopal G, Krishnamoorthy K, Mohan R, Kim S-J. An investigation of the electrical properties of graphene oxide films. Mater. Chem. Phys. 2012; 132: 29-33. DOI: 10.1016/j.matchemphys.2011.10.040.

[15] Song T-B, Chen Q, Zhou H, Jiang C, Wang H-H, Yang YM, Liu Y, You J, Yang Y. Perovskite solar cells:

film formation and properties. J Mater Chem A. 2015. DOI: 10.1039/C4TA05246C.

[16] Urbina A. The balance between efficiency, stability and environmental impacts in perovskite solar cells: a review. J Phys Energy 2020;2. DOI: 10.1088/2515-7655/ab5eee.

[17] Zhou H, Chen Q, Li G, Luo S, Song T-b, Duan H-S, Hong Z, You J, Liu Y, Yang Y. Interface engineering of highly efficient perovskite solar cells. New York; 2014. Contract No: 6196.

[18] Ameen S, Akhtar MS, Seo H-K, Nazeeruddin MK, Shin H-S. An Insight into Atmospheric Plasma Jet Modified ZnO Quantum Dots Thin Film for Flexible Perovskite Solar Cell: Optoelectronic Transient and Charge Trapping Studies. J Phys Chem C 2015;119: 10379-10390.

[19] Thuat NT, Thoa BB, Tran NB, Tu Nm, Minh Nn, Huong Hnl, Trang PT, Van P.V.T, Tu TT, Linh DT. Fabrication of organolead iodide perovskite solar cells with niobium-doped titanium dioxide as compact layer. Comm. Phys. 2017; 27,121-130. DOI: 10.15625/0868-3166/27/2/9811.

[20] Mali SS, Patil JV, Kim H, Hong CK. Synthesis of SnO_2 nanofibers and nanobelts electron transporting layer for efficient perovskite solar cells. Nanoscale. 2018. DOI:10.1039/C8NR00695D.

[21] Zhou D, Zhou T, Tian Y, Zhu X, Tu Y. Perovskite-Based Solar Cells: Materials, Methods, and Future Perspectives. J Nanometer. 2018;2018. DOI:10.1155/2018/8148072.

[22] Vivo P, Salunke JK, Priimagi A. Hole-Transporting Materials for Printable Perovskite Solar Cells. Materials 2017;10: 1087. DOI:10.3390/ma10091087.

[23] Mahmood T, Aslam M, Naeem A. Graphene/Metal Oxide Nanocomposite Usage as Photoanode in Dye-Sensitized and Perovskite Solar Cells. In: editor. Reconfigurable Materials. United Kingdom: intechopen; 2020.

[24] Habisreutinger SN, Leijtens T, Eperon GE, Stranks SD, Nicholas RJ, Snaith HJ. Carbon Nanotube/Polymer Composites as a Highly Stable Hole Collection Layer in Perovskite Solar Cells. Nano Lett. 2014;14: 5561-5568.

[25] Chen Y, Zhang L, Zhang Y, Gao H, Yan H. Large-area perovskite solar cells – a review of recent progress and issues RSC Adv. 2018; 8: 10489-10508 DOI:10.1039/C8RA00384J.

[26] Fan J, Jia B, Gu M. Perovskite-based low-cost and high-efficiency hybrid halide solar cells. Photon Res. 2014; 2: 111-120. DOI: 10.1364/PRJ.2.000111.

[27] Schoonman J. Organic–inorganic lead halide perovskite solar cell materials: A possible stability problem. Chem Phys Lett. 2015; 619: 193-195. DOI: 10.1016/j.cplett.2014.11.063.

[28] Shi Z, Jayatissa AH. Perovskites-Based Solar Cells: A Review of Recent Progress, Materials and Processing Methods. Materials (Basel). 2018; 11: 1-34. DOI:10.3390/ma11050729.

[29] Ava T, Al Mamun A, Marsillac S, Namkoong G. A Review: Thermal Stability of Methylammonium Lead Halide Based Perovskite Solar Cells. Applied Sciences. 2019; 9:188. DOI: 10.3390/app9010188

[30] Jiang Q, Zhang X, You J. SnO_2: A Wonderful Electron Transport Layer for Perovskite Solar Cells. 2018. DOI: 10.1002/smll.201801154.

[31] Tan H, Jain A, Voznyy O, Lan X, Arquer FPGd, Fan JZ, Quintero-Bermudez R, Yuan M, Zhang B,

Zhao Y, Fan F, Li P, Quan LN, Zhao Y, Lu Z-H, Yang Z, Hoogland S, Sargent EH. Efficient and stable solution-processed planar perovskite solar cells via contact passivation. Science. 2017; 355: 722-726.

[32] Dkhissi Y, Huang F, Rubanov S, Xiao M, Bach U, Spiccia L, Caruso RA, Cheng Y-B. Low temperature processing of flexible planar perovskite solar cells with efficiency over 10%. J Power Sour. 2015;278: 325-331. DOI: 10.1016/j. jpowsour.2014.12.104.

[33] Diao X-F, Tang Y-l, Tang T-Y, Xie Q, Xiang K, Liu G-f. Study on the stability of organic–inorganic perovskite solar cell materials based on first principle. Molecul Phys. 2019. DOI:10.1080/00268 976.2019.1665200

[34] Boix PP, Nonomura K, Mathews N, Mhaisalkar SG. Current progress and future perspectives for organic/inorganic perovskite solar cells. Mater Today 2014; 17: 16-23. DOI: 10.1016/j.mattod.2013.12.002.

[35] Green MA, Ho-Baillie A, Snaith HJ. The emergence of perovskite solar cells Nat Photon. 2014 DOI:10.1038/ nphoton.2014.134

[36] Park N-G. Perovskite solar cells: an emerging photovoltaic technology. Mater Today. 2015;18 DOI: 10.1016/j. mattod. 2014.07.007

[37] Song Z, Watthage SC, Phillips AB, Heben MJ. Pathways toward high-performance perovskite solar cells: review of recent advances in organo-metal halide perovskites for photovoltaic applications. J Photon Energy. 2016; 6. DOI: 10.1117/1.JPE.6.022001

[38] Dymshits A, Henning A, Segev G, Rosenwaks Y, Etgar L. The electronic structure of metal oxide/ organo metal halide perovskite junctions in perovskite based solar cells. Sci Rep. 2015;5 1-6. DOI:10.1038/srep08704

[39] Li D, Shi J, Xu Y, Luo Y, Wu H, Meng Q. Inorganic–organic halide perovskites for new photovoltaic technology. Nat Sci Review 2018; 5: 559- 576. DOI: 10.1093/nsr/nwx100

[40] Mitzi DB, Prikas MT, Chondroudis K. Thin Film Deposition of Organic-Inorganic Hybrid Materials Using a Single Source Thermal Ablation Technique. Chem Mater 1999; 11: 542-544.

[41] Burschka J, Pellet N, Moon SJ, Humphry-Baker R, Gao P, Nazeeruddin MK, Gratzel M. Sequential deposition as a route to high-performance perovskite-sensitized solar cells. Nature. 2013; 499: 316-319. DOI: 10.1038/nature12340

[42] Chen Q, Zhou H, Hong Z, Luo S, Duan H-S, Wang H-H, Liu Y, Li G, Yang Y. Planar Heterojunction Perovskite Solar Cells via Vapor-Assisted Solution Process. J Am Chem Soc. 2014; 136: 622-625. DOI:10.1021/ja411509g

[43] Era M, Morimoto S, Tsutsui T, Saito S. Organic inorganic heterostructure electroluminescent device using a layered perovskite semiconductor $(C_6H_5C_2H_4NH_3)_2$ PbI_4 Appl Phys Lett. 1994; 65. DOI: 10.1063/1.112265

[44] Liu M, Johnston MB, Snaith HJ. Efficient planar heterojunction perovskite solar cells by vapour deposition. Nature. 2013; 501: 395-398. DOI: 10.1038/nature12509

[45] Leyden MR, Jiang Y, Qi Y. Chemical Vapor Deposition Grown Formamidinium Perovskite Solar Modules with High Steady State Power and Thermal Stability J Mater Chem A. 2016. DOI: 10.1039/C6TA04267H

[46] Banerjee D, Chattopadhyay KK. Hybrid Inorganic Organic Perovskites: A Low-Cost-Efficient Optoelectronic Material In: editor. Perovskite

Photovoltaics. Elsevier; 2018. 123-162. DOI: 10.1016/B978-0-12-812915-9.00005-8.

[47] Giacomo FD, Razza S, Matteocci F, D'Epifanio A, Licoccia S, Brown TM, Carlo AD. High efficiency CH3NH3PbI $(_{3\ x})Cl_x$ perovskite solar cells with poly(3-hexylthiophene) hole transport layer. J Power Sour. 2014; 251: 152-156.

[48] He M, Zheng D, Wang M, Lin C, Lin Z. High Efficiency Perovskite Solar Cells: From Complex Nanostructure to Planar Heterojunction. J Mater Chem A, 2013. DOI: 10.1039/C3TA14160H

[49] Aharon S, Cohen BE, Etgar L. Hybrid Lead Halide Iodide and Lead Halide Bromide in Efficient Hole Conductor Free Perovskite Solar Cell. J Phys Chem C 2014; 118:17160-17165.

[50] Yang G, Chen C, Yao F, Chen Z, Zhang Q, Zheng X, Ma J, Lei H, Qin P, Xiong L, Ke W, Li G, Yan Y, Fang G. Effective Carrier Concentration Tuning of SnO_2 Quantum Dot Electron Selective Layers for High-Performance Planar Perovskite Solar Cells. Adv Mater. 2018. DOI:10.1002/adma.201706023.

[51] Anaraki EH, Kermanpur A, Steier L, Domanski K, Matsui T, Tress W, Saliba M, Abate A, Gratzel M, Hagfeldt A, Correa-Baena J-P. Highly efficient and stable planar perovskite solar cells by solution-processed tin oxide. Energy Environ Sci. 2016. DOI:10.1039/C6EE02390H.

[52] Xiong L, Qin M, Chen C, Wen J, Yang G, Guo Y, Ma J, Zhang Q, Qin P, Li S, Fang G. Fully High-Temperature-Processed SnO_2 as Blocking Layer and Scaffold for Efficient, Stable, and Hysteresis-Free Mesoporous Perovskite Solar Cells Adv Funct Mater. 2018. DOI:10.1002/adfm.201706276.

[53] Bi D, Yi C, Luo J, Décoppet J-D, Zhang F, Zakeeruddin SM, Li X, Hagfeldt A, Grätzel M. Polymer-templated nucleation and crystal growth of perovskite films for solar cells with efficiency greater than 21%. Nat Energy. 2016. DOI: 10.1038/nenergy.2016.142.

[54] Chung I, Lee B, He J, Chang RPH, Kanatzidis MG. All-solid-state dye-sensitized solar cells with high efficiency. Nature. 2012; 486-490. DOI:10.1038/nature11067.

[55] Choi H, Jeong J, Kim H-B, Kim S, Walker B, Kim G-H, Kim JY. Cesium-doped methylammonium lead iodide perovskite light absorber for hybrid solar cells. Nano Energy. 2014; 7: 80-85.

[56] Saliba M, Matsui T, Seo J-Y, Domanski K, Correa-Baena J-P, Nazeeruddin MK, Zakeeruddin SM, Tress W, Abate A, Hagfeldt A, Gratzel M. Cesium-containing triple cation perovskite solar cells: improved stability, reproducibility and high efficiency. Energy Environ Sci. 2016; 9: 1989-1997. DOI:10.1039/C5EE03874J

[57] Song Z, Wang C, Phillips AB, Grice CR, Zhao D, Yu Y, Chen C, Li C, Yin X, Ellingson RJ, Heben MJ, Yan Y. Probing the Origins of Photodegradation in Organic-Inorganic Metal Halide Perovskites with Time-Resolved Mass Spectrometry. Sustain Energy Fuel. 2018. DOI:10.1039/C8SE00358K.

[58] Chen Z, Wang JJ, Ren Y, Yu C, Shum K. Schottky solar cells based on $CsSnI_3$ thin-films. Phys Lett. 2012; 101. DOI:10.1063/1.4748888.

[59] Lv X, Xiao G, Feng X, Liu J, Yao X, Cao J. Acyl hydrazone-based porphyrin derivative as hole transport material for efficient and thermally stable perovskite solar cells Dyes and Pigments. 2018. DOI: 10.1016/j.dyepig.2018.08.056.

[60] Jiang J, Wang Q, Jin Z, Zhang X, Lei J, Bin H, Zhang Z-G, Li Y, Liu SF. Polymer Doping for High-Efficiency Perovskite Solar Cells with Improved

Moisture Stability. Adv Energy Mater. 2017. DOI: 10.1002/aenm.201701757.

[61] Zhang J, Gao X, Deng Y, Zha Y, Yuan C. Comparison of life cycle environmental impacts of different perovskite solar cell systems. Sol Energy Mater & Solar Cell 2017;166: 9-17. DOI: 10.1016/j.solmat.2017.03.008.

[62] Zhang S, Han G. Intrinsic and environmental stability issues of perovskite photovoltaics. Prog Energy 2020 2:1-44. DOI:10.1088/2516-1083/ab70d9.

[63] Schileo G, Grancini G. Halide perovskites: current issues and new strategies to push material and device stability. J Phys: Energy 021005. 2020; 2: 1-12. DOI:10.1088/2515-7655/ab6cc4

[64] Szostak R, Silva JC, Turren-Cruz S-H, Soares MM, Freitas Ro, Hagfeldt A, Tolentino HCN, Nogueira AF. Nanoscale mapping of chemical composition in organic-inorganic hybrid perovskite films. Sci Adv. 2019; 5: 1-7.

[65] Wang Q, Chen B, Liu Y, Deng Y, Bai Y, Dong Q, Huang J. Scaling Behavior of Moisture-induced Grain Degradation in Polycrystalline Hybrid Perovskite Thin Films. Energy Environ Sci. 2016. DOI:10.1039/C6EE02941H.

[66] Smith IC, Hoke ET, Solis-Ibarra D, McGehee MD, Karunadasa HI. A Layered Hybrid Perovskite Solar-Cell Absorber with Enhanced Moisture Stability. Angew Chem Int Ed. 2014;53: 1-5. DOI: 10.1002/anie.201406466

[67] Bryant D, Aristidou N, Pont S, Sanchez-Molina I, Chotchunangatchaval T, Wheeler S, Durrant JR, Haque SA. Light and oxygen induced degradation limits the operational stability of methylammonium lead triiodide perovskite solar cells. Energy Environ Sci. 2016. DOI:10.1039/C6EE00409A

[68] Motti SG, Meggiolaro D, Barker AJ, Perini CAR, Ball JM, Angelis FD, Petrozza A, Gandini M, Mosconi E, Kim M. Controlling competing photochemical reactions stabilizes perovskite solar cells. Nature Photonics. 2019. DOI: 10.1038/s41566-019-0435-1

[69] Juarez-Perez EJ, Hawash Z, Raga SR, Ono LK, Qi Y. Thermal degradation of CH3NH3PbI3 perovskite into NH3 and CH3I gases observed by coupled thermogravimetry–mass spectrometry analysis† Energy Environ Sci. 2016;9. DOI: 10.1039/c6ee02016j

[70] JuarezPerez EJ, Ono LK, Maeda M, Jiang Y, Hawash Z, Qi Y. Photo-, Thermal-decomposition in Methylammonium Halide Lead Perovskites and inferred design principles to increase photovoltaic device stability. J Mater Chem A. 2018. DOI: 10.1039/C8TA03501F

[71] Yang J, Hong Q, Yuan Z, Xu R, Guo X, Xiong S, Liu X, Braun S, Li Y, Tang J, Duan C, Fahlman M, Bao Q. Unraveling Photostability of Mixed Cation Perovskite Films in Extreme Environment. Adv Optical Mater 2018. DOI: 10.1002/adom.201800262.

[72] Conings B, Drijkoningen J, Gauquelin N, Babayigit A, D'Haen J, D'Olieslaeger L, Ethirajan A, Verbeeck J, Manca J, Mosconi E, Angelis FD, Boyen H-G. Intrinsic Thermal Instability of Methylammonium Lead Trihalide Perovskite. Adv Energy Mater. 2015; 5. DOI: 10.1002/aenm.201500477.

[73] Saliba M, Matsui T, Domanski K, Seo J-Y, Ummadisingu A, Zakeeruddin SM, Correa-Baena J-P, Tress WR, Abate A, Hagfeldt A, Grätzel M. Incorporation of rubidium cations into perovskite solar cells improves photovoltaic performance. Science. 2016; 354: 206-209.

[74] Grancini G, Rolda'n-Carmona C, Zimmermann I, Mosconi E, X.Lee,

Martineau D, Narbey S, Oswald F, Angelis FD, Graetzel M, Nazeeruddin MK. One-Year stable perovskite solar cells by 2D/3D interface engineering. Nature Communications 2017; 8. DOI:10.1038/ncomms15684.

[75] Lin Y, Bai Y, Fang Y, Chen Z, Yang S, Zheng X, Shi Tang, Liu Y, Zhao J, Huang J. Enhanced Thermal Stability in Perovskite Solar Cells by Assembling 2D/3D Stacking Structures. J Phys Chem Lett 2018; 9: 654-658. DOI: 10.1021/acs.jpclett.7b02679.

[76] Jung M, Shin TJ, Seo J, Kim G, Seok SI. Structural features and their functions in surfactant-armoured methylammonium lead iodide perovskites for highly efficient and stable solar cells. Energy Environ Sci. 2018. DOI: 10.1039/C8EE00995C.

[77] Du J, Wang Y, Zhang Y, Zhao G, Jia Y, Zhang X, Liu Y. Ionic Liquid-Assisted Improvements in the Thermal Stability of CH3NH3PbI3 Perovskite Photovoltaics. Phys Status Solidi RRL 2018. DOI:10.1002/pssr.201800130

[78] Xia R, Fei Z, Drigo N, Bobbink FD, Huang Z, Jasiu–nas R, Franckevicius M, Gulbinas V, Mensi M, Fang X, Roldán-Carmona C, Nazeeruddin MK, Dyson PJ. Retarding Thermal Degradation in Hybrid Perovskites by Ionic Liquid Additives. Adv Funct. Mater. 2019. DOI:10.1002/adfm.201902021

[79] Zou J, Liu W, Deng W, Lei G, Zeng S, Xiong J, Gu H, Hu Z, Wang X, Li J. An efficient guanidinium isothiocyanate additive for improving the photovoltaic performances and thermal stability of perovskite solar cells. Electrochim. Acta 2018. DOI: 10.1016/j.electacta.2018.08.117.

8

Static and Dynamic Analysis of Piezoelectric Laminated Composite Beams and Plates

Chung Nguyen Thai, Thinh Tran Ich and Thuy Le Xuan

Abstract

In this chapter, the mechanical behavior analysis of piezoelectric laminated composite beams and plates is influenced subjected to static, dynamic, and aerodynamic loads. Algorithm for dynamic, stability problem analysis and vibration control of laminated composite beams and plates with piezoelectric layers is presented. In addition, numerical calculations, considering the effect of factors on static, dynamic, and stability response of piezoelectric laminated composite beams and plates are also clearly presented. The content of this chapter can equip readers with the knowledge used to calculate the static, dynamic, and vibration control of composite beams, panels made of piezoelectric layers applied in the field different techniques.

Keywords: beams, plates, static, dynamic, piezoelectric, composite, stiffened

1. Introduction

The content of this chapter is the inheritance and development of the research results of the authors and other authors by published scientific works on composite materials, piezoelectric and structural calculation by piezoelectric composite materials.

2. Electromechanical interaction of piezoelectric materials

2.1 Mechanical-electrical behavior relations

Let us consider a block of elastic material in an environment with an electric field of zero, the relationship between stress and strain is followed Hooke's law, and written as follows [1, 2]:

$$\{\sigma\} = [c]\{\varepsilon\}, \tag{1}$$

where $\{\sigma\}$ is the mechanical stress vector, $\{\varepsilon\}$ is the mechanical strain vector, and $[c]$ is the material stiffness matrix of beam.

Mechanical-electrical relations in piezoelectric materials have an interactive relationship, strain $\{\varepsilon\}$ will produce $e\varepsilon$ - polarization, where e is the voltage stress factor when there is no mechanical strain. The imposed electric field E produces the -eE stress in the piezoelectric material according to the reverse voltage effect. Therefore, we have a mathematical model that describes the mechanical-electrical interaction relationship in piezoelectric materials as follows [3–6]:

$$\{\sigma\} = [c]\{\varepsilon\} - [e]\{E\}, \tag{2}$$

$$\{D\} = [e]^T\{\varepsilon\} + [p]\{E\}, \tag{3}$$

$$\text{or } \{D\} = [d]^T\{\sigma\} + [p]\{E\}, \tag{4}$$

where [e] is the piezoelectric stress coefficient matrix, [p] is the dielectric constant matrix, {E} is the vector of applied electric field (V/m), and {D} is the vector of electric displacement (C/m^2).

For the linear problem and small strain, strain vector in the piezoelectric structures can be defined as follows:

$$\{\varepsilon\} = [s]\{\sigma\} + [d]\{E\}, \tag{5}$$

in which [s] is the matrix of compliance coefficients (m^2/N), [d] is the matrix of piezoelectric strain constants (m/V).

In the field of engineering, piezoelectric materials are used by two types. The first type, the piezoelectric layers or the piezoelectric patches act as actuators, called the piezoelectric actuators. In this case, the piezoelectric layers are strained when imposing an electric field on it. The second type, the piezoelectric layers or piezoelectric patches act as sensors, called piezoelectric sensors. In this case, the voltage is generated in piezoelectric layers when there is mechanical strain.

2.2 Piezoelectric actuators and sensors

2.2.1 Piezoelectric actuators

Eq. (5) can be written in the matrix form as follows [4, 6]:

$$\{\varepsilon\} = \begin{Bmatrix} \varepsilon_{11} \\ \varepsilon_{22} \\ \varepsilon_{33} \\ \gamma_{23} \\ \gamma_{13} \\ \gamma_{12} \end{Bmatrix} = \begin{bmatrix} s_{11} & s_{12} & s_{13} & s_{14} & s_{15} & s_{16} \\ s_{21} & s_{22} & s_{23} & s_{24} & s_{25} & s_{26} \\ s_{31} & s_{32} & s_{33} & s_{34} & s_{35} & s_{36} \\ s_{41} & s_{42} & s_{43} & s_{44} & s_{45} & s_{46} \\ s_{51} & s_{52} & s_{53} & s_{54} & s_{55} & s_{56} \\ s_{61} & s_{62} & s_{63} & s_{64} & s_{65} & s_{66} \end{bmatrix} \begin{Bmatrix} \sigma_{11} \\ \sigma_{22} \\ \sigma_{33} \\ \tau_{23} \\ \tau_{13} \\ \tau_{12} \end{Bmatrix} + \begin{bmatrix} d_{11} & d_{21} & d_{31} \\ d_{12} & d_{22} & d_{32} \\ d_{13} & d_{23} & d_{33} \\ d_{14} & d_{24} & d_{34} \\ d_{15} & d_{25} & d_{35} \\ d_{16} & d_{26} & d_{36} \end{bmatrix} \begin{Bmatrix} E_1 \\ E_2 \\ E_3 \end{Bmatrix}, \tag{6}$$

Assuming that the device is pulled along the axis 3, and viewing the piezoelectric material as a transversely isotropic material, which is true for piezoelectric ceramics, many of the parameters in the above matrices will be either zero, or can be expressed through each other. In particular, the non-zero compliance coeffi-cients are $s_{11}, s_{12}, s_{13}, s_{21}, s_{22}, s_{23}, s_{31}, s_{32}, s_{33}, s_{44}, s_{55}, s_{66}$, in which $s_{12} = s_{21}, s_{13} = s_{31}$, $s_{23} = s_{32}$, $s_{44} = s_{55}$, $s_{66} = 2(s_{11} - s_{12})$.

Finally, Eq. (6) becomes:

$$\{\varepsilon\} = \begin{Bmatrix} \varepsilon_{11} \\ \varepsilon_{22} \\ \varepsilon_{33} \\ \gamma_{23} \\ \gamma_{13} \\ \gamma_{12} \end{Bmatrix} = \begin{bmatrix} s_{11} & s_{12} & s_{13} & 0 & 0 & 0 \\ s_{12} & s_{22} & s_{23} & 0 & 0 & 0 \\ s_{13} & s_{23} & s_{33} & 0 & 0 & 0 \\ 0 & 0 & 0 & s_{44} & 0 & 0 \\ 0 & 0 & 0 & 0 & s_{55} & 0 \\ 0 & 0 & 0 & 0 & 0 & s_{66} \end{bmatrix} \begin{Bmatrix} \sigma_{11} \\ \sigma_{22} \\ \sigma_{33} \\ \tau_{23} \\ \tau_{13} \\ \tau_{12} \end{Bmatrix} + \begin{bmatrix} 0 & 0 & d_{31} \\ 0 & 0 & d_{32} \\ 0 & 0 & d_{33} \\ 0 & d_{24} & 0 \\ d_{15} & 0 & 0 \\ 0 & 0 & 0 \end{bmatrix} \begin{Bmatrix} E_1 \\ E_2 \\ E_3 \end{Bmatrix}, \tag{7}$$

where E_1, E_2, and E_3 are electric fields in the 1, 2, and 3 directions, respectively.

2.2.2 Piezoelectric sensors

The induction charge equation of piezoelectric sensor layers is derived from Eq. (4) can be written in the matrix form as [4, 6, 7]:

$$\{D\} = \begin{Bmatrix} D_1 \\ D_2 \\ D_3 \end{Bmatrix} = \begin{bmatrix} d_{11} & d_{12} & d_{13} & d_{14} & d_{15} & d_{16} \\ d_{21} & d_{22} & d_{23} & d_{24} & d_{25} & d_{26} \\ d_{31} & d_{32} & d_{33} & d_{34} & d_{35} & d_{36} \end{bmatrix} \begin{Bmatrix} \sigma_{11} \\ \sigma_{22} \\ \sigma_{33} \\ \tau_{23} \\ \tau_{13} \\ \tau_{12} \end{Bmatrix} + \begin{bmatrix} p_{11} & p_{12} & p_{13} \\ p_{21} & p_{22} & p_{23} \\ p_{31} & p_{32} & p_{33} \end{bmatrix} \begin{Bmatrix} E_1 \\ E_2 \\ E_3 \end{Bmatrix}, \tag{8}$$

The non-zero piezoelectric strain constants are d_{31}, d_{32}, d_{15}, d_{24}, and d_{33}, in which $d_{31} = d_{32}$, $d_{15} = d_{24}$. And the non-zero dielectric coefficients are p_{11}, p_{22}, and p_{33}, where $p_{11} = p_{22}$. Eq. (8) becomes:

$$\{D\} = \begin{Bmatrix} D_1 \\ D_2 \\ D_3 \end{Bmatrix} = \begin{bmatrix} 0 & 0 & 0 & 0 & d_{15} & 0 \\ 0 & 0 & 0 & d_{24} & 0 & 0 \\ d_{31} & d_{32} & d_{33} & 0 & 0 & 0 \end{bmatrix} \begin{Bmatrix} \sigma_{11} \\ \sigma_{22} \\ \sigma_{33} \\ \tau_{23} \\ \tau_{13} \\ \tau_{12} \end{Bmatrix} + \begin{bmatrix} p_{11} & 0 & 0 \\ 0 & p_{22} & 0 \\ 0 & 0 & p_{33} \end{bmatrix} \begin{Bmatrix} E_1 \\ E_2 \\ E_3 \end{Bmatrix}, \tag{9}$$

where D_1, D_2, D_3, p_{11}, p_{22}, and p_{33} are the displacement charge, dielectric constant in the 1, 2, and 3 directions, respectively.

Normally, the voltage is transmitted through the thickness of the actuator layers.

3. Static and dynamic analysis of laminated composite beams with piezoelectric layers

3.1 Displacement and strain

Based on the first-order shear deformation theory (FSDT), the displacement field at any point of the beam is defined as [1, 2]:

$$\begin{aligned} u(x,z) &= u_0(x) + z\theta_y(x), \\ w(x,z) &= w_0(x), \end{aligned} \tag{10}$$

where u, w denotes the displacements of a point (x, z) in the beam; u_0, w_0 are the displacements of a point at the beam neutral axis, and θ_y is the rotation of the transverse normal about the y axis. The bending and shear strains associated with the displacement field in Eq. (10) are defined as:

$$\{\varepsilon\} = \begin{Bmatrix} \varepsilon_x \\ \gamma_{xz} \end{Bmatrix} = \begin{Bmatrix} \dfrac{du}{dx} \\ \dfrac{du}{dz} + \dfrac{dw}{dx} \end{Bmatrix} = \begin{Bmatrix} \dfrac{du_0}{dx} + z\dfrac{d\theta_z}{dx} \\ \theta_y + \dfrac{dw_0}{dx} \end{Bmatrix} = \begin{bmatrix} \dfrac{d}{dx} & 0 & z\dfrac{d}{dx} \\ 0 & \dfrac{d}{dx} & 1 \end{bmatrix} \begin{Bmatrix} u_0 \\ w_0 \\ \theta_z \end{Bmatrix}, \tag{11}$$

in which ε_x, γ_{xz} are the normal strain, and shear strain, respectively.

Using finite element method, we consider 2-node bending elements with 3 degrees of freedom per node (**Figure 1**).

The displacements of the beam neutral axis are expressed in local coordinate system in the form:

$$\{d_0\} = \begin{Bmatrix} u_0 \\ v_0 \\ \theta_z \end{Bmatrix} = \begin{Bmatrix} [N^u]\{q^u\} \\ [N^v]\{q^v\} \\ [N^{\theta_z}]\{q^{\theta_z}\} \end{Bmatrix} = [N^M]\{q_b\}_e, \tag{12}$$

where $\{q_b\}_e$ is the *vector of vector of nodal displacements* of element, $[N_M]$ is the matrix mechanical shape functions:

$$\{q_b\}_e = \{q_1\ q_2\ q_3\ q_4\ q_5\ q_6\}^T, \tag{13}$$

$$\underbrace{[N^M]}_{3\times 6} = \begin{bmatrix} [N^u] & 0 & 0 \\ 0 & [N^v] & 0 \\ 0 & 0 & [N^{\theta_z}] \end{bmatrix}, \tag{14}$$

in which $[N^u]$, $[N^v]$, $[N^{\theta z}]$ are, in this order, the row vectors of longitudinal, transverse along *y*, and rotation about *z* shape functions.

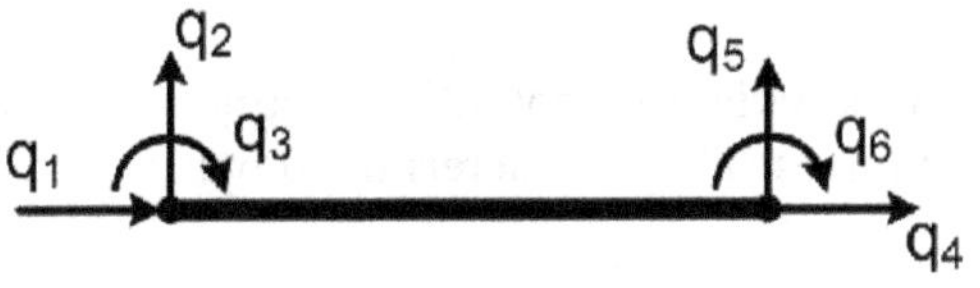

Figure 1.
Two noded beam element.

Substituting Eq. (12) into Eq. (11), we obtain:

$$\underbrace{\{\varepsilon\}}_{2\times1} = \underbrace{[B_b]}_{2\times6}\underbrace{\{q_b\}_e}_{6\times1}, \tag{15}$$

$$\text{where } [B_b] = \begin{bmatrix} \frac{d}{dx} & 0 & z\frac{d}{dx} \\ 0 & \frac{d}{dx} & 1 \end{bmatrix}[N^M]. \tag{16}$$

The electric potential is constant over the element surface:

$$\phi_k = \sum_{i=1}^{n} N_i\phi_i, \tag{17}$$

where n is the element node number.

A voltage ϕ is applied across an actuator of layer thickness t_p generates an electric field vector {E}, such that [4, 8–10]:

$$\{E_k\} = -\nabla\phi_k = \{\,0 \quad 0 \quad E_k^z\,\}, \tag{18}$$

in which

$$E_k^z = -\frac{\phi_k}{t_{pk}} = [B_\phi]\{\phi\} = \begin{bmatrix} 0 & 0 & \frac{1}{t_{p1}} & 0 & 0 & 0 \\ 0 & 0 & 0 & 0 & 0 & \frac{1}{t_{p2}} \end{bmatrix}^T \begin{Bmatrix} \phi_1 \\ \phi_2 \end{Bmatrix}, \tag{19}$$

where t_{pk} is the thickness of the kth piezoelectric layer.

Substituting Eq. (19) into Eq. (18), the electric field vector {E} can also be defined in terms of nodal variables as:

$$\{E\} = -[B_\phi]\{\phi\}_e, \tag{20}$$

Using Eqs. (15), and (20), the linear piezoelectric constitutive equations coupling the elastic and electric fields will be completely determined by Eqs. (2) and (3).

3.2 Finite element equations

Using Hamilton's principle, we have [11–13]:

$$\int_{t_1}^{t_2} (T^e - U^e - W^e)dt = 0, \tag{21}$$

where T^e, U^e are the kinetic and potential energy, respectively and W^e is the work done by external forces. They are determined by:

$$T^e = \frac{1}{2}\int_{V_e} \rho\{\dot{q}\}_e^T\{q\}_e dV, \tag{22}$$

$$U^e = \frac{1}{2}\int_{V_e} \{\varepsilon\}_e^T \{\sigma\}_e dV, \tag{23}$$

$$W^e = \int_{V_e} \{q\}_e^T \{f_b\}_e dV + \int_{S_e} \{q\}_e^T \{f_s\}_e dS + \{q\}_e^T \{f_c\}_e, \tag{24}$$

in which $\{f_b\}_e, \{f_s\}_e, \{f_c\}_e$ are the body, surface, and concentrated forces acting on the element, respectively. V_e and S_e are elemental volume and area.

Substituting Eqs. (15), (2), (20), (22), (23), and (24) into Eq. (21), one obtains:

$$[M_{bb}^e]\{\ddot{q}\}_e + [K_{bb}^e]\{q\}_e + \left[K_{b\phi}^e\right]\{\phi\}_e = \{f\}_e, \tag{25}$$

$$\left[K_{\phi b}^e\right]\{q\}_e - \left[K_{\phi\phi}^e\right]\{\phi\}_e = \{Q\}_e, \tag{26}$$

where

$$\text{Element mass matrix: } [M_{bb}^e] = \int_{V_e} \rho [N^M]^T [N^M] dV, \tag{27}$$

$$\text{Element mechanical stiffness matrix: } [K_{bb}^e] = \int_{S_e} [B_b]^T [H][B_b] dS, \tag{28}$$

Element mechanical-electrical coupling stiffness matrix:

$$\left[K_{b\phi}^e\right] = \int_{S_e} [B_b]^T [\overline{e}]\,[B_\phi] dS, \tag{29}$$

Element electrical-mechanical coupling stiffness matrix:

$$\left[K_{\phi b}^e\right] = \left[K_{b\phi}^e\right]^T, \tag{30}$$

Element piezoelectric permittivity matrix:

$$\left[K_{\phi\phi}^e\right] = -\int_{S_e} [B_\phi]^T [\overline{p}]\,[B_\phi] dS, \tag{31}$$

$$\text{where } [H] = \begin{bmatrix} c_{11} & 0 \\ 0 & c_{22} \end{bmatrix}, [\overline{e}] = \begin{bmatrix} e_{11} & e_{12} \\ e_{21} & e_{22} \end{bmatrix}, [\overline{p}] = \begin{bmatrix} t_{p1}p_{11} & 0 \\ 0 & t_{p2}p_{22} \end{bmatrix}, \tag{32}$$

$\{f\}_e$, $\{Q\}_e$ are the applied external load and charge, respectively.

3.2.1 Static analysis

In the case of beams subjected to static loads, zero acceleration, from Eqs. (25) and (26), we obtain the static equations of the beam as follows:

$$[K_{bb}^e]\{q\}_e + \left[K_{b\phi}^e\right]\{\phi\}_e = \{f\}_e, \tag{33}$$

$$\left[K^{e}_{\phi b}\right]\{q\}_e - \left[K^{e}_{\phi\phi}\right]\{\phi\}_e = \{Q\}_e, \tag{34}$$

Assembling the element equations yields general static equation:

$$[K_{bb}]\{q\} + \left[K_{b\phi}\right]\{\phi\} = \{f\}, \tag{35}$$

$$\left[K_{\phi b}\right]\{q\} - \left[K_{\phi\phi}\right]\{\phi\} = \{Q\}. \tag{36}$$

where $[K_{bb}]$, $[K_{\phi\phi}]$ are the overall mechanical stiffness and piezoelectric permittivity matrices respectively; $[K_{b\phi}]$ and $[K_{\phi b}]$ are the overall mechanical - electrical and electrical - mechanical coupling stiffness matrices, respectively, and $\{q\}$, $\{\phi\}$ are respectively the overall mechanical displacement, and electric potential vector.

Substituting Eq. (36) into Eq. (35) yields:

$$\left([K_{bb}] + \left[K_{b\phi}\right]\left[K_{\phi\phi}\right]^{-1}\left[K_{\phi b}\right]\right)\{q\} = \{f\} + \left[K_{b\phi}\right]\left[K_{\phi\phi}\right]^{-1}\{Q\}, \tag{37}$$

Substituting $\{q\}$ from Eq. (37) into Eq. (36), we obtain the vector $\{\phi\}$.

3.2.2 Dynamic analysis

From Eqs. (25) and (26), assembling the element equations yields general dynamic equation of motion:

$$[M_{bb}]\{\ddot{q}\} + [K_{bb}]\{q\} + \left[K_{b\phi}\right]\{\phi\} = \{f\}, \tag{38}$$

$$\left[K_{\phi b}\right]\{q\} - \left[K_{\phi\phi}\right]\{\phi\} = \{Q\}, \tag{39}$$

Substituting Eq. (39) into Eq. (38), we obtain:

$$[M_{bb}]\{\ddot{q}\} + \left([K_{bb}] + \left[K_{b\phi}\right]\left[K_{\phi\phi}\right]^{-1}\left[K_{\phi b}\right]\right)\{q\} = \{f\} + \left[K_{b\phi}\right]\left[K_{\phi\phi}\right]^{-1}\{Q\}, \tag{40}$$

3.2.3 Free vibration analysis

For free vibrations, from Eq. (40), the governing equation is:

$$[M_{bb}]\{\ddot{q}\} + \left([K_{bb}] + \left[K_{b\phi}\right]\left[K_{\phi\phi}\right]^{-1}\left[K_{\phi b}\right]\right)\{q\} = \{0\}. \tag{41}$$

The beam vibrations induce charges and electric potentials in sensor layers. Therefore, the control system allows current to flow and feeds back to the actuators. In this case, if we apply no external charge Q to a sensor, from Eq. (39), we will have:

$$\left[K_{\phi\phi}\right]_s^{-1}\left[K_{\phi b}\right]_s\{q\}_s = \{\phi\}_s. \tag{42}$$

and $\{Q\}_s = \left[K_{\phi b}\right]_s\{q\}_s$ is the induced charge due to strain.

The operation of the amplified control loop implies, the actuating voltage is determined by the following relationship [1, 10, 14]:

$$\{\phi\}_a = G_d\{\phi\}_s + G_v\{\dot{\phi}\}_s, \tag{43}$$

where G_d and G_v are the feedback control gains for displacement and velocity.

Substituting Eq. (43) into Eq. (39), the charge in the actuator due to actuator strain in response to the beam vibration modified by control system feedback is:

$$\left[K_{\phi b}\right]_a\{q\}_a - \left[K_{\phi\phi}\right]_a\left(G_d\{\phi\}_s + G_v\{\dot{\phi}\}_s\right) = \{Q\}_a. \quad (44)$$

Substituting (42) into (44) leads to:

$$\{Q\}_a = \left[K_{\phi b}\right]_a\{q\}_a - G_d\left[K_{\phi\phi}\right]_a\left[K_{\phi\phi}\right]_s^{-1}\left[K_{\phi b}\right]_s\{q\}_s - G_v\left[K_{\phi\phi}\right]_a\left[K_{\phi\phi}\right]_s^{-1}\left[K_{\phi b}\right]_s\{\dot{q}\}_s. \quad (45)$$

Substituting Eq. (45) into (40), we obtain:

$$\begin{aligned}&[M_{bb}]\{\ddot{q}\} + \left([K_{bb}] + \left[K_{b\phi}\right]\left[K_{\phi\phi}\right]^{-1}\left[K_{\phi b}\right]\right)\{q\} = \{f\}+\\ &+\left[K_{b\phi}\right]\left[K_{\phi\phi}\right]^{-1}\begin{pmatrix}\left[K_{\phi b}\right]_a\{q\}_a - G_v\left[K_{\phi\phi}\right]_a\left[K_{\phi\phi}\right]_s^{-1}\left[K_{\phi b}\right]_s\{\dot{q}\}_s -\\ -G_d\left[K_{\phi\phi}\right]_a\left[K_{\phi\phi}\right]_s^{-1}\left[K_{\phi b}\right]_s\{q\}_s\end{pmatrix},\end{aligned} \quad (46)$$

in which $\{q\}_s \equiv \{q\}_a \equiv \{q\}$ is the beam displacement vector, $[K_{\phi\phi}]_a$ = $[K_{\phi\phi}]_s$ = $[K_{\phi\phi}]$ is the piezoelectric permittivity matrix, and $[K_{\phi b}]_a$ = $[K_{\phi b}]_s$ = $[K_{\phi b}]$ is the mechanical-electrical coupling stiffness matrix.

Therefore, Eq. (46) becomes:

$$\begin{aligned}&[M_{bb}]\{\ddot{q}\} + [K_{bb}]\{q\} + G_v\left[K_{b\phi}\right]\left[K_{\phi\phi}\right]^{-1}\left[K_{\phi\phi}\right]\left[K_{\phi\phi}\right]^{-1}\left[K_{\phi b}\right]\{\dot{q}\}+\\ &+G_d\left[K_{b\phi}\right]\left[K_{\phi\phi}\right]^{-1}\left[K_{\phi\phi}\right]\left[K_{\phi\phi}\right]^{-1}\left[K_{\phi b}\right]\{q\} = \{f\}.\end{aligned} \quad (47)$$

In the case of considering the structural damping, the equation of motion of the beam is:

$$[M_{bb}]\{\ddot{q}\} + ([C_A] + [C_R])\{\dot{q}\} + K^*\{q\} = \{f\}, \quad (48)$$

where $[C_A] = G_v\left[K_{b\phi}\right]\left[K_{\phi\phi}\right]^{-1}\left[K_{\phi\phi}\right]\left[K_{\phi\phi}\right]^{-1}\left[K_{\phi b}\right]$ is the active damping matrix, $[K^*] = \left([K_{bb}] + G_d\left[K_{b\phi}\right]\left[K_{\phi\phi}\right]^{-1}\left[K_{\phi\phi}\right]\left[K_{\phi\phi}\right]^{-1}\left[K_{\phi b}\right]\right)$ is the total of mechanical stiffness matrix and piezoelectric, $[C_R] = \alpha_R[M_{bb}] + \beta_R[K_{bb}]$ is the overall structural damping matrix, α_R, and β_R are respectively the Rayleigh damping coefficients, which are generally determined by the first and second natural frequencies (ω_1, ω_2) and ratio of damping ξ, *{f}* is the overall mechanical force vector.

Eq. (48) can be solved by the direct integration Newmark's method.

3.3 Numerical analysis

An example for free vibration of laminated beam affected by piezoelectric layers is presented here. The beam is made of four layers symmetrically (0°/90°/90°/0°) of epoxy-T300/976 graphite material with 2.5 mm thickness per layer, and with one layer piezo ceramic materials bonded to the top and bottom surfaces, 2.0 mm thickness per layer as shown in **Figure 2** is considered (a = 0.254 m, b = 0.0254 m). The material properties of the piezo ceramic layers and graphite-epoxy are shown in **Table 1**.

The direct integration Newmark's method is used with parameters α_R = 0.5, β_R = 0.25; integral time step Δt = 0.005 s with total time calculated t = 15 s.

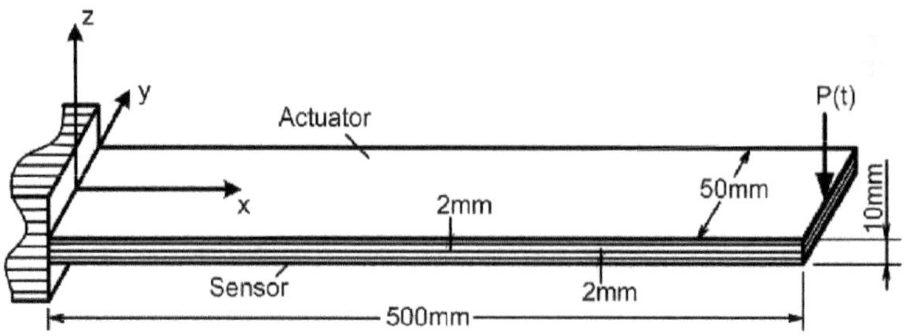

Figure 2.
Piezoelectric composite cantilever beam.

Properties	PZT G1195 N	T300/976
E_{11} [N/cm^2]	0.63×10^6	1.50×10^6
$E_{22} = E_{33}$ [N/cm^2]	0.63×10^6	0.09×10^6
$\nu_{12} = \nu_{13} = \nu_{23}$	0.3	0.3
$G_{12} = G_{13}$ [N/cm^2]	0.242×10^6	0.071×10^6
G_{23} [N/cm^2]	0.242×10^6	0.025×10^6
ρ [kg/m^3]	7600	1600
$d_{31} = d_{32}$ (m/V)	254×10^{-12}	—
$p_{11} = p_{22}$ (F/m)	15.3×10^{-9}	—
p_{33} (F/m)	15.0×10^{-9}	—

Table 1.
Relevant mechanical properties of respective materials.

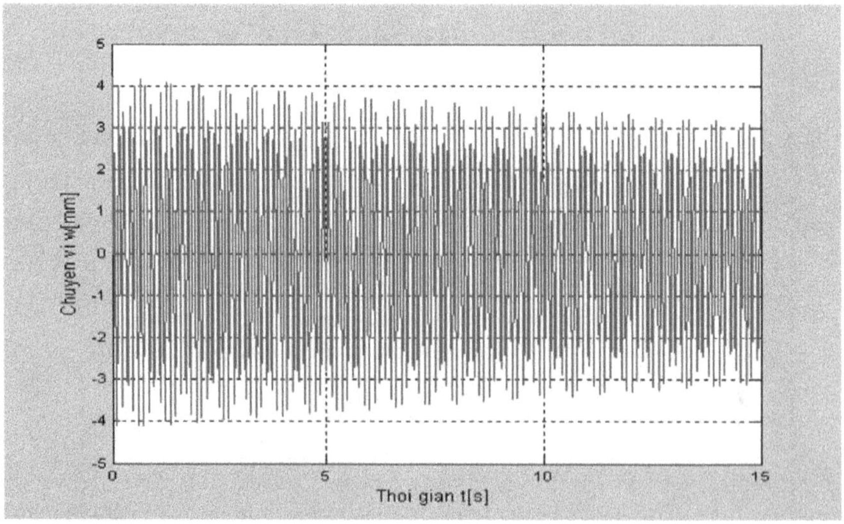

Figure 3.
Vertical displacement response ($G_v = 0$, $G_d = 0$ – Case 1).

Figures 3 and **4** illustrate the vertical displacement w at the free end of the beam for two cases:

Case 1: With structural damping, and without piezoelectric damping ($G_v = 0$, $G_d = 0$).

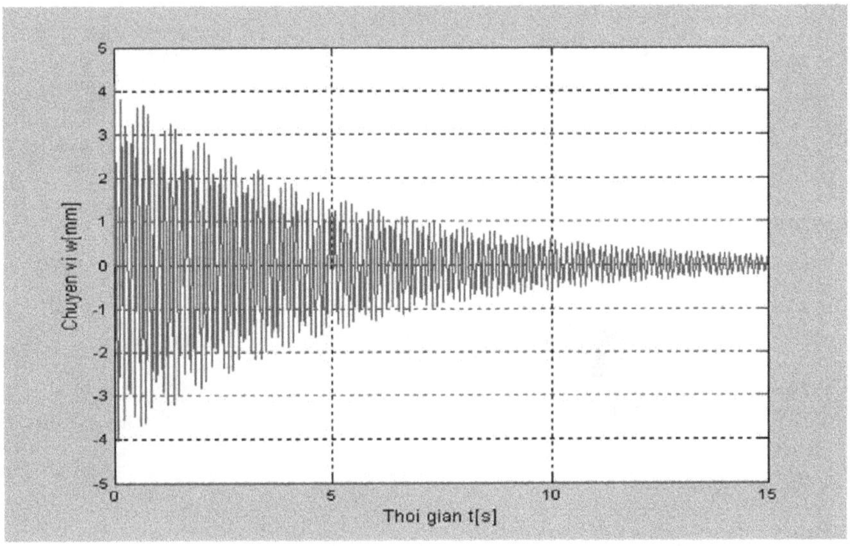

Figure 4.
Vertical displacement response (G_v = 0.5, G_d = 30 – Case 2).

Case 2: With structural damping, with piezoelectric damping (G_v = 0.5, G_d = 30).

4. Dynamic analysis of laminated piezoelectric composite plates

4.1 The electromechanical behavioral relations in the plate

Consider laminated composite plates with general coordinate system (x, y, z), in which the x, y plane coincides with the neutral plane of the plate. The top and bottom surfaces of the plate are bonded to the piezoelectric patches or piezoelectric layers (actuator and sensor). The plate under the load acting on its neutral plane has any temporal variation rule (**Figure 5**).

Hypothesis: The piezoelectric composite plate corresponds with Reissner-Mindlin theory. The material layers are arranged symmetrically through the neutral plane of the plate, ideally adhesive with each other.

4.1.1 Strain - displacement relations

Based on the first-order shear deformation theory, the displacement fields at any point in the plate are [7, 8]:

$$\begin{aligned} u(x,y,z,t) &= u_0(x,y,t) + z\theta_y(x,y,t), \\ v(x,y,z,t) &= v_0(x,y,t) - z\theta_x(x,y,t), \\ w(x,y,z,t) &= w_0(x,y,t), \end{aligned} \tag{49}$$

where u, v and w are the displacements of a general point (x, y, z) in the laminate along x, y and z directions, respectively. u_0, v_0, w_0, θ_x and θ_y are the displacements and rotations of a midplane transverse normal about the y-and x-axes respectively.

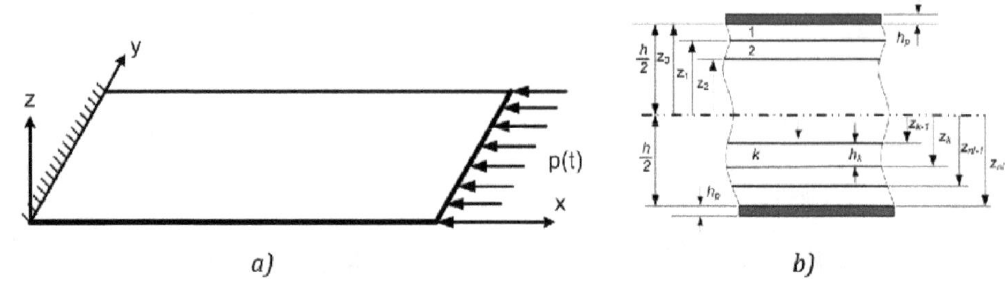

Figure 5.
Piezoelectric composite plate and coordinate system of the plate (a), and lamina details (b).

The components of the strain vector corresponding to the displacement field (49) are defined as:

For the linear strain:

$$\begin{aligned} &\varepsilon_x = \frac{\partial u}{\partial x} = \frac{\partial u_0}{\partial x} + z\frac{\partial \theta_y}{\partial x}, \varepsilon_y = \frac{\partial v}{\partial y} = \frac{\partial v_0}{\partial y} - z\frac{\partial \theta_x}{\partial y}, \\ &\gamma_{xy} = \left(\frac{\partial u}{\partial y} + \frac{\partial v}{\partial x}\right) + \frac{\partial w}{\partial x}\cdot\frac{\partial w}{\partial y} = \left(\frac{\partial u_0}{\partial y} + \frac{\partial v_0}{\partial x}\right) + z\left(\frac{\partial \theta_y}{\partial x} - \frac{\partial \theta_x}{\partial y}\right), \\ &\gamma_{xz} = \frac{\partial u}{\partial z} + \frac{\partial w}{\partial x} = \frac{\partial w_0}{\partial x} + \theta_y, \gamma_{yz} = \frac{\partial v}{\partial z} + \frac{\partial w}{\partial y} = \frac{\partial w_0}{\partial y} - \theta_x, \end{aligned} \tag{50}$$

or in the vector form:

$$\begin{aligned} &\begin{Bmatrix} \varepsilon_x \\ \varepsilon_y \\ \gamma_{xy} \end{Bmatrix} = \begin{Bmatrix} \varepsilon_x^o \\ \varepsilon_y^o \\ \gamma_{xy}^o \end{Bmatrix} + z\begin{Bmatrix} \kappa_x \\ \kappa_y \\ \kappa_{xy} \end{Bmatrix} = \begin{bmatrix} \frac{\partial}{\partial x} & 0 \\ 0 & \frac{\partial}{\partial y} \\ \frac{\partial}{\partial y} & \frac{\partial}{\partial x} \end{bmatrix} \begin{Bmatrix} u_0 \\ v_0 \end{Bmatrix} + z\begin{bmatrix} -\frac{\partial}{\partial y} & 0 \\ 0 & -\frac{\partial}{\partial x} \\ -\frac{\partial}{\partial y} & \frac{\partial}{\partial x} \end{bmatrix} \begin{Bmatrix} \theta_x \\ \theta_y \end{Bmatrix} = \\ &= [D_\varepsilon]\begin{Bmatrix} u_0 \\ v_0 \end{Bmatrix} + [D_\kappa]\begin{Bmatrix} \theta_x \\ \theta_y \end{Bmatrix} = \{\varepsilon_0\} + z\{\kappa\} = \{\varepsilon_b^L\}, \end{aligned} \tag{51}$$

$$\begin{Bmatrix} \gamma_{xz} \\ \gamma_{yz} \end{Bmatrix} = \begin{bmatrix} \frac{\partial}{\partial x} & 0 & 1 \\ \frac{\partial}{\partial y} & -1 & 0 \end{bmatrix} \begin{Bmatrix} w^o \\ \theta_x \\ \theta_y \end{Bmatrix} = \left[\, \{^w D\} \quad -[I_s] \,\right] \begin{Bmatrix} w_0 \\ \theta_x \\ \theta_y \end{Bmatrix} = \{\varepsilon_s\}. \tag{52}$$

and for the nonlinear strain:

$$\begin{Bmatrix} \varepsilon_x \\ \varepsilon_y \\ \gamma_{xy} \end{Bmatrix} = \{\varepsilon_b^L\} + \{\varepsilon^N\} = \{\varepsilon_b^N\}, \tag{53}$$

$$\begin{Bmatrix} \gamma_{xz} \\ \gamma_{yz} \end{Bmatrix} = \{\varepsilon_s\}, \tag{54}$$

where $\{\varepsilon^N\} = \frac{1}{2}\begin{bmatrix} \frac{\partial w_0}{\partial x} & 0 \\ 0 & \frac{\partial w_0}{\partial y} \\ \frac{\partial w_0}{\partial y} & \frac{\partial w_0}{\partial x} \end{bmatrix}\begin{Bmatrix} \frac{\partial}{\partial x} \\ \frac{\partial}{\partial y} \end{Bmatrix} w_0$ is the non-linear strain vector, $\{\varepsilon_b^L\}$ is the linear strain vector, $\{\varepsilon_s\}$ is the shear strain vector.

4.1.2 Stress-strain relations

The equation system describing the stress-strain relations and mechanical-electrical quantities is respectively written as [8, 14]:

$$\begin{aligned} \{\sigma_b\} &= [Q]\{\varepsilon_b^N\} - [e]\{E\}, \\ \{\tau_b\} &= [Q_s]\{\varepsilon_s\}, \end{aligned} \tag{55}$$

$$\{D\} = [e]\{\varepsilon_b^N\} + [p]\{E\}, \tag{56}$$

where $\{\sigma_b\} = \{\sigma_x \quad \sigma_y \quad \tau_{xy}\}^T$ is the plane stress vector, $\{\tau_b\} = \{\tau_{yz} \quad \tau_{xz}\}^T$ is the shear stress vector, [Q] is the ply in-plane stiffness coefficient matrix in the structural coordinate system, $[Q_s]$ is the ply out-of-plane shear stiffness coefficient matrix in the structural coordinate system. Notice that $\{\tau_b\}$ is free from piezoelectric effects.

The in-plane force vector at the state pre-buckling:

$$\{N^0\} = \{N_x^0 \quad N_y^0 \quad N_{xy}^0\}^T = \sum_{k=1}^{n} \int_{h_{k-1}}^{h_k} \begin{Bmatrix} \sigma_x^0 \\ \sigma_y^0 \\ \tau_{xy}^0 \end{Bmatrix}_k dz. \tag{57}$$

4.1.3 Total potential energy

The total potential energy of the system is given by:

$$\Pi = \frac{1}{2}\int_{V_p} \{\varepsilon_b^N\}^T\{\sigma_b\}dV + \frac{1}{2}\int_{V_p} \{\varepsilon_s\}^T\{\tau_b\}dV - \frac{1}{2}\int_{V_p} \{E\}^T\{D\}dV - W, \tag{58}$$

where W is the energy of external forces, V_p is the entire domain including composite and piezoelectric materials.

Introducing [A], [B], [D], $[A_s]$, and vectors $\{N_p\}$, $\{M_p\}$ as [8]:

$$\begin{aligned} ([A],[B],[D]) &= \int_{-h/2}^{h/2} (1,z,z^2)[Q]dz, \\ [A_s] &= \int_{-h/2}^{h/2} [Q_s]dz, \; (\{N_p\},\{M_p\}) = \int_{-h/2}^{h/2} (1,z)[e]\{E\}dz, \end{aligned} \tag{59}$$

where h is the total laminated thickness and combining with (5), (6) the total potential energy equation (8) can be written

$$\Pi = \frac{1}{2}\int_{\Omega}\{\varepsilon_0\}^T[A]\{\varepsilon_0\}d\Omega + \frac{1}{2}\int_{\Omega}\{\kappa\}^T[D]\{\kappa\}d\Omega + \frac{1}{2}\int_{\Omega}\{\varepsilon_s\}^T[A_s]\{\varepsilon_s\}d\Omega + \\ + \int_{\Omega}\{\varepsilon^N\}^T([A]\{\varepsilon_0\} - [N_p])d\Omega - \int_{\Omega}\{\varepsilon_0\}^T[N_p]d\Omega - \int_{\Omega}\{\kappa\}^T[M_p]d\Omega - W, \tag{60}$$

where Ω is the plane xy domain of the plate.

4.2 Dynamic stability analysis of laminated composite plate with piezoelectric layers

4.2.1 Finite element models

Nine-node Lagrangian finite elements are used with the displacement and strain fields represented by Eqs. (49), (53), and (54). In the developed models, there is one electric potential degree of freedom for each piezoelectric layer to represent the piezoelectric behavior and thus the vector of electrical degrees of freedom is [6, 14]:

$$\{\phi^e\} = \{\cdot \quad \cdot \quad \phi_j^e \quad \cdot \quad \cdot\}^T, \; j = 1, \ldots, NPL^e, \tag{61}$$

in which NPL^e is the number of piezoelectric layers in a given element.
The vector of degrees of freedom for the element $\{q^e\}$ is:

$$\{q^e\} = \{\,\{q_1^e\} \quad \{q_2^e\} \quad \ldots \quad \{q_9^e\} \quad \phi^e\,\}^T, \tag{62}$$

where $\{q_i^e\} = \{\,u_i \quad v_i \quad w_i \quad \theta_{x_i} \quad \theta_{y_i}\,\}$ is the mechanical displacement vector for node i.

4.2.2 Dynamic equations

The dynamic equations of piezoelectric composite plate can be derived by using Hamilton's principle, accordingly, the vibration equation of the membrane (without damping) with in-plane loads is:

$$[M_{ss}]\{\ddot{q}_{ss}\} + [K_{ss}]\{q_{ss}\} = \{F(t)\}. \tag{63}$$

The equation of bending vibrations with out-of-plane loads is:

$$\begin{bmatrix} [M_{bb}] & [0] \\ [0] & [0] \end{bmatrix}\begin{Bmatrix} \{\ddot{q}_{bb}\} \\ \{\ddot{\phi}\} \end{Bmatrix} + \begin{bmatrix} [C_R] & [0] \\ [0] & [0] \end{bmatrix}\begin{Bmatrix} \{\dot{q}_{bb}\} \\ \{\dot{\phi}\} \end{Bmatrix} \\ + \begin{bmatrix} [K_{bb}] + [K_G] & [K_{b\phi}] \\ [K_{\phi b}] & -[K_{\phi\phi}] \end{bmatrix}\begin{Bmatrix} \{q_{bb}\} \\ \{\phi\} \end{Bmatrix} \\ = \begin{Bmatrix} \{R\} \\ \{Q_{el}\} \end{Bmatrix}, \tag{64}$$

where $[M_{ss}]$, $[K_{ss}]$ are the overall mass, membrane elastic stiffness matrix respectively, and $\{q_{ss}\}, \{\dot{q}_{ss}\}, \{\ddot{q}_{ss}\}$ are respectively the membrane displacement, velocity, acceleration vector. $[M_{bb}]$, $[K_{bb}]$ and $\{q_{bb}\}, \{\dot{q}_{bb}\}, \{\ddot{q}_{bb}\}$ are the overall mass, bending elastic stiffness matrix and the bending displacement, velocity, acceleration vector; $[K_G]$ is the overall geometric stiffness matrix; ($[K_G]$ is a function of external in-plane loads); {F(t)} is the in-plane load vector, {R} is the normal load vector, $\{Q_{el}\}$ is the vector containing the nodal charges and in-balance charges.

The element coefficient matrices are:

$$[K_G^e] = [K_{Gx}^e] + [K_{Gy}^e] + [K_{Gxy}^e], \tag{65}$$

$$[K_{Gx}^e] = \int_{A_e} N_x^0 [N_x'] [N_x']^T dA_e,$$

$$\text{where } [K_{Gy}^e] = \int_{A_e} N_y^0 [N_y'] [N_y']^T dA_e, \tag{66}$$

$$[K_{Gxy}^e] = \int_{A_e} N_{xy}^0 [N_x'] [N_y']^T dA_e,$$

$$\text{in which } [N_x'] = \frac{\partial}{\partial x}[N(x,y)],\ [N_y'] = \frac{\partial}{\partial y}[N(x,y)], \tag{67}$$

$$\frac{\partial \mathrm{w}}{\partial \mathrm{x}} = \left[\frac{\partial \mathrm{N}}{\partial \mathrm{x}}\right]\{q_{bb}^e\} = [N_x']\{q_{bb}^e\},\ \frac{\partial \mathrm{w}}{\partial y} = \left[\frac{\partial \mathrm{N}}{\partial y}\right]\{q_{bb}^e\} = [N_y']\{q_{bb}^e\} \tag{68}$$

$$[K_G] = \sum_{ne} [K_G^e] \tag{69}$$

4.2.3 Dynamic stability analysis

When the plate is subjected to in-plane loads only ({R} = {0}), the in-plane stresses can lead to buckling, from Eqs. (63) and (64) the governing differential equations of motion of the damped system may be written as:

$$\begin{aligned} &[M_{ss}]\{\ddot{q}_{ss}\} + [K_{ss}]\{q_{ss}\} = \{F(t)\}, \\ &[M_{bb}]\{\ddot{q}_{bb}\} + [C_R]\{\dot{q}_{bb}\} + ([K_{bb}] + [K_G])\{q_{bb}\} + [K_{b\phi}]\{\phi\} = \{0\}, \\ &[K_{\phi b}]\{q_{bb}\} - [K_{\phi\phi}]\{\phi\} = \{Q_{el}\}. \end{aligned} \tag{70}$$

Eq. (70) is rewritten as:

$$\begin{aligned} &[M_{ss}]\{\ddot{q}_{ss}\} + [K_{ss}]\{q_{ss}\} = \{F(t)\}, \\ &[M_{bb}]\{\ddot{q}_{bb}\} + ([C_A] + [C_R])\{\dot{q}_{bb}\} + ([K^*] + [K_G])\{q_{bb}\} = \{0\}. \end{aligned} \tag{71}$$

The overall geometric stiffness matrix $[K_G]$ is defined as follows:

- In the case of only tensile or compression plates (w = 0): Solving Eq. (71) helps us to present unknown displacement vector $\{q_{ss}\}$, and then stress vector:

$$\{\sigma_{ss}\} = [A_s][B_s]\{q_{ss}\}, \tag{72}$$

where $[A_s]$ and $[B_s]$ are the stiffness coefficient matrix and strain-displacement matrix of the plane problem.

- In the case of bending plate (w $\neq$ 0), the stress vector is:

$$\begin{aligned} &\{\sigma_{sb}\} = \{\sigma_{ss}\} + \{\sigma_{bb}\}, \\ &\{\sigma_{bb}\} = [A_b][B_b]\{q_{bb}\}, \end{aligned} \tag{73}$$

where $[A_b]$ and $[B_s]$ are the stiffness coefficient matrix and strain-displacement matrix of the plane bending problem.

Stability criteria [14]:

- In the case of plate subjected to periodic in-plane loads and without damping, the elastic stability problems become simple only by solving the linear equations to determine the eigenvalues.

- In case of the plate under any in-plane dynamic load and with damping, the elastic stability problems become very complex. This iterative method can be proved effectively and the following dynamic stability criteria are used:

    - *Plate is considered to be stable if the maximum bending deflection is three times smaller than the plate 's thickness: Eq. (71) has the solution $(w_i)_{max}$ satisfying the condition $0 \leq |w_i|_{max} < 3h$, where w_i is the deflection of the plate at node number i.*

    - *Plate is called to be in critical status if the maximum bending deflection of the plate is three times equal to the plate's thickness. Eq. (71) has the solution $(w_i)_{max}$ satisfying the condition $|w_i|_{max} = 3h$.*

    - *Plate is called to be at buckling if the maximum deflection of the plate is three times larger than the plate's thickness: Eq. (71) has the solution $(w_i)_{max}$ satisfying the condition $|w_i|_{max} > 3h$.*

The identification of critical forces is carried out by the iterative method.

4.2.4 Iterative algorithm

Step 1. Defining the matrices, the external load vector and errors of load iterations.

Step 2. Solving Eq. (71) to present unknown displacement vector, $\{q_{ss}\}$ and the stress vector is defined by (72), updating the geometric stiffness matrix $[K_G]$.

Step 3. Solving Eq. (71) to present unknown bending displacement vector $\{q_{bb}\}$, and then testing stability conditions.

– If for all $|w_i| = 0$: increase load, recalculate from step 2;

– If at least one value $|w_i| \neq 0$:

+ In case: $0 < |w_i|_{max} < 3h$: Define stress vector by Eq. (73), update the geometric stiffness matrix $[K_G]$. Increase load, recalculate from step 2;

+ In case: $0 \leq \frac{||w_i|_{max} - 3h|}{|w_i|_{max}} \leq \varepsilon_D$: Critical load $p = p_{cr}$. End.

4.2.5 Numerical analysis

Stability analysis of piezoelectric composite plate with dimensions $a \times b \times h$, where $a = 0.25$ m, $b = 0.30$ m, $h = 0.002$ m. Piezoelectric composite plate is composed of three layers, in which two layers of piezoelectric PZT-5A at its top and bottom are considered, each layer thickness $h_p = 0.00075$ m; the middle layer material is Graphite/Epoxy material, with thickness $h_1 = 0.0005$ m. The material properties for graphite/epoxy and PZT-5A are shown in Section 5.1 above. One short edge of the plate is clamped, the other three edges are free. The in-plane half-sine load is evenly distributed on the short edge of the plate: $p(t) = p_0 \sin(2\pi ft)$, where p_0 is the amplitude of load, $f = 1/T = 1/0.01 = 100$ Hz ($0 \le t \le T/2 = 0.005$ s) is the excitation frequency, voltage applied V = 50 V. The iterative error of the load $\varepsilon_D = 0.02\%$ is chosen.

Consider two cases: with damping ($\xi = 0.05$, $G_v = 0.5$, $G_d = 15$) and without damping.,($\xi = 0\ 0$ $G_v = 0\ 0$ $G_d = 15$).,. The response of vertical displacement at the plate centroid over the plate thickness for the two cases is shown in **Figure 6**.

The results show that the critical load of the plate with damping is larger than that without damping. In the two cases above, the critical load rises by 6.8%.

Analyze the stability of the plate with damping when a voltage of −200, −150, −100, −50, 0, 50, 100, 150 and 200 V is applied to the actuator layer of the piezoelectric composite plate.

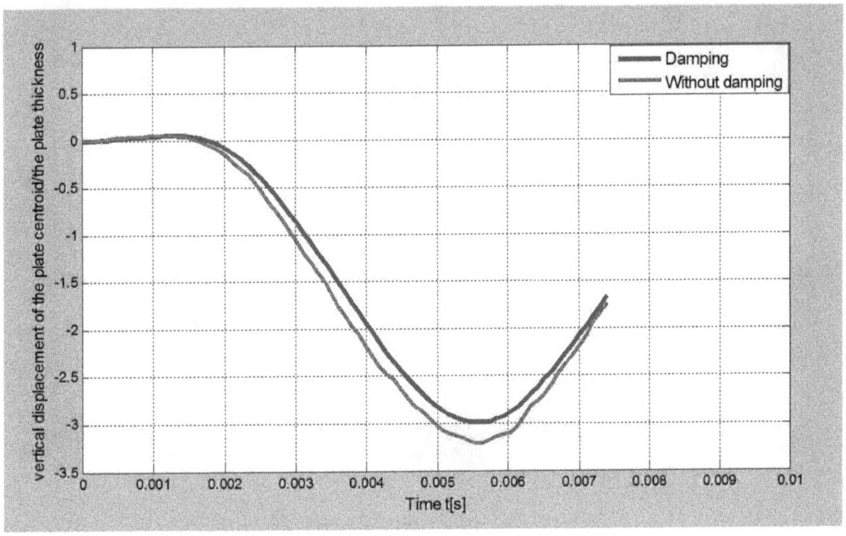

Figure 6.
Vertical displacement response at the plate centroid over the plate thickness.

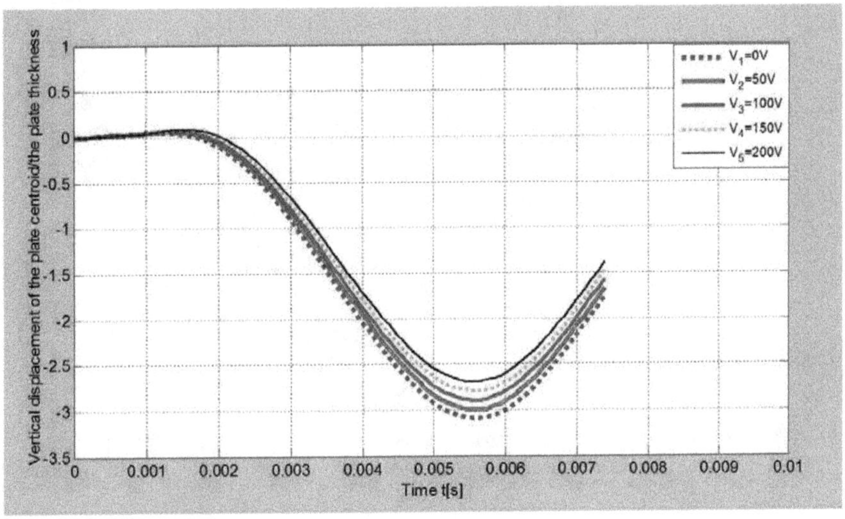

Figure 7.
Vertical displacement response at the plate centroid over the plate thickness.

Figure 7 shows the time history of the vertical displacement at the plate centroid over the plate thickness when a voltage of 0, 50, 100, 150 and 200 V is applied. The relation between critical load and voltages is shown in **Figure 8**.

The results show that the voltage applied to the piezoelectric layers affects the stability of the plate. As the voltage increases, the critical load of the plate also increases.

When the amplitude of the load changes from $0.25p_{cr}$ to $1.5p_{cr}$ (where p_{cr} is the amplitude of the critical load), a voltage of 50 V is applied to the actuator layer of the plate.

The results show the time history response of the vertical displacement at the plate centroid over the plate thickness as seen in **Figure 9**.

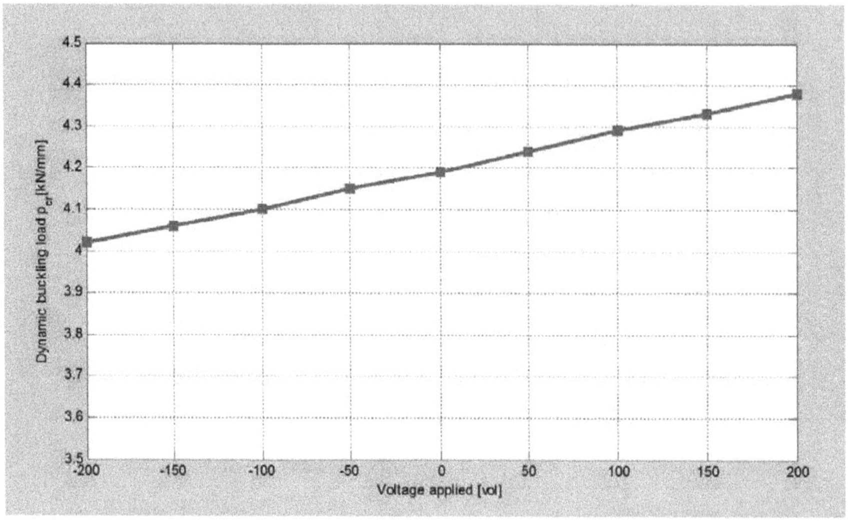

Figure 8.
Critical load-voltage relation.

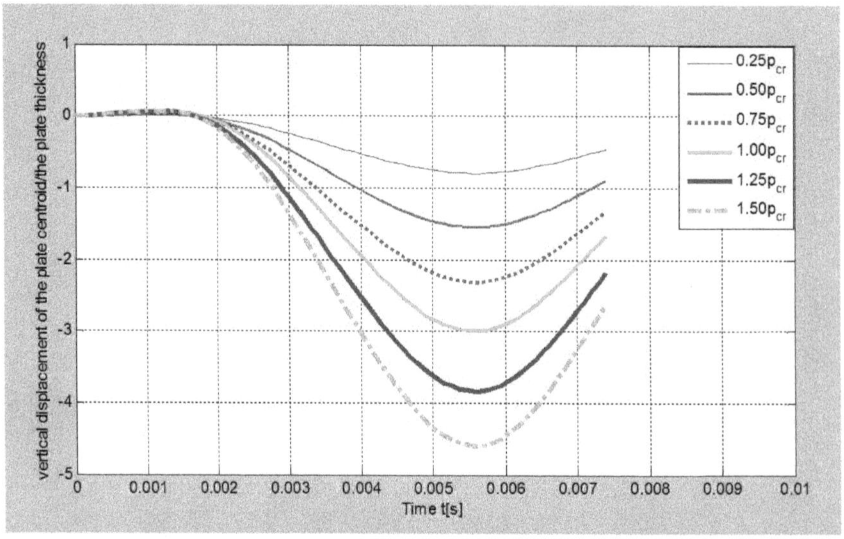

Figure 9.
Time history of the vertical displacement at the plate centroid over the plate thickness when $p_o = 0.25p_{cr}$, $0.5p_{cr}$, $0.75p_{cr}$, $1.0p_{cr}$, $1.25p_{cr}$, *and* $1.5p_{cr}$.

4.3 Dynamic analysis of piezoelectric stiffened composite plates subjected to airflow

Consider isoparametric piezoelectric laminated stiffened plate with the general coordinate system (x, y, z), in which the x, y plane coincides with the neutral plane of the plate. The top surface and lower surface of the plate are bonded to the piezoelectric patches (actuator and sensor). The plate subjected to the airflow load acting (**Figure 10**).

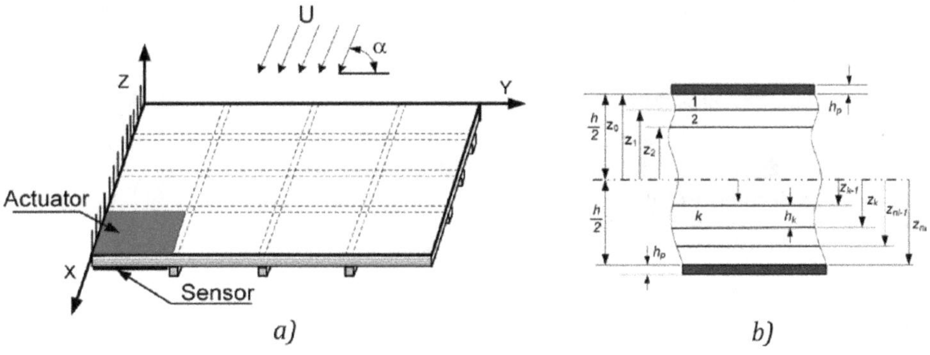

Figure 10.
Smart stiffened plate subjected to airflow. (a) Smart stiffened plate and coordinate system and (b) Lamina details.

The dynamic equations of a finite smart composite plate are written as follows:

$$[M]_e\{\ddot{u}\}_e + [C_A]_e\{\dot{u}\}_e + \left(\left[K_{bb}^{\ln}\right]_e + \left[K_{bb}^{nl}\right]_e + [K_A]_e\right)\{u\}_e = \{f\}_e^m, \tag{74}$$

where $\left[K_{bb}^{\ln}\right]_e = \int\limits_{V_e} \left[B_b^{\ln}\right]^T [Q] \left[B_b^{\ln}\right] dV$, and $\left[K_{bb}^{nl}\right]_e = \int\limits_{V_e} \left[B_b^{nl}\right]^T [Q] \left[B_b^{nl}\right] dV$ are the element linear mechanical stiffness and nonlinear mechanical stiffness respectively, $\{f\}_e^m$ is element external mechanical force vector.

4.3.1 Formulation of Stiffener:

4.3.1.1 Formulation of x-Stiffener

$$\begin{aligned} U_{xs}(x,z) &= u_0(x) + z\theta_{xs}(x), \\ W_{xs}(x,z) &= w_{xs}(x). \end{aligned} \tag{75}$$

where x-axis is taken along the stiffener centerline and the z-axis is its upward normal. The plate and stiffener element shown in **Figure 11**.

If we consider that the x-stiffener is attached to the lower side of the plate, conditions of displacement compatibility along their line of connection can be written as:

$$u_p\big|_{z=-t_p/2} = u_{xs}\big|_{z=t_{xs}/2}, \theta_{xp}\big|_{z=-t_p/2} = \theta_{xs}\big|_{z=t_{xs}/2}, w_p\big|_{z=-t_p/2} = w_{xs}\big|_{z=t_{xs}/2}, \tag{76}$$

where t_p is the plate thickness and t_{xs} is the x-stiffener depth.

The element stiffness and mass matrices are defined as follows [2, 15]:

$$[K_{xs}]_e = \int\limits_{l_e} [B_{xs}]^T [D_{xs}][B_{xs}] dx, \tag{77}$$

$$[M_{xs}]_e = \int_{A_e} \left[P\left([N_{u^0}]^T [N_{u^0}] + [N_w]^T [N_w] \right) + I_y \left([N_{\theta_x}]^T [N_{\theta_x}] \right) \right] dA, \qquad (78)$$

with $[B_{xs}]$ is the strain-displacement relations matrix, $[D_{xs}]$ is the stress-strain relations matrix and l_e is the element length, $[N_{u^0}]$, $[N_w]$ *and* $[N_{\theta_x}]$ are the shape function matrices relating the primary variables u_0, w, x, in terms of nodal unknowns, I_y is the area moment of inertia related to the y-axis and $P = \sum_{k=1}^{n} \int_{h_{k-1}}^{h_k} \rho_k dz$, with ρ_k is density of kth layer.

4.3.1.2 Formulation of y-Stiffener

The same as for x-stiffener, the element stiffness and mass matrices of the y-stiffener are defined as follows:

$$[K_{ys}]_e = \int_{l_e} [B_{ys}]^T [D_{ys}] [B_{ys}] dy, \qquad (79)$$

$$[M_{ys}]_e = \int_{A_e} \left[P\left([N_{u^0}]^T [N_{u^0}] + [N_w]^T [N_w] \right) + I_x \left([N_{\theta_y}]^T [N_{\theta_y}] \right) \right] dA, \qquad (80)$$

4.3.2 Modeling the effect of aerodynamic pressure and motion equations of the smart composite plate-stiffeners element

Based on the first order theory, the aerodynamic pressure l_h and moment m_θ, can be described as [15–17]:

$$l_w = \frac{1}{2}\rho_a (U\cos\alpha)^2 B \left[kH_1^* \frac{\dot{w}}{U\cos\alpha} + kH_2^* \frac{B\dot{\theta}}{U\cos\alpha} + k^2 H_3^* \theta \right] + \frac{1}{2} C_p \rho_a (U\sin\alpha)^2,$$
$$m_\theta = \frac{1}{2}\rho_a (U\cos\alpha)^2 B^2 \left[kA_1^* \frac{\dot{w}}{U\cos\alpha} + kA_2^* \frac{B\dot{\theta}}{U\cos\alpha} + k^2 A_3^* \theta \right], \qquad (81)$$

where $k = b\omega/U$ is defined as the reduced frequency, ω is the circular frequency of oscillation of the airfoil, U is the wind velocity, B is the half-chord length of the airfoil or half-width of the plate, ρ_a is the air density and α is the angle of attack.

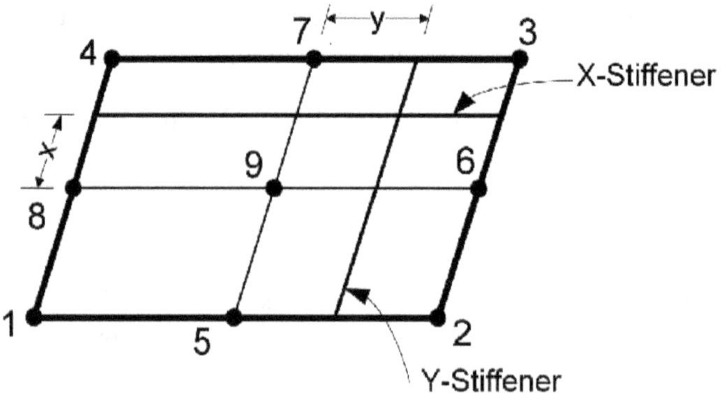

Figure 11.
Modeling of plate and stiffener element.

The functions $A_i^*(K), H_i^*(K)$ are defined as follows:

$$H_1^*(K) = -\frac{\pi}{k}F(k), H_2^*(K) = -\frac{\pi}{4k}\left[1 + F(k) + \frac{2G(k)}{k}\right],$$
$$H_3^*(K) = -\frac{\pi}{2k^2}\left[F(k) - \frac{kG(k)}{2}\right], A_1^*(K) = \frac{\pi}{4k}F(k), \tag{82}$$
$$A_2^*(K) = -\frac{\pi}{16k}\left[1 - F(k) - \frac{2G(k)}{k}\right], A_3^*(K) = \frac{\pi}{8k^2}\left[\frac{k^2}{8} + F(k) - \frac{kG(k)}{2}\right],$$

where F(k) and G(k) are defined as:

$$F(k) = \frac{0.500502k^3 + 0.512607k^2 + 0.2104k + 0.021573}{k^3 + 1.035378k^2 + 0.251293k + 0.021508},$$
$$G(k) = -\frac{0.000146k^3 + 0.122397k^2 + 0.327214k + 0.001995}{k^3 + 2.481481k^2 + 0.93453k + 0.089318}. \tag{83}$$

Using finite element method, aerodynamic force vector can be described as:

$$\{f\}_e^{air} = -[K^{air}]_e\{u\}_e - [C^{air}]_e\{\dot{u}\}_e + \{f\}_e^n, \tag{84}$$

with $[K^{air}]_e$, $[C^{air}]_e$ and $\{f\}_e^n$ are the aerodynamic stiffness, damping matrices and lift force vector, respectively

$$[K_e^{air}] = \rho_a(U\cos\alpha)^2Bk^2\int_{A_e}\left[H_3^*(k)[N_w]^T[N_{\theta x}] + BA_3^*(k)\left[\frac{\partial N_{\theta y}}{\partial x}\right]^T[N_{\theta x}]\right]dA, \tag{85}$$

$$[C_e^{air}] = \rho_a(U\cos\alpha)Bk\left[\begin{array}{l}\int_{A_e}\left(H_1^*(k)[N_w]^T[N_w] + BH_2^*(k)[N_w]^T[N_{\theta x}]dA\right)\\ +\int_{A_e}\left(BA_1^*(k)\left[\frac{\partial N_{\theta y}}{\partial x}\right]^T[N_w] + B^2A_2^*(k)\left[\frac{\partial N_{\theta y}}{\partial x}\right]^T[N_{\theta x}]dA\right)\end{array}\right], \tag{86}$$

$$\{f\}_e^n = C_p\rho_a(U\sin\alpha)^2\int_{A_e}[N_w]^TdA, \tag{87}$$

where A_e is the element area, $[N_w]$, $[N_\theta]$ are the shape functions.

From Eqs. (74) and (84), the governing equations of motion of the smart composite plate-stiffeners element subjected to an aerodynamic force without damping can be derived as:

$$[M^*]_e\{\ddot{u}\}_e + [C_A]_e\{\dot{u}\}_e + \left([K^*]_e + [K_A]_e + [K^{air}]_e\right)\{u\}_e = \{f^*\}_e^m, \tag{88}$$

where $[M^*]_e = [M]_e + [M_{xs}]_e + [M_{ys}]_e$, $[K^*]_e = [K_{bb}^{\ln}]_e + [K_{bb}^{nl}]_e + [K_{xs}]_e + [K_{ys}]_e$, $\{f^*\}_e^m = \{f\}_e^m + \{f\}_e^n$.

4.3.3 Governing differential equations for total system

Finally, the elemental equations of motion are assembled to obtain the open-loop global equation of motion of the overall stiffened composite plate with the PZT patches as follows:

$$[M^*]\{\ddot{u}\} + ([C_R] + [C_A])\{\dot{u}\} + ([K^*] + [K_A] + [K^{air}])\{u\} = \{f^*\}^m, \quad (89)$$

where $[C_R] = \alpha_R[M_{bb}] + \beta_R([K_{bb}^{\text{ln}}] + [K_{bb}^{nl}])$.

The solution of nonlinear Eq. (89) is carried out by using Newmark direct and Newton-Raphson iteration method.

4.3.4 Numerical applications

A rectangle cantilever laminated composite plate is assumed to be $[0°/90°]_s$ with total thickness 4 mm, length of 600 mm and width of 400 mm with three stiffeners along each direction x and y. The geometrical dimension of the stiffener is 5 mm of high and 10 mm of width . The plate and stiffeners are made of graphite/epoxy with mechanical properties: E_{11} = 181 GPa, E_{22} = E_{33} = 10.3 GPa, E_{12} = 7.17 GPa, ν_{12} = 0.35, ν_{23} = ν_{32} = 0.38, ρ = 1600 kg·m^{-3}. Material properties for piezoelectric layer made of PZT-5A are: d_{31} = d_{32} = -171×10^{-12} m/V, d_{33} = 374×10^{-12} m/V, d_{15} = d_{24} = -584×10^{-12} m/V, G_{12} = 7.17 GPa, G_{23} = 2.87 GPa, G_{32} = 7.17 GPa, ν_{PZT} = 0.3, ρ_{PZT} = 7600 kg·m^{-3} and thickness t_{PZT} = 0.15876 mm, ξ = 0.05, G_v = 0.5, G_d = 15. The effects of the excitation frequency and location of the actuators are presented through a parametric study to examine the vibration shape of the composite plate activated by the surface bonded piezoelectric actuators. The iterative

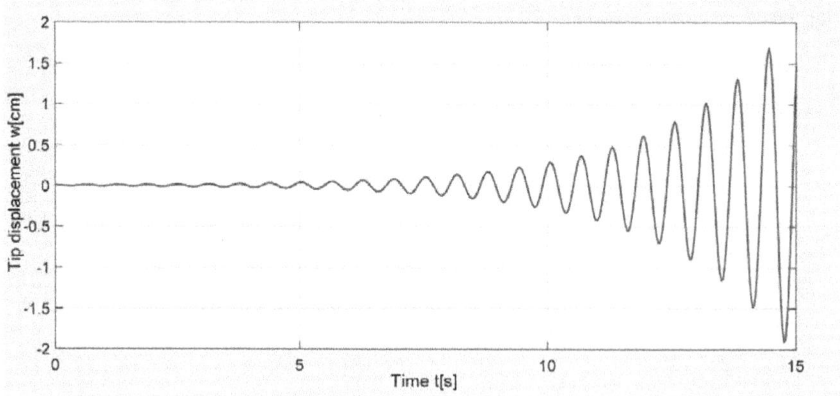

a)

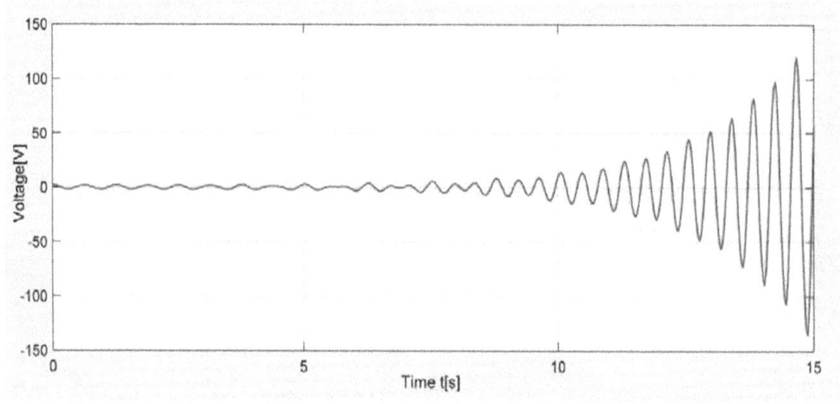

b)

Figure 12.
History of the plate at a critical airflow velocity U_{cr} = 30.5 m/s. (a) Displacement response and (b) Piezoelectric voltage response.

error of the load ε_D = 0.02% is chosen. The piezoelectric stiffened composite plate is subjected to the airflow in the positive x direction as shown in **Figure 10a**.

Dynamic response of the piezoelectric stiffened composite plate is shown in **Figure 12**.

Author details

Chung Nguyen Thai[1*], Thinh Tran Ich[2] and Thuy Le Xuan[1]

1 Le Quy Don Technical University, Hanoi, Vietnam

2 Hanoi University of Science and Technology, Hanoi, Vietnam

*Address all correspondence to: chungnt@mta.edu.vn

References

[1] Chung NT. Basic Finite Element Method and Programming ANSYS in Mechanical Engineering. Publishing house of the Vietnam Ministry of Defence; 2016. ISBN: 978-604-51-1959-4

[2] Bathe KJ, Wilson EL. Numerical Method in Finite Method Analysis. New Delhi: Prentice Hall of India Private Limited; 1978

[3] Haskins JF, Walsh JL. Vibration of ferroelectric cylindrical shells with transverse isotropy. The Journal of the Acoustical Society of America. 1957; **29**:729

[4] Tiersten HF. Linear Piezoelectric Plate Vibrations. New York: Plenum Press; 1969

[5] Adelman NT, Stavsky Y. Vibrations of radially polarized composite piezoceramic cylinders and disks. Journal of Sound and Vibration. 1975;**43**:37

[6] Yang J. The Mechanics of Piezoelectric Structures. World Scientific Publishing Co. Pte. Ltd.; 2006

[7] Vinson JR. Plate and Panel Structures of Isotropic, Composite and Piezoelectric Materials, Including Sandwich Construction. Springer; 2005. ISBN: 1-4020-3110-6 (HB)

[8] Reddy JN. Mechanics of Laminated Composite Plates and Shells Theory and Analysis. 2nd ed. CRC Press; 2004

[9] Toupin RA. Piezoelectric relation and the radial deformation of polarized spherical shell. The Journal of the Acoustical Society of America. 1959; **31**:315

[10] Thinh TI, Ngoc LK. Static behavior and vibration control of piezoelectric cantilever composite plates and comparison with experiments. Computational Materials Science. 2010; **49**:S276-S280

[11] Meirovitch L. Methods of Analytical Dynamics. Vol. 2003. New York: Dover Publications Inc.; 2003

[12] Chopra AK. Dynamics of Structures Theory and Applications to Earthquake Engineering. 2nd ed. Pearson Education Asia Limited and Tsinghua University Press; 2004

[13] Advanced Dynamic of Structures, NTUST - CT 6006. 2006

[14] Chung NT, Luong HX, Xuan NTT. Dynamic stability analysis of laminated composite plate with piezoelectric layers. Vietnam Journal of Mechanics. 2014;**36**(02):95-107

[15] Chung NT, Thuy NN, Thu DTN, Chau LH. Numerical and experimental analysis of the dynamic behavior of piezoelectric stiffened composite plates subjected to airflow. Mathematical Problems in Engineering. 2019;**2019**: 2697242. DOI: 10.1155/2019/2697242

[16] John DH. Wind Loading of Structures. New York, NY: Simultaneously Published in the USA and Canada; 2003. p. 10001

[17] Dowell EH. Aeroelasticity of Plates and Shells. Leyden: Noordhoff International Publishing; 1975. ISBN: 90-286-0404-9

9

Structural Phase Transitions of Hybrid Perovskites $CH_3NH_3PbX_3$ (X = Br, Cl) from Synchrotron and Neutron Diffraction Data

Carlos Alberto López, María Consuelo Alvarez-Galván, Carmen Abia, María Teresa Fernández-Díaz and José Antonio Alonso

Abstract

Methylammonium (MA) lead trihalide perovskites, that is, $CH_3NH_3PbX_3$ (X = I, Br, Cl), have emerged as a new class of light-absorbing materials for photovoltaic applications. Indeed, since their implementation in solar-cell heterojunctions, they reached efficiencies above 23%. From a crystallographic point of view, there are many open questions that should be addressed, including the role of the internal motion of methylammonium groups within PbX_6 lattice under extreme conditions, such as low/high temperature or high pressure. For instance, in $MAPbBr_3$ perovskites, the octahedral tilting can be induced upon cooling, lowering the space group from the aristotype $Pm\bar{3}m$ to $I4/mcm$ and $Pnma$. The band gap engineering brought about by the chemical management of $MAPb(Br,Cl)_3$ perovskites has been controllably tuned: the gap progressively increases with the concentration of Cl ions from 2.1 to 2.9 eV. In this chapter, we review recent structural studies by state-of-the-art techniques, relevant to the crystallographic characterization of these materials, in close relationship with their light-absorption properties.

Keywords: methyl ammonium (MA) lead trihalide perovskite, phase transition, octahedral tilting, MA orientation, Fourier synthesis, H location

1. Introduction

Organic-inorganic hybrid perovskites have risen as promising materials for a new generation of solar cells because of their ease of manufacturing and good performance, competing with the best photovoltaic devices based on silicon [1–5]. The introduction of $CH_3NH_3PbX_3$ (X = Br and I) as the absorber material in an electrolyte-based solar cell structure established the beginning of perovskite-based photovoltaics [2]. However, the power conversion efficiency (PCE) and cell stability were low due to the corrosion of the perovskites by the liquid electrolyte. The replacement of the liquid electrolyte with a solid hole-transporting material led to a key progress in 2012, resulting in both higher PCE and cell stability [3]. Subsequently, great efforts have been devoted to the improvement of these hybrid perovskite-based cells, including the development of a wide range of architectures, which allowed the PCEs to increase up to 23% within the past years [6]. These halide perovskites have been shown to be finer and to exhibit more suitable optical and electrical properties, like the adaptability of the band gap by introducing different halides (i.e., bromide, chloride, iodide) and its relative proportion in the perovskite [7–9], the high absorption coefficient [10], and the long lifetime of the photogenerated species [11].

$CH_3NH_3PbI_3$-based perovskite solar cells have been more extensively investigated, due to their fast charge extraction rates and their near-complete visible light absorption, related to a relatively low band gap, around 1.6 eV [11, 12]. However, the poor stability of $CH_3NH_3PbI_3$ and rapid degradation in humidity has remained a drawback for its practical application, as a commercialized product [13–16]. $CH_3NH_3PbBr_3$ constitutes a promising alternative, which presents a good charge transport in devices due to its long exciton diffusion length [17]. In addition, its cubic phase and low ionic mobility lead to a better stability under air and moist conditions, compared to the pseudo-cubic $CH_3NH_3PbI_3$ phase, [7, 17–19]. Nevertheless, there are some undesirable features to be concerned about in these bromide-based perovskites; it is the case of a larger band gap (2.2 eV), which decreases the solar light absorption [20, 21], the relatively large exciton binding energy and the reduced light absorption beyond its band edge at 550 nm (linked to its, previously indicated, larger band gap), associated with more limited efficiencies of $CH_3NH_3PbBr_3$ solar cells [7, 18, 19, 22–24].

In parallel with the evaluation of the influence of a particular chemical composition of the perovskite, it is mandatory to determine the crystallographic structure under conditions in which the sample will be used. As other perovskite structures, $CH_3NH_3PbBr_3$ consists of a framework constituted by corner-sharing $PbBr_6$ octahedra, determining large cages where $CH_3NH_3^+$ units are located. In order to correlate the crystal structure of the perovskites with their electro-optical properties, it is necessary to exhaustively study the structural details, including the orientation of the methylammonium (MA) units within the perovskite cage in the course of phase transitions. Although synchrotron XRD data are essential, this can only be fully accomplished using neutrons as a probe due to the presence of protons. Usually, the MA configuration is linked to the rotation or tilting of the PbX_6 octahedra; the more symmetric the octahedral framework is, the more delocalized appears the organic unit inside the inorganic cage. This information is essential to establish relationships between these structures and the macroscopic phenomenology (optical and physi-cal properties when these compounds are used as optoelectronic materials).

$MAPbBr_3$ was previously studied by diffraction techniques in single crystal form by X-ray or in deuterated samples by neutron beams [25–27]. On the other hand, $CH_3NH_3PbCl_3$ is also an alternative material that presents a wider band gap (3.1 eV), being also sensitive to the UV region. Furthermore, this compound exhibits a fast photoresponse and long-term photostability, having a charge carrier concentration, mobility, and diffusion length comparable with the best-developed crystal structures of $CH_3NH_3PbI_3$ and $CH_3NH_3PbBr_3$ [28–30]. However, both $CH_3NH_3PbCl_3$ and the mixed anion $CH_3NH_3Pb(Br_{1-x}Cl_x)_3$ have been less investigated [31].

In this chapter, we review our previous work on the $CH_3NH_3Pb(Br_{1-x}Cl_x)_3$ system, where we have investigated the crystallographic features in powdered, non-deuterated samples, using the mentioned state of the art techniques, neutron and synchrotron X-ray diffraction [32–34]. The crystal structure of $MAPbBr_3$ has been determined and refined at different temperatures, describing the evolution of the orientation of MA group in the 120–295 K temperature range. We found a partial delocalization in the cubic phase (at room temperature), where C and N atoms present large multiplicity positions, becoming progressively localized across the sequence cubic-tetragonal and finally orthorhombic, at 120 K, with MA units fully oriented in the (101) plane [32]. Regarding to Cl-containing phases, a systematic study of the structural properties of hybrid mixed perovskites $CH_3NH_3Pb(Br_{1-x}Cl_x)_3$ by combining synchrotron X-ray diffraction and UV-vis spectroscopic analyses reveals that the band gap can be chemically tuned according to the Br/Cl ratio. The orientation of the organic MA units may also play an important role in the optoelectronic properties of these materials. By neutron powder diffraction, we found at RT three different orientations depending on the chlorine content and, therefore, on the unit-cell size. At lower temperatures, we unveiled that the halide disorder prevents the cooperative rearrangements needed to drive the octahedral PbX_6 tiltings in intermediate Br/Cl ratios; only $CH_3NH_3PbCl_3$ underwent conspicuous phase transitions (cubic at room temperature, evolving to tetragonal and orthorhombic at 120 K) [33]. H-bond interactions with the halide ions stabilize these conformations, in accordance to reported theoretical calculations [34].

2. Crystal growth

The crystal growth of $CH_3NH_3Pb(Br_{1-x}Cl_x)_3$ (x = 0, 0.33, 0.5, 0.67 and 1) [32–34] was made from stoichiometric amounts of CH_3NH_3X and PbX_2 (X = Cl, Br). Previously, the methyl ammonium bromide and methyl ammonium chloride were synthesized from methyl amine (CH_3NH_2) and the corresponding acid HBr and HCl, respectively, according to the following reaction:

$$CH_3NH_2 + HX \rightarrow CH_3NH_3X \text{ (from X = Br to Cl)} \quad (1)$$

Then, the obtained methyl ammonium halides were reacted with the lead halide in stoichiometric amounts in dimethyl formamide (DMF) according to the following reaction:

$$CH_3NH_3X + PbX_2 \rightarrow CH_3NH_3PbX_3 \text{ (from X = Br to Cl)} \quad (2)$$

From this procedure, the mixed halide perovskites were obtained as well crystallized materials, showing crystals of variable sizes and colors, varying from orange for $CH_3NH_3PbBr_3$ to white for $CH_3NH_3PbCl_3$, adopting progressively paler hues of yellow as Cl content increases, as shown in the optical microscope images included in **Figure 1**. The effect of halide composition on the morphology and structure of

Figure 1.
SEM images of the mixed perovskites and optical microscope images of as-grown $CH_3NH_3PbX_3$ (from X = Br to Cl) perovskites. The insets show the color variation as Cl is introduced.

crystals was observed in SEM images as is also shown in **Figure 1**. In all cases, the obtained perovskites show cuboid-type microcrystals. The content of chloride induces a decrease in the size of the crystals of the mixed perovskites.

3. Synchrotron X-ray diffraction studies

Synchrotron X-ray diffraction (SXRD) technique provides an extreme angular resolution of the patterns, useful to define the symmetry of the different phases and to determine its evolution below room temperature (RT). The crystal structures at RT are cubic, and they are well defined in the space group *Pm* $\bar{3}$ *m*. In this model, the lead and bromine atoms are placed in 1*a* (0,0,0) and 3*d* (1/2,0,0) Wyckoff sites, respectively; and the organic unit is positioned in 1*b* (1/2,1/2,1/2). **Figure 2a** shows selected reflections of the Rietveld refinements corresponding to x = 0 in comparison with x = 0.33, 0.5, 0.67, and 1 members at RT. **Figure 2b** and **c** plot the unit-cell parameter variation and the anisotropic atomic displacement parameters (ADPs) of X site, respectively. The unit-cell parameters exhibit an expected reduction as the amount of Cl increases, but this change is not linear. As shown in the **Figure 2c**, the disks perpendicular to the Pb–X–Pb bonds exhibit an oblate shape, meaning the ADPs of X atoms are considerably anisotropic. Since the thermal vibrations in this direction are allowed in perovskites, this behavior is not surprising. However, for the intermediate mixed halide phases (x = 0.33, 0.5, and 0.67), the ADPs show a non-monotonic variation compared to both end members (x = 0 and 1), although the difference does not overcome two times the standard deviations and is less significant.

These anomalies were assigned to the structural disorder introduced by the mixture of halides, for x = 0.33, 0.5, and 0.67. The ADPs should account for the structural

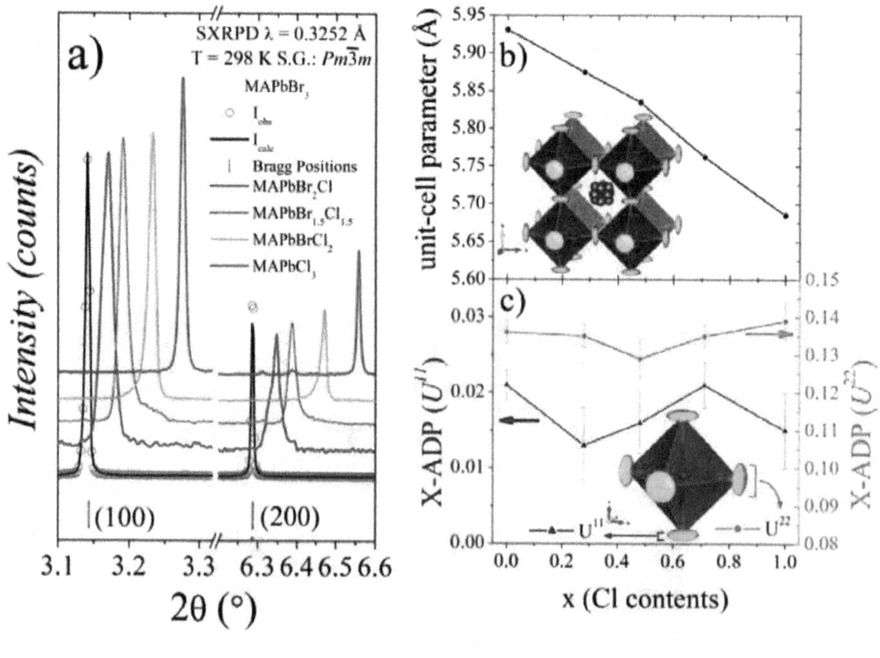

Figure 2.
(a) SXRD profiles for $CH_3NH_3PbBr_3$ at RT, after a pattern matching showing the characteristic perovskite peaks and the absence of impurities. Red circles are the experimental points, the black full line is the calculated profile. The green vertical marks represent the allowed Bragg positions in the Pm $\bar{3}$*m space group. The Cl-doped patterns are added in this plot to compare with the bromide parent. (b) Unit-cell parameters evolution and (c) variation of the anisotropic atomic displacement parameters (ADP) of X site with the Cl contents.*

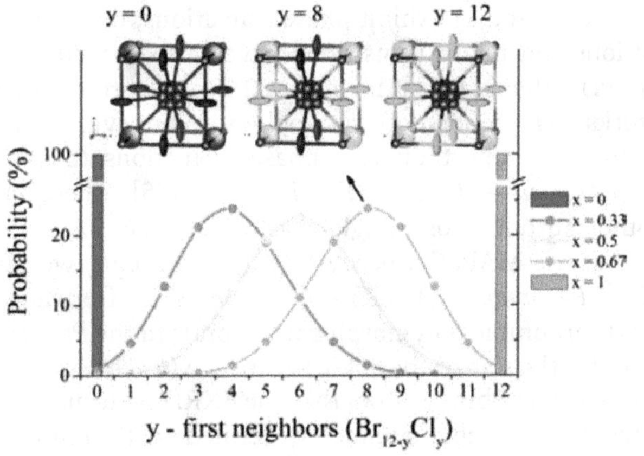

Figure 3.
Statistical probability distribution of the halide environment in $CH_3NH_3Pb(Cl_{1-x}Br_x)_3$ (x = 0, 0.33, 0.5, 0.67, and 1).

disorder, which can also generate a perturbation in the interactions between the inorganic PbX_6 skeleton and the methyl-ammonium units. This perturbation is absent in $MAPbBr_3$ and $MAPbCl_3$, containing single halide ions. In **Figure 3**, a statistical probability distribution of the halide environment of MA groups is plotted, showing this behavior for the different samples ($CH_3NH_3Pb(Cl_{1-x}Br_x)_3$, x = 0, 0.33, 0.5, 0.67, and 1). It is remarkable that for $MAPbBr_3$ and $MAPbCl_3$ the probability is 100%, because all of their MA units are coordinated to 12 Br or Cl atoms; this state contrasts with the mixed halide situations (x = 0.33, 0.5, and 0.67). These distributions reveal the high structural disorder given in mixed situations, in contrast to both end members. These probably induce tensions in the lattice preventing a linear behavior between the pure bromine and chlorine compounds. The inserts in **Figure 3** include illustrative schemes of the extreme situations in comparison with an intermediate case where the MA is coordinated to eight chlorides and four bromides, y = 8.

The thermal variation of the crystallographic structures was followed between 120 K and RT. **Figure 4** shows the temperature evolution of selected diffraction

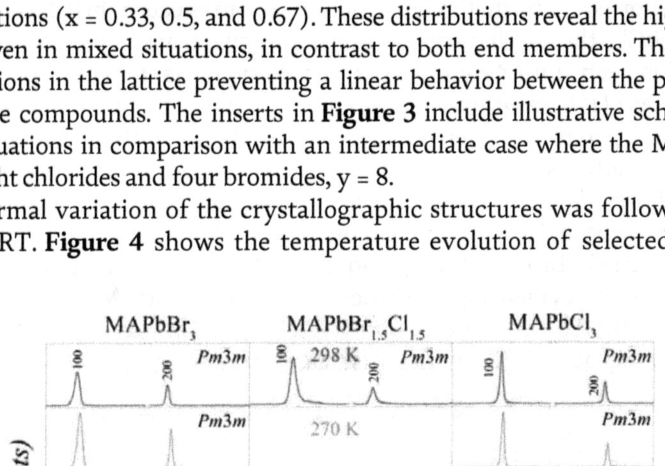

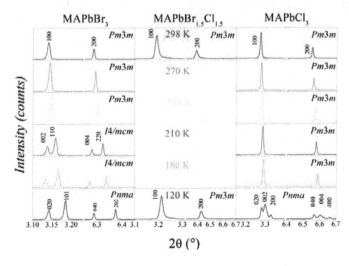

Figure 4.
Thermal evolution of selected diffraction lines in which the phase transitions are evidenced, from SXRD data, collected at MSPD diffractometer at ALBA synchrotron (Spain) [32–34].

lines for all members of the series. In this temperature range, only both end members of the series (x = 0 and 1) exhibit phase transitions; in contrast, for the mixed halide compositions, no phase transitions have been detected, observing cubic structures ($Pm\bar{3}m$) either at RT or down to 120 K, as illustrated in **Figure 4**.

Previous works also report on the polymorphic evolution of Br and Cl phases. Swainson et al. report on two phase transitions for $MAPbBr_3$: $Pnma \leftarrow (\approx 150\ K) \rightarrow I4/mcm \leftarrow (\approx 223\ K) \rightarrow Pm\bar{3}m$ [25]. These phases were also described by other authors from single crystal data [26, 27]. It was reported by Poglitsch et al. that the $MAPbCl_3$ perovskite goes through two phase transitions: $P222_1 \leftarrow (173\ K) \rightarrow P4/mmm \leftarrow (179\ K) \rightarrow Pm\bar{3}m$ [35]. Afterward, Chi et al. stated that the orthorhombic polymorph corresponds to the *Pnma* space group, but with a unit-cell twice the size of the cubic aristotype ($a \approx b \approx c \approx 2a_p$) [36].

Our work on $CH_3NH_3PbBr_3$ shows that the SXRD patterns collected at RT, 270 and 240 K correspond to cubic symmetry, defined in the $Pm\bar{3}m$ space group; at 210 and 180 K to tetragonal symmetry in the $I4/mcm$ space group; and at 120 K to orthorhombic symmetry, defined in the *Pnma* space group [32]. On the other hand, $CH_3NH_3PbCl_3$ remains stable as cubic down to 180 K; however, the SXRD patterns at 120 and 150 K exhibit a conspicuous splitting of some reflections [33]. According to the model proposed by Chi et al. [36]; this splitting (at 120 and 150 K) is possible. Nevertheless, some extra lines during the preliminary refinements seem to indicate that there is another phase, orthorhombic (*Pnma*), resembling the one observed in $MAPbBr_3$ ($a \approx \sqrt{2}a_p$; $b \approx 2a_p$; $c \approx \sqrt{2}a_p$). If the coexistence of both phases mentioned is taken into consideration, a satisfactory refinement at 150 and 120 K can be completed, with only slight differences in the peak width; being wider for the case of the conventional *Pnma* phase compared to the doubled *Pnma* structure. A possible explanation could be that microstructural features cause this phase mixture.

For $MAPbCl_3$, the mechanism or transient state from cubic to orthorhombic symmetry has been until now only partially known. As indicated previously, the tetragonal phase ($P4/mmm$) was observed in a very narrow range of temperature, 172.9–178.8 K. Up to now, no new reports have appeared on this transient tetragonal phase. Several patterns were collected sequentially [33] in this narrow temperature range, with intervals of 2.5 K, as illustrated in **Figure 5a**, where a different phase is evidenced between 169 and 164 K. This phase, corresponding to the pattern at 167.2 K, was initially fitted to the tetragonal model; however, additional diffraction lines were observed evidencing an orthorhombic symmetry. Matching the patterns observed at 120, 150, and 160 K, *Pnma* ($a \approx \sqrt{2}a_p$; $b \approx 2a_p$; $c \approx \sqrt{2}a_p$) provided a satisfactory fit at 167.2 K. In this last case, at low temperatures the unit-cell

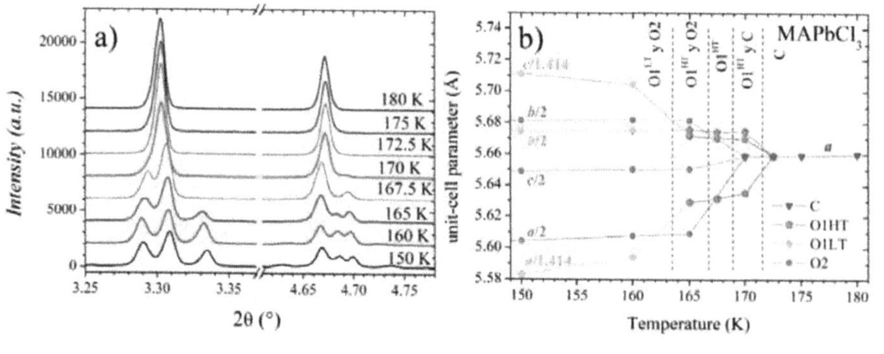

Figure 5.
(a) Selected angular region of the SXRD patterns of $MAPbCl_3$ at increasing temperatures (from 150 to 180 K), illustrating the evolution and splitting of some characteristic reflections. (b) Thermal evolution of the unit-cell parameters of the $MAPbCl_3$ perovskite in the same temperature range [33].

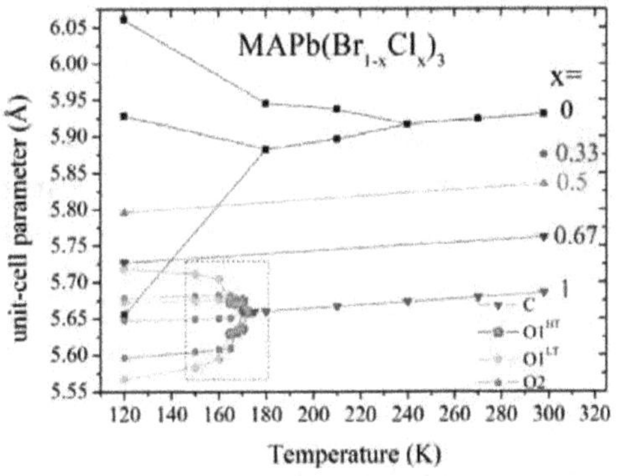

Figure 6.
Thermal evolution of the unit-cell parameters of $MAPb(Br_{1-x}Cl_x)_3$ perovskites in the 120–300 K temperature range [33].

parameters tend to be considerably less split. In **Figure 5b**, the unit-cell parameters variation as a function of temperature is displayed in this narrow temperature range.

From these facts, it is clear that $MAPbCl_3$ goes through a more complex transition compared to the ones already reported. To summarize, there are three phases: cubic, orthorhombic ($a \approx \sqrt{2}a_p$; $b \approx 2a_p$; $c \approx \sqrt{2}a_p$), and a second orthorhombic ($a \approx 2a_p$; $b \approx 2a_p$; $c \approx 2a_p$), as indicated in **Figure 5b** as C, O1, and O2, respectively. Moreover, the O1 phase can be separated into two states with different distorted degrees: $O1^{HT}$ and $O1^{LT}$. In the first case, the unit-cell parameters splitting with respect to cubic a_p is substantially lower than in the second case. Moreover, an inversion between $b/2$ and $c/\sqrt{2}$ values is observed in **Figure 5b**. Additionally, a high tetragonality is observed in $O1^{HT}$ where the following relationship between the unit-cell parameters is observed: $a/\sqrt{2} < b/2 \approx c/\sqrt{2}$. This fact can explain the previous description in the tetragonal symmetry in this short temperature range.

As reported in [32, 33], $MAPbCl_3$ and $MAPbBr_3$ are stable above 179 and 237 K respectively; therefore, it is relevant to mention the lack of any other phase transition down to 120 K in mixed halide perovskites. The anion disorder in the mixed-halide phases could explain this surprising behavior. As it was previously mentioned, the interactions between the organic and inorganic parts differ from one sample to another within the present series (see **Figure 3**). The phase transitions are prevented due to the halide disorder, because it does not allow the rearrangements that are required in terms of octahedral tilting. Finally, SXRD measurements and their analysis (Rietveld refinements) provide new evidence that allows to complete the polymorphic evolution in the $MAPb(Br_{1-x}Cl_x)_3$ family, as displayed in **Figure 6**.

4. Neutron diffraction studies

The neutron powder diffraction (NPD) investigation is essential to obtain a detailed description of these hybrid materials. The NPD data were collected at the D2B diffractometer (ILL, France) [32, 34]. In particular, the distribution of MA groups can be elucidated taking advantage of the high contrast in the coherent scattering lengths of Pb, Br, Cl, C, N, and H, of 9.405, 6.795, 9.577, 6.646, 9.36, and −3.739 fm, respectively.

However, there is an issue in the resolution of the structure that needs to be considered. The incoherent background in a powder experiment is considerably large due to the presence of hydrogen, which has an important inelastic component

(25,274 barns) that gives rise to a significant incoherent background. This is not an obstacle for the crystallographic Rietveld refinement if sufficient statistic is achieved during the measurement.

On the other hand, due to the negative contribution of hydrogen to the scattering, some strategies are necessary to find the adequate distribution of MA in the inorganic PbX_6 framework. First, the instrumental and the unit-cell parameters were refined using the Le Bail method. The structure of the octahedral PbX_6 framework was eventually considered in the model, placing the lead atoms at 1*a* (0,0,0) sites and bromine/chlorine atoms at 3*d* (1/2,0,0) positions in the $Pm\bar{3}m$ space group. Later on, Difference Fourier Maps (DFM) were acquired from the observed and calculated patterns, unveiling the missing negative and positive nuclear density (in scattering length) around A site of the perovskite structure (1*b* (1/2,1/2,1/2)).

This analysis in $MAPbBr_3$ [32] yields the positive (represented in yellow) and negative (represented in blue) isosurfaces shown in **Figure 7**a, which correspond to the C/N and H positions, respectively. The negative scattering regions are due to the negative scattering of protons while the positive zone corresponds to C/N atoms. These nuclear densities support that the MA units are delocalized in the A site of the perovskite. Considering the positive density, the C/N atoms are located at 24*i* (1/2,*y*,*y*) positions. Then, the H positions can be elucidated from the geometric shape of methylammonium group. The observed geometry can be satisfied with two hydrogen atoms located at 24*l* (1/2,*y*,*z*) and 24*m* (*x*,*x*,*z*) Wyckoff sites. This analysis in $MAPbBr_3$reveals that the MA group is delocalized at room temperature along the [110] direction, involving six possible orientations.

In contrast, the DFM for $MAPbBr_2Cl$ show that the positive and negative densities match exactly with those expected for the MA cations oriented along [111] directions. It is possible to deduce that the C/N and H atoms are placed at 8*g* (*x*, *x*, *x*) and 24*m* (*x*, *x*, *z*) Wyckoff positions, respectively (**Figure** 7b). $MAPbBr_{1.5}Cl_{1.5}$ exhibits a peculiar negative distribution (see **Figure 7c**), where there are four negative zones along the [001] directions, suggesting that the MA units are along this direction. However, along [111] directions, there also appears a non-negligible density, which is unrealistic. This dichotomy is resolved by analyzing the C/N density from the positive surface; it unveils that the C/N atoms are indeed delocalized along the [100]

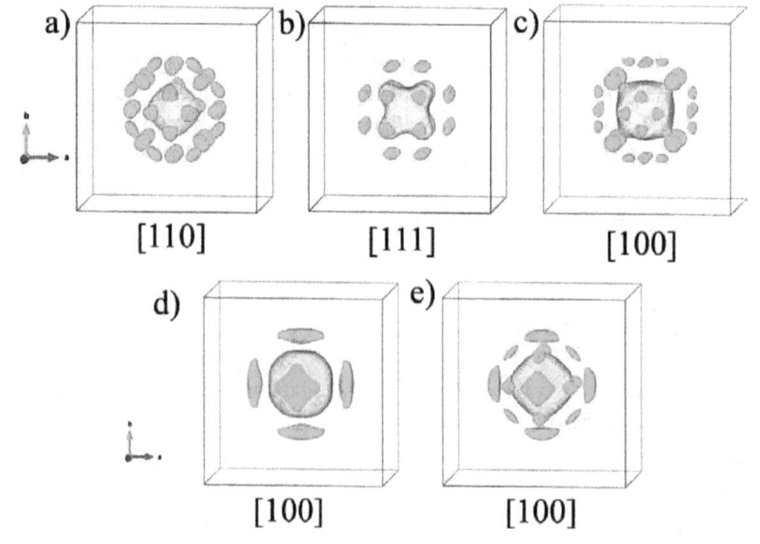

Figure 7.
DFM isosurfaces of $MAPb(Br_{1-x}Cl_x)_3$ *series for x = 0 (a), 0.33 (b), 0.5 (c), 0.67 (d), and 1 (e), from NPD data [32, 34].*

directions. The same procedure in $MAPbBrCl_2$ and $MAPbCl_3$, see **Figure 7d** and **e**, shows that the MA groups are oriented along [100] directions, in the same way as $MAPbBr_{1.5}Cl_{1.5}$. From this result, which only can be obtained from NPD data, a clear restriction in the MA delocalization is unveiled, evolving from [110] to [111] and finally to [100] directions, while X is progressively enriched in Cl^- anions.

From the Fourier synthesis maps, a crystallographic model can be built with the MA configurations, to start the structural Rietveld refinements from the NPD data. Additionally, the MA displacement toward $-NH_3$ group also can be considered, as it was proposed from theoretical calculations [37, 38]. This last approach corresponds to the plausible chemical interactions between the organic cation and inorganic framework; the greater electronegativity of nitrogen produces H-bond interactions shorter than C–H···X. Considering this last fact and the DFM results, the Rietveld refinement of NPD data was made. Minor amounts of MACl were identified and included in the refinements as second phase in the chlorine phase. The Rietveld fits for $MAPbBr_3$ and $MAPbCl_3$ are illustrated in **Figure 8**. The crystal structure data after these refinements are listed in **Table 1** for all the compositions. This table shows the three different combinations of C/N and H positions considering three possibilities for the MA delocalization in the PbX_6 framework observed in the $MAPb(Br_{1-x}Cl_x)_3$ series. The variation of the unit-cell parameter with the Cldoping is illustrated in **Figure 9**; here, a change of slope is observed in the perovskite with an equimolar amount of Br and Cl atoms, according to the anomaly obtained from synchrotron data.

For x = 0.5, 0.67, and 1, the unit cell experiences a contraction, keeping a constant slope as Cl is introduced, with the MA oriented in the same direction. On the other hand, for x = 0 and 0.33, the unit cell parameters do not follow this trend, the MA orientation is different.

A feature to highlight is the difference in negative density (arising from H positions) observed between [111] and [100] orientations. While along the [111] direction, the three terminal H atoms of the $H_3C\text{-}NH_3^+$ units are fixed, as it is shown in **Figure 7b**. From the negative density displayed in **Figure 7c–e**, it is noticeable that four H atoms are comprised in the (100) plane, suggesting that the MA units occupy equivalent positions, so that each H points to each of the four halides in the edges of the perovskite unit cell, thus manifesting the importance of H···X hydrogen bonds. This bonded H is located in H11 site and the other two H of the amine group are located in the H12 site. Similarly, the methyl group is formed by C2, H21, and H22 atoms.

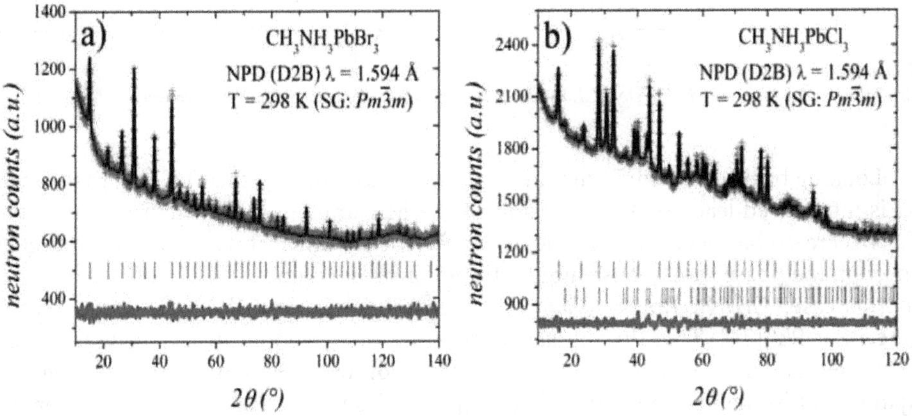

Figure 8.
Rietveld plots of a) $MAPbBr_3$ and b) $PbPbCl_3$ from NPD data, at RT.

$MAPb(Br_{1-x}Cl_x)_3$	x = 0 [32]	x = 0.33 [34]	x = 0.5 [34]	x = 0.67 [34]	x = 1 [34]
a (Å)	5.9259(4)	5.8454(8)	5.80459(6)	5.76955(7)	5.6813(5)
Pb (1*a*) (0,0,0)					
$U^{11} = U^{22} = U^{33}$	0.025(2)	0.048(3)	0.0313(3)	0.031(2)	0.0239(4)
$Br_{1-x}Cl_x$ (3*d*) (0.5,0,0)					
x	0	0.38(3)	0.55(5)	0.67(8)	1
U^{11}	0.018(5)	0.028(5)	0.0147(5)	0.026(4)	0.0471(8)
$U^{22} = U^{33}$	0.138(5)	0.130(5)	0.1318(6)	0.104(4)	0.1151(7)
N/C	C/N (0.5,*y*,*y*) 0.407(2)	N1(*x*,*x*,*x*) 0.5809(5)	N1(0.5,0.5,*z*) 0.6290(4)	N1(0.5,0.5,*z*) 0.6294(5)	N1(0.5,0.5,*z*) 0.6594(5)
		C2(*x*,*x*,*x*) 0.4378(5)	C2(0.5,0.5,*z*) 0.3777(6)	C2(0.5,0.5,*z*) 0.3782(6)	C2(0.5,0.5,*z*) 0.4113(5)
U_{iso}*/U_{eq}	0.039(7)	0.061(1)*	0.051*	0.051*	0.039(1)*
H	H1(0.5,*y*,*x*) 0.414(9) 0.247(9)	H11(*x*,*y*,*x*) 0.5839(6) 0.7363(9)	H11(0.5,*y*,*z*) 0.664(1) 0.702(3)	H11(0.5,*y*,*z*) 0.665(1) 0.702(3)	H11(0.5,*y*,*z*) 0.670(1) 0.720(2)
	H2(*x*,*x*,*z*) 0.326(3) 0.448(4)	H12(*x*,*y*,*x*) 0.4416(7) 0.2841(8)	H12(*x*,*y*,*z*) 0.640(1) 0.419(1) 0.702(3)	H12(*x*,*y*,*z*) 0.641(1) 0.418(1) 0.702(3)	H12(*x*,*y*,*z*) 0.6445(9) 0.4166(9) 0.720(2)
			H21(0.5,*y*,*z*) 0.331(1) 0.305(3)	H21(0.5,*y*,*z*) 0.331(1) 0.306(1)	H21(0.5,*y*,*z*) 0.326(1) 0.351(2)
			H22(*x*,*y*,*z*) 0.357(1) 0.583(1) 0.305(3)	H22(*x*,*y*,*z*) 0.356(1) 0.583(1) 0.306(3)	H22(*x*,*y*,*z*) 0.354(1) 0.585(1) 0.351(2)
U_{iso}*/U_{eq}	0.06(1)	0.063*	0.063*	0.063*	0.059(2)*
R_p(%)	1.3	1.3	1.0	1.2	0.8
R_{wp}(%)	1.6	1.7	1.3	1.5	0.9
χ^2	1.1	1.1	1.3	1.1	1.2
R_{Bragg}(%)	3.9	10.8	6.5	6.5	6.4

Table 1.
Crystallographic data for $MAPb(Br_{1-x}Cl_x)_3$ with x = 0, 0.33, 0.5, 0.67, and 1 at room temperature from NPD.

Looking back at the unit-cell variation, the contraction of the PbX_6 octahedron as Cl is introduced leads to a localization of the MA units, limiting their orienta-tion. This decrease has been deeply studied below room temperature, as the crystal goes through the phase transitions: cubic-tetragonal-orthorhombic [37, 39, 40]; however, in this case, $CH_3NH_3Pb(Br_{1-x}Cl_x)_3$, the behavior is observed in the same crystal system at RT [34]. This finding shows that the MA freedom degree reduction can occur either by a decrease in the symmetry or by size reduction (keeping the symmetry) in the PbX_6 network. These discoveries renew the question whether the inorganic framework deformation occurs and then the MA units accommodate in the preferred directions, or the reduction in the MA freedom degree enables the

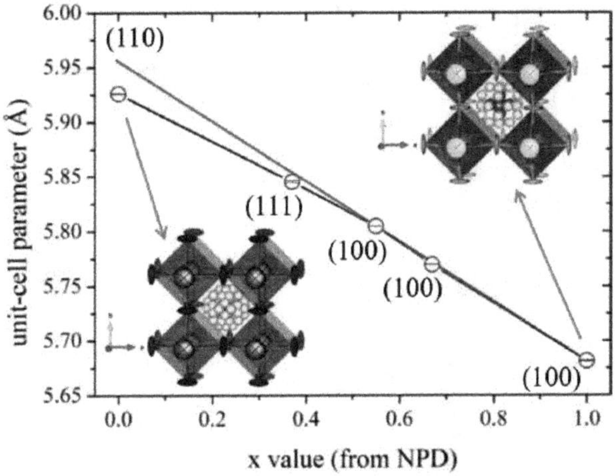

Figure 9.
Unit-cell parameters variation for $MAPb(Br_{1-x}Cl_x)_3$ from NPD data, at RT.

inorganic framework to adopt a low symmetry. This dichotomy is as the *chicken and egg* paradox, which has previously been analyzed from DFT calculations, concluding that phase transitions occur as a synergic effect between both organic and inorganic components [38]. In the present case, the MA restrictions occur without symmetry reduction in the inorganic framework, but as a consequence of a contraction in the unit-cell size, so it is possible to think that the PbX_6 lattice drives the MA freedom degree. However, it is only an appreciation and it is not enough to resolve this dichotomy.

On the other hand, the role of H···X interactions in the MA conformations is not unknown [41, 42] but shows some controversy among the different results. Theoretical studies from DFT indicate that the energy difference between the low index [100], [110], and [111] orientations of the MA units is similar, but with a subtle preference for the [100] orientation in the tetragonal $MAPbI_3$ [40]. However, the ab initio simulations performed by Shimamura et al. [43] unveiled that [110] was the preferred orientation due to the H···X interactions in $MAPbI_3$. Differently, Li et al. identify [111] and [100] for the $MAPbI_3$ and $MAPbCl_3$ perovskites as the favorable orientations [38]. Afterward, Varadwaj et al. reinforced this last idea by also stating that these orientations were the most energetically favorable [37]. Although there is significant theoretical work regarding the MA units orientation in cubic symmetry, experimental evidence is lacking [25, 27, 36]. It is the case of the Baikie et al. [27] report that focuses on the three structures for X = I, Br, and Cl from X-ray and neutron diffraction data. In all the cases, the MA components were refined in the [110] direction. In this context of theoretical results, the considerations made in the $CH_3NH_3Pb(Br_{1-x}Cl_x)_3$ refinements offer new possibilities to achieve a detailed analysis of the crystal studies in comparison with these theoretical works. Once the atomic positions are refined, a "deconvolution" of the MA units orientation can be done for the directions [111] and [100]. The H···X interactions obtained can be compared with those predicted theoretically.

The MA units within the cubic unit cell are presented in **Figure 10**, the three alignment possibilities ([110], [111], and [100]) are shown. The MA in $MAPbBr_3$ is aligned along [110] and, as can be seen in **Figure 10a** and **d**, in this orientation the N–H bonds point toward face centers or cube corners forming smalls angles N–H···Br; hence, this configuration is not suitable for hydrogen bonding. On the other hand, for x = 0.33 (along [111] direction), the H–C and H–N interact with a

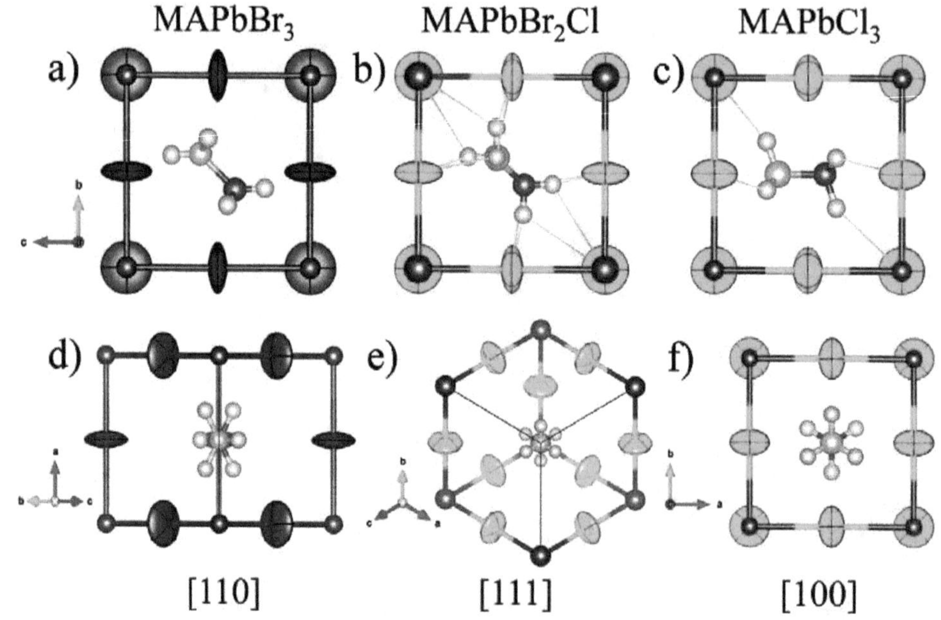

Figure 10.
Methyl-ammonium conformations in the MAPbBr3 (a) and (d), MAPbBr2Cl (b) and (e) and MAPbCl3 (c) and (f) perovskites, from NPD data at RT. The lowest figures (d), (e) and (f) are shown along the MA direction.

unique halide (X = $Cl_{0.33}Br_{0.67}$) with a distance of 2.921(5) and 3.088(5) Å, respectively (**Figure 10b** and **e**). As mentioned, the MA conformation is slightly displaced, the (C)H···X distances are larger than (N)H···X, which could be explained by the bigger electronegativity of N compared to C.

Varadwaj et al. reported on DFT calculations for the $MAPbBr_3$ perovskite [37]; the distances were determined, being ≈2.44 and ≈3.47 Å for (N)H···X and (C)H···X, respectively, in agreement with the NPD experimental results. Carrying out the same procedure for x = 0.5, 0.67 and 1, we can determine the orientation of the MA units, which resulted to be [100]. The MA units can be rotated in four positions along the three [100] directions giving 12 possible conformations. With the help of DFT calculations, the orientation of the MA unit for bromine and chlorine phases can be estimated, with the additional fact that the (N)H···X distances are shorter that the (C)H···X ones [37, 38]. **Figure 10c** and **f** illustrate a MA unit from the obtained C, N, and H atomic positions for $MAPbCl_3$; two different types of H-bonds can be distinguished: normal (N1–H11···X) and bifurcated (N1–H12···X). Similarly, the methyl group shows the same interactions through H21 and H22 atoms. The obtained distances for x= 1 phase are 2.459(8) and 2.615(8) Å for N1–H11···X and N1–H12···X, respectively [38].

The structural refinement from NPD data provide, for the first time, experimental evidence of the MA alignments, confirming the theoretical results reported up to now.

5. Optoelectronic properties

The optoelectronic properties are characterized using different techniques such as diffuse reflectance UV-vis spectra and building a photodetector device where the current-voltage (IV) curves are measured under illuminations. This device was fabricated by drop-casting the perovskite solution onto Au/Cr pre-patterned electrodes.

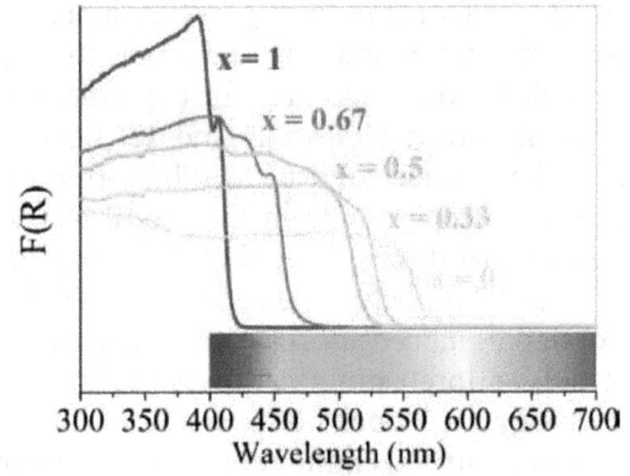

Figure 11.
UV-vis absorption spectra for $MAPb(Br_{1-x}Cl_x)_3$. The colored line corresponds to the band gap energy.

The UV-vis spectra, illustrated in **Figure 11**, are used to calculate the optical absorption coefficient (α) according to the Kubelka-Munk equation:

$$F(R) = \alpha = (1 - R)2/(2R) \quad (3)$$

where R is the reflectance (%). The optical band gap is determined from the extrapolation of the linear part of the transformed Kubelka-Munk spectrum with the hν axis. Absorbance vs. wavelength of the incident radiation for $MAPb(Br_{1-x}Cl_x)_3$ series is illustrated in **Figure 11**. In this plot, the included color chart allows accounting for the sample color shown in **Figure 1**.

The absorption edge presents a gradual evolution, which is consistent with the continuous structural changes, mainly with the unit-cell parameter evolution. The energy of the gap experiences an increment as Br is progressively replaced with Cl. This is, in principle, expected as a consequence of the smaller covalency of the lead-halide bonds within the PbX_6 (X = Cl, Br) octahedra [31]. Recent studies by density functional theory (DFT) show that the edges of the valence band and conduction band of $CH_3NH_3X_3$ perovskites are principally made up of p_X and p_{Pb} states, respectively [44]. It is shown that near the gap, the predominant contribution is brought about by Pb-Pb transitions, involving *s*-type orbitals. The characteristic absorption band above 3 eV is fundamentally driven by s_{Pb}- p_{Pb} interactions, tuned by the unit-cell

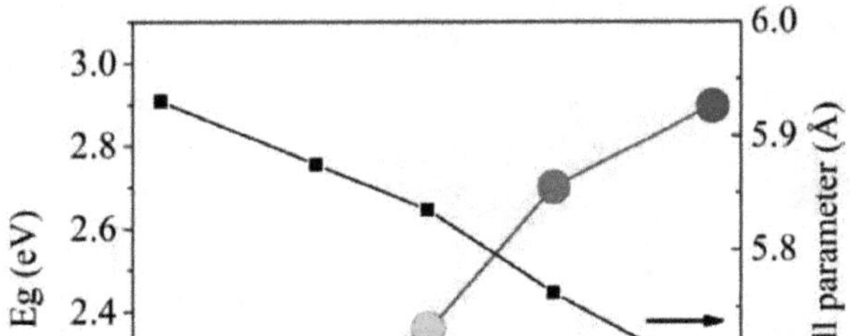

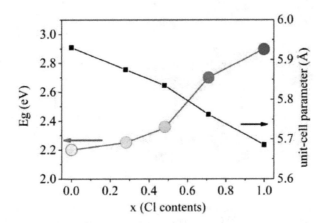

Figure 12.
Comparison between the band gap energy and unit-cell parameter in $MAPb(Br_{1-x}Cl_x)_3$, as a function of the Cl content, x. The color in the circles corresponds to the band gap energy.

size, with a lower contribution by p_{Pb}-p_X and s_X-p_X transitions. These latter contributions decrease from iodine to bromine and then to chlorine, that is, when increasing the electronegativity of the halide, which accounts for the variation of the gap with the chemical nature of X, as shown in **Figures 11** and **12**. Endres et al. [45] found a remarkably low DOS at the valence band of $MAPbX_3$, from either ultraviolet and inverse photoemission spectroscopies or from theoretical densities of states (DOS), calculated via density functional theory. They found a strong band dispersion at the valence band due to the coupling between halide p and Pb s antibonding orbitals, observing the absence of significant densities of states tailing into the perovskite gaps.

As already mentioned, an interesting feature is that the compositional evolution of the band gap energy (Eg) does not possesses a linear behavior; it seems that the deviation from linearity is reminiscent of that unveiled for the variation of the unit-cell parameters at RT (see **Figure 12**). By additionally contrasting the Eg value with the Pb–X bond lengths, a non-linear behavior is apparent, suggesting that the evolution of Eg is not only related to the inorganic features. From this point of view, it is interesting to consider that, although all these studies are based on the nominally simple cubic perovskite structure, these compounds are in fact very complex. For example, in $(CH_3NH_3)PbI_3$, the dynamics associ-ated with the $(CH_3NH_3)^+$ ions are still not fully understood, although ab initio calculations show [46] that at room and higher temperature, the rotation of CH_3NH_3 molecules can be viewed as effectively giving local structures that are cubic and tetragonal-like from the point of view of the PbI_3framework, though

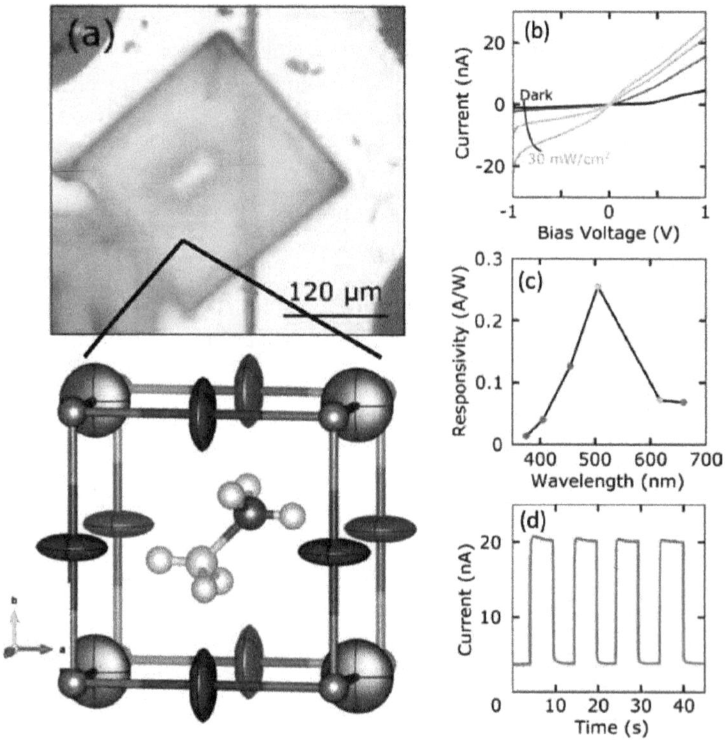

Figure 13.
(a) View of the photodetector device, including a crystal of $MAPbBr_3$grown between two pre-patterned electrodes, separated apart by 10 μm. (b) Current-voltage (I-V) curves obtained with 505-nm illumination, by changing the incident optical power densities. (c) Spectrum displaying the responsivity for different wavelengths. (d) Generated photocurrent vs. time, when a modulated illumination is applied with a fixed voltage [32].

in fact having lower symmetry. These arrangements are locally polar, with sizable polarization, ~10 μC/cm^2 due to the dipoles on the organic part [46]. The structural transitions are thus analogous to the transitions between two ferroelectric structures, where there is strong screening of charged defects that can lead to enhanced mobility and charge collection.

This observation seems to confirm that the increase in the Eg with the amount of Cl^- is closely correlated with the effect of the MA and its freedom degree in the inorganic framework. The change in the MA delocalization upon the incorporation of smaller Cl^- anions is an additional parameter to be considered in the tuning of the optical properties of these hybrid materials.

Furthermore, the potential of these materials in optoelectronic applications can be exemplified for $MAPbBr_3$ measuring the current-voltage (I-V) in a photodetector device that is illustrated in **Figure 13a** [32]. The effect of the optical power densities under illumination at 505 nm in the I-V curves is plotted in **Figure 13b**. This device shows a photoresponse among the best ones reported for other perovskite-based photodetectors, even considering different geometries: 2D-$MAPbI_3$ [47, 48], thin films [49, 50], nanowires [51, 52], and networks [53]. Starting from I-V measurements using different wavelengths from alternative light sources, always keeping the same illumination power of 1 mW/cm^2, it is possible to calculate the wavelength responsivity spectrum of the photodetector, by using the next formula:

$$R = I_{ph}/P \tag{4}$$

with R being the responsivity, I_{ph} the generated photocurrent, and P the illumination power on the device. These data are illustrated in **Figure 13c**, they show a maximum in the responsivity (0.26 A/W) for light with wavelength of 505 nm, which is near the wavelength with maximum solar irradiance. **Figure 13d** plots the current through the device as a function of time with a fixed voltage under pulsed illumination. This illumination mode allows to characterize the response time of the photodetector, in this device <100 ms (limited by the experimental setup), which is comparable with values previously reported [47, 48, 54, 55].

6. Conclusions

This chapter reviews recent structural results on the $CH_3NH_3(Br_{1-x}Cl_x)_3$ series. The combination of SXRD and NPD has permitted to address issues related to phase transitions below RT, and the conformation of the $CH_3NH_3^+$ units within the inorganic cages, formed by corner-sharing PBX_6 octahedra. Interestingly, it was observed that the progressive localization of the organic units is not only achieved through a reduction in symmetry (from cubic to tetragonal, and finally to orthorhombic), but also simply as a consequence of the contraction of the cubic unit cells as the Cl con-tents increase. This can be interpreted as an evolution of the X—H hydrogen bonds, as previously observed from DFT calculations. In summary, we have contributed with novel experimental evidence that facilitates a more reasonable design of hybrid halide perovskites with tunable properties for solar cell technologies.

Acknowledgements

This work was supported by the Spanish MINECO for funding MAT2017-84496-R. The authors thank ALBA and ILL for making the facilities available. CAL acknowledges ANPCyT and UNSL for financial support (projects PICT2017-1842 and PROICO 2-2016), Argentine.

We acknowledge support for the publication fee by the CSIC Open Access Publication Support Initiative through its Unit of Information Resources for Research (URICI).

Author details

Carlos Alberto López[1,2*], María Consuelo Alvarez-Galván[3], Carmen Abia[1,4], María Teresa Fernández-Díaz[4] and José Antonio Alonso[1*]

1 Instituto de Ciencia de Materiales de Madrid, CSIC, Madrid, Spain

2 Instituto de Investigaciones en Tecnología Química (INTEQUI), UNSL, CONICET, Facultad de Química, Bioquímica y Farmacia, UNSL, San Luis, Argentina

3 Instituto de Catálisis y Petroleoquímica, CSIC, Madrid, Spain

4 Institut Laue Langevin, Grenoble, France

*Address all correspondence to: calopez@unsl.edu.ar and ja.alonso@icmm.csic.es

References

[1] Lee MM, Teuscher J, Miyasaka T, et al. Efficient hybrid solar cells based on meso-superstructured organometal halide perovskites. Science (80-). 2012;**338**:643-647

[2] Kojima A, Teshima K, Shirai Y, et al. Organometal halide perovskites as visible-light sensitizers for photovoltaic cells. Journal of the American Chemical Society. 2009;**131**:6050-6051

[3] Kim H-S, Lee C-R, Im J-H, et al. Lead iodide perovskite sensitized all-solid-state submicron thin film mesoscopic solar cell with efficiency exceeding 9%. Scientific Reports. 2012;**2**:591

[4] Ye M, Hong X, Zhang F, et al. Recent advancements in perovskite solar cells: Flexibility, stability and large scale. Journal of Materials Chemistry A. 2016;**4**:6755-6771

[5] Fan Z, Sun K, Wang J. Perovskites for photovoltaics: A combined review of organic–inorganic halide perovskites and ferroelectric oxide perovskites. Journal of Materials Chemistry A. 2015;**3**:18809-18828

[6] Green MA, Emery K, Hishikawa Y, et al. Solar cell efficiency tables (version 48). Progress in Photovoltaics: Research and Applications. 2016;**24**:905-913

[7] Noh JH, Im SH, Heo JH, et al. Chemical management for colorful, efficient, and stable inorganic–organic hybrid nanostructured solar cells. Nano Letters. 2013;**13**:1764-1769

[8] Sadhanala A, Deschler F, Thomas TH, et al. Preparation of single-phase films of CH3NH3Pb(I1–xBrx)3 with sharp optical band edges. Journal of Physical Chemistry Letters. 2014;**5**:2501-2505

[9] Eperon GE, Stranks SD, Menelaou C, et al. Formamidinium lead trihalide: A broadly tunable perovskite for efficient planar heterojunction solar cells. Energy & Environmentaal Science. 2014;**7**:982

[10] De Wolf S, Holovsky J, Moon S-J, et al. Organometallic halide perovskites: Sharp optical absorption edge and its relation to photovoltaic performance. Journal of Physical Chemistry Letters. 2014;**5**:1035-1039

[11] Stranks SD, Eperon GE, Grancini G, et al. Electron-hole diffusion lengths exceeding 1 micrometer in an organometal trihalide perovskite absorber. Science (80-). 2013;**342**: 341-344

[12] Xing G, Mathews N, Sun S, et al. Long-range balanced electron- and hole-transport lengths in organic-inorganic CH3NH3PbI3. Science (80-). 2013;**342**:344-347

[13] Han Y, Meyer S, Dkhissi Y, et al. Degradation observations of encapsulated planar $CH_3NH_3PbI_3$ perovskite solar cells at high temperatures and humidity. Journal of Materials Chemistry A. 2015;**3**:8139-8147

[14] Yang J, Siempelkamp BD, Liu D, et al. Investigation of $CH_3NH_3PbI_3$ degradation rates and mechanisms in controlled humidity environments using in situ techniques. ACS Nano. 2015;**9**:1955-1963

[15] Akbulatov AF, Frolova LA, Dremova NN, et al. Light or heat: What is killing lead halide perovskites under solar cell operation conditions? Journal of Physical Chemistry Letters. 2020;**11**:333-339

[16] Senocrate A, Kim GY, Grätzel M, et al. Thermochemical stability of hybrid halide perovskites. ACS Energy Letters. 2019;**4**:2859-2870

[17] Kedem N, Brenner TM, Kulbak M, et al. Light-induced increase of electron

diffusion length in a p–n junction type CH3NH3PbBr3 perovskite solar cell. Journal of Physical Chemistry Letters. 2015;**6**:2469-2476

[18] Talbert EM, Zarick HF, Orfield NJ, et al. Interplay of structural and compositional effects on carrier recombination in mixed-halide perovskites. RSC Advances. 2016;**6**:86947-86954

[19] Sheng R, Ho-Baillie A, Huang S, et al. Methylammonium lead bromide perovskite-based solar cells by vapor-assisted deposition. Journal of Physical Chemistry C. 2015;**119**:3545-3549

[20] Zheng X, Chen B, Wu C, et al. Room temperature fabrication of CH3NH3PbBr3 by anti-solvent assisted crystallization approach for perovskite solar cells with fast response and small J–V hysteresis. Nano Energy. 2015;**17**:269-278

[21] Arora N, Orlandi S, Dar MI, et al. High open-circuit voltage: Fabrication of formamidinium lead bromide perovskite solar cells using fluorene–dithiophene derivatives as hole-transporting materials. ACS Energy Letters. 2016;**1**:107-112

[22] Edri E, Kirmayer S, Cahen D, et al. High open-circuit voltage solar cells based on organic–inorganic lead bromide perovskite. Journal of Physical Chemistry Letters. 2013;**4**:897-902

[23] Heo JH, Song DH, Im SH. Planar $CH_3NH_3PbBr_3$ hybrid solar cells with 10.4% power conversion efficiency, fabricated by controlled crystallization in the spin-coating process. Advanced Materials. 2014;**26**:8179-8183

[24] Zarick HF, Boulesbaa A, Puretzky AA, et al. Ultrafast carrier dynamics in bimetallic nanostructure-enhanced methylammonium lead bromide perovskites. Nanoscale.2017;**9**:1475-1483

[25] Swainson IP, Hammond RP, Soullière C, et al. Phase transitions in the perovskite methylammonium lead bromide, $CH_3ND_3PbBr_3$. Journal of Solid State Chemistry. 2003;**176**:97-104

[26] Mashiyama H, Kawamura Y, Kasano H, et al. Disordered configuration of methylammonium of $CH_3NH_3PbBr_3$ determined by single crystal neutron diffractometry. Ferroelectrics. 2007;**348**: 182-186

[27] Baikie T, Barrow NS, Fang Y, et al. A combined single crystal neutron/X-ray diffraction and solid-state nuclear magnetic resonance study of the hybrid perovskites $CH_3NH_3PbX_3$ (X = I, Br and Cl). Journal of Materials Chemistry A. 2015;**3**:9298-9307

[28] Maculan G, Sheikh AD, Abdelhady AL, et al. $CH_3NH_3PbCl_3$ single crystals: Inverse temperature crystallization and visible-blind UV-photodetector. Journal of Physical Chemistry Letters. 2015;**6**:3781-3786

[29] Rana S, Awasthi K, Bhosale SS, et al. Temperature-dependent electroabsorption and electrophoto-luminescence and exciton binding energy in $MAPbBr_3$ perovskite quantum dots. Journal of Physical Chemistry C. 2019;**123**:19927-19937

[30] Gao Z-R, Sun X-F, Wu Y-Y, et al. Ferroelectricity of the orthorhombic and tetragonal $MAPbBr_3$ single crystal. Journal of Physical Chemistry Letters. 2019;**10**:2522-2527

[31] Comin R, Walters G, Thibau ES, et al. Structural, optical, and electronic studies of wide-bandgap lead halide perovskites. Journal of Materials Chemistry C. 2015;**3**:8839-8843

[32] López CA, Martínez-Huerta MV, Alvarez-Galván MC, et al. Elucidating the methylammonium (MA) conformation in $MAPbBr_3$ perovskite with application in solar cells. Inorganic Chemistry. 2017;**56**:14214-14219

[33] Alvarez-Galván MC, Alonso JA, López CA, et al. Crystal growth, structural phase transitions, and optical gap evolution of $CH_3NH_3Pb(Br_1–xClx)_3$ perovskites. Crystal Growth & Design. 2018;**19**(2):918-924. DOI: 10.1021/acs.cgd.8b01463

[34] López CA, Álvarez-Galván MC, Martínez-Huerta MV, et al. Dynamic disorder restriction of methylammonium (MA) groups in chloride–doped MAPbBr3 hybrid perovskites: A neutron powder diffraction study. Chemistry: A European Journal. 2019;**25**:4496-4500

[35] Poglitsch A, Weber D. Dynamic disorder in methylammonium-trihalogenoplumbates (II) observed by millimeter-wave spectroscopy. The Journal of Chemical Physics. 1987;**87**:6373-6378

[36] Chi L, Swainson I, Cranswick L, et al. The ordered phase of methylammonium lead chloride $CH_3ND_3PbCl_3$. Journal of Solid State Chemistry. 2005;**178**:1376-1385

[37] Varadwaj A, Varadwaj PR, Marques HM, et al. Halogen in materials design: Revealing the nature of hydrogen bonding and other non-covalent interactions in the polymorphic transformations of methylammonium lead tribromide perovskite. Materials Today Chemistry. 2018;**9**:1-16

[38] Li J, Rinke P. Atomic structure of metal-halide perovskites from first principles: The chicken-and-egg paradox of the organic-inorganic interaction. Physical Review B. 2016;**94**: 045201

[39] Leguy AMA, Goñi AR, Frost JM, et al. Dynamic disorder, phonon lifetimes, and the assignment of modes to the vibrational spectra of methylammonium lead halide perovskites. Physical Chemistry Chemical Physics. 2016;**18**:27051-27066

[40] Frost JM, Walsh A. What is moving in hybrid halide perovskite solar cells? Accounts of Chemical Research. 2016;**49**:528-535

[41] Yin T, Fang Y, Fan X, et al. Hydrogen-bonding evolution during the polymorphic transformations in $CH_3NH_3PbBr_3$: Experiment and theory. Chemistry of Materials. 2017;**29**:5974-5981

[42] Bernasconi A, Page K, Dai Z, et al. Ubiquitous short-range distortion of hybrid perovskites and hydrogen-bonding role: The $MAPbCl_3$ case. Journal of Physical Chemistry C. 2018;**122**:28265-28272

[43] Shimamura K, Hakamata T, Shimojo F, et al. Rotation mechanism of methylammonium molecules in organometal halide perovskite in cubic phase: An ab initio molecular dynamics study. The Journal of Chemical Physics. 2016;**145**:224503

[44] Tablero CC. The effect of the halide anion on the optical properties of lead halide perovskites. Solar Energy Materials & Solar Cells. 2019;**195**: 269-273

[45] Endres J, Egger DA, Kulbak M, et al. Valence and conduction band densities of states of metal halide perovskites: A combined experimental-theoretical study. Journal of Physical Chemistry Letters. 2016;**7**:2722-2729

[46] Ong KP, Wu S, Nguyen TH, et al. Multi band gap electronic structure in $CH_3NH_3PbI_3$. Scientific Reports. 2019;**9**:2144

[47] Wang G, Li D, Cheng H-C, et al. Wafer-scale growth of large arrays of perovskite microplate crystals for functional electronics and optoelectronics. Science Advances. 2015;**1**:e1500613

[48] Liu J, Xue Y, Wang Z, et al. Two-dimensional $CH_3NH_3PbI_3$ perovskite:

Synthesis and optoelectronic application. ACS Nano. 2016;**10**:3536-3542

[49] Zhang Y, Du J, Wu X, et al. Ultrasensitive photodetectors based on island-structured $CH_3NH_3PbI_3$ thin films. ACS Applied Materials & Interfaces. 2015;**7**:21634-21638

[50] Lu H, Tian W, Cao F, et al. A self-powered and stable all-perovskite photodetector-solar cell nanosystem. Advanced Functional Materials. 2016;**26**:1296-1302

[51] Deng H, Dong D, Qiao K, et al. Growth, patterning and alignment of organolead iodide perovskite nanowires for optoelectronic devices. Nanoscale. 2015;**7**:4163-4170

[52] Zhuo S, Zhang J, Shi Y, et al. Self-template-directed synthesis of porous perovskite nanowires at room temperature for high-performance visible-light photodetectors. Angewandte Chemie International Edition. 2015;**54**:5693-5696

[53] Deng H, Yang X, Dong D, et al. Flexible and semitransparent organolead triiodide perovskite network photodetector arrays with high stability. Nano Letters. 2015;**15**:7963-7969

[54] Li D, Dong G, Li W, et al. High performance organic-inorganic perovskite-optocoupler based on low-voltage and fast response perovskite compound photodetector. Scientific Reports. 2015;**5**:7902

[55] Dong R, Fang Y, Chae J, et al. High-gain and low-driving-voltage photodetectors based on organolead triiodide perovskites. Advanced Materials. 2015;**27**:1912-1918

10

Effect of Piezoelectric Filed on the Optical Properties of (311) A and (311) B Oriented InAlAs/InP Heterostructures

Badreddine Smiri, Faouzi Saidi and Hassen Maaref

Abstract

InAlAs alloy was grown by MOCVD on an InP (311) substrate with different polarities. Measurements of photoluminescence (PL) and photoreflectance (PR) were performed to study the impact of the V/III flux ratio. It is discovered that the PL line was shifted to a greater energy side with the increasing excitation power density, and no saturation was observed of its related PL intensity. It is a fingerprint of type II transition emission. However, the recombination of the type II interface showed a powerful dependence on AsH3 overpressure and substrate polarity. In fact, we have noted an opposite behavior of type II energy transition shift from A to B polarity substrate in respect to V/III ratio variation. PR signals corresponding to Franz-Keldysh Oscillation (FKO) were observed. The analysis of their period has allowed one to assess the value of the PZ field in the samples. PL-luminescence measurements were performed out as a function of temperature. PL peak energy, PL intensity, and half maximum full width show anomalous behaviors. Indicating the existence of localized carriers, they were ascribed to the energy potential modulation associated with the indium cluster formation and PZ field.

Keywords: V/III flux ratio, substrate polarity, piezoelectric field, Franz-Keldysh oscillation, photoreflectance, photoluminescence

1. Introduction

Recently, scientists have focused their interest in InAlAs/InP grown on non-conventional (n11) planes. For example, (311) A and (311) B are not acquired as compared to the used (001) surface due to their remarkable characteristics [1]. In addition, InAlAs semiconductor layers grown on (311) A/B-oriented InP substrates give several unique characteristics compared to those grown on InP (100). Indeed, in (311) plane, the strain and hydrostatic deformations are discovered to be improved compared to those on (100) plane [1–3]. The primary reasons for this are: (i) the presence of a built-in electric field, produced through the piezoelectric effect in the layer [1, 4, 5] and (ii) the difference in arsenic segregation at the inverse interface. It is expected that these factors will be heavily dependent on growth conditions such as substrate orientation, V/III ratio.

InAlAs-InP materials have attracted tremendous interest over the past decades due to a variety of potential applications such as optical, optoelectronic and electronic devices [6, 7] due to its large direct band-gap energy, high electron mobility and the type II nature of the interface [8]. These advancement efforts were appointed by the fabrication and commercialization of a variety of devices such as Quantum cascade lasers (QCLs) [9, 10], Avalanche Photodiodes (APDs) [11] and high-electron-mobility transistor (HEMT) [12 – 14]. The heterostructures of $In_xAl_{1-x}As/InP$ have a type II transition [1, 15] which becomes a promising contender for the optical telecommunication light source. Different techniques such as Molecular Beam Epitaxy (MBE) and Metal-Organic Chemical Deposition (MOCVD) have developed this type of structure. However, the InAlAs material itself suffered from a large density of hetero-epitaxy-inherent defects [1, 6]: I Al content, (ii) phase separation, and (iii) InAlAs growth spinodal decomposition. Despite all this, the full potential of devices based on InAlAs/InP has still been obtained.

These issues are expected to be highly dependent on growth conditions (substrate polarity, V/III ratio, etc.) due to the large difference in bond energy between Al-As and In-As [15]. To date, most study work on the optical and electrical properties of InAlAs was performed on the conventional (100) planes, but little is known for the non-conventional (n11) planes. Different substrate orientations show various surface states, which are expected to influence the growth mode and even the optical and electrical properties of epilayers.

The existence of aluminum in the InAlAs layer, therefore, prompts the existence of the In- and Al-rich clusters, which is the consequence of the non-uniformity in the alloy composition. As a result, it contributes to the undulation of the InAlAs bandgap from which the localized energy level is present. In addition, the substrate polarity [A or B] in our nanostructures alters the containment of electrons-holes by changing the strain and the existence of piezoelectric (PZ) field within the structure. Similar heterostructures such as In0.21Ga0.79As/GaAs (311) A MQWs [16] and heterostructures AlGaAs/GaAs grown on (100), (311) A and (311) B-oriented substrates [17] have seen the carrier localization phenomenon. The previous investigation will be constrained to those samples implanted at high index (11N). We have shown in our latest study [18] that the presence of localized carriers has been attributed to the energy potential modulation related to the existence of Indium clusters and PZ-field.

The aim of our chapter is to study the effect of PZ-field on the optical properties of InAlAs/InP (311) with different substrate polarity, elaborated by MOCVD. The research of their optical characteristics by PR and PL spectroscopy is a significant step to demonstrate the possibility of incorporating our structure into optoelectronic applications such as 1.55 μm devices.

2. Experimental details

The studies are conducted on InP/InAlAs/InP (311) double heterostructures, marked as S1, S1′, S1″, S2, S2′ and S2″, which are cultivated at different V/III ratios by low-pressure metal-organic chemical vapor deposition (MOCVD). More information about development is summarized in **Table 1**. The source materials for the growth process are trimethylindium (TMIn), trimethylaluminum (TMAl) and(AsH3). At a substrate temperature of 600°C, an InP layer of 100 nm thickness was developed. The growth rate of InP is approximately 0.17 nm/s. A 270 nm thick layer of $In_xAl_{1-x}As$ was subsequently deposed. Each sample was finally capped with an InP layer of 10 nm.

The source of excitation is the 514.5 nm line of the continuous-wave Ar^+ laser with an excitation density of 80 W/cm^2 in PL measurements. Spectral analysis of

Samples	Substrate orientation	V/III ratio molar
S1	(311) B	25
S1′	(311) B	50
S1″	(311) B	125
S2	(311) A	25
S2′	(311) A	50
S2″	(311) B	125

Table 1.
Growth conditions of the samples S1, S1′, S1″, S2, S2′ and S2″.

the luminescence measurements was performed out using JOBIN YVON HRD1 monochromatic and identified by a cooled Ge diode detector with a built-in amplifier. PR measurements were performed using a standard setup with the 514.5 nm line of Ar^+ laser as the pump light, which was mechanically chopped at 970 Hz. The probe light was acquired from a 250 W tungsten halogen lamp dispersed with a 275 mm focal length monochromatic. The reflectance signal is detected by an InGaAs and silicon photodiodes.

3. Results and discussions

3.1 Photoluminescence study

3.1.1 Effect of substrate polarity: PZ-field

PL spectra were registered at low temperature to confirm the effect of PZ-field on the optical properties of InAlAs/InP (311) A/B:

Figure 1 illustrates the 10K-PL spectra of InAlAs/InP samples grown on (311) B and (311) A, respectively denoted S1 and S2. A higher energy side, both emissions at around 1.13 and 1.23 eV, for the samples S1 and S2, can be related to the interfacial defects between the InAlAs and InP layers [1, 19]. Both PL bands occur at about 0.8 and 1.03 eV for S1 and S2, respectively, on the lower energy side. A gradual $InAs_xP_{1-x}$ layer formation at the interface between InP and InAlAs (see **Figure 2**) was explained by Hallara et al. [6]. For sample S1, an emission situated at around 1.27 eV may be related to acceptor-band recombination [1].

To analyze the origin of the inverted interface (InP/InAlAs), we suggested a model based on arsenic segregation (some atomic monolayers) and linked the theoretical calculations with the experimental results.

We can conclude that the radiative recombination around the inverse interface with emission 1.03 eV for the polarity A is due to the appearance of a gradual layer for 3ML of $InAs_xP_{1-x}$. In this case, the arsenic content of xAs is about 40%, but in the inverted transition for polarity B, it is in the order of 70%. It is possible to estimate the band offset between InP and In0.513Al0.487As layers based on this reference [20]. In fact, the interface between InP and $InAs_xP_{1-x}$ layers is type I, although it is type II for In0.513Al0.487As and $InAs_xP_{1-x}$, where the xAs content ranged from 0 to 0.78.

There are two Gaussian peaks in the PL spectrum (see **Figure 1**). For the type II transition, an asymmetric band tail appeared in both S1 and S2 samples, resulting from unintentional thin strained InAs layers created at the InAlAs-InP interface [1, 18]. To explain more, this layer's smaller band gap can conduct a quantum well

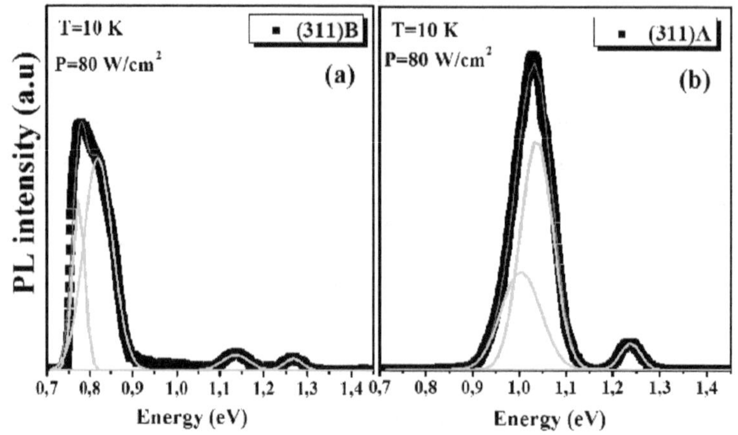

Figure 1.
(a) and (b) Low temperature PL spectra of InAlAs on InP (311) A and on InP (311) B, respectively. The green solid line is the Gaussian-fitting curve.

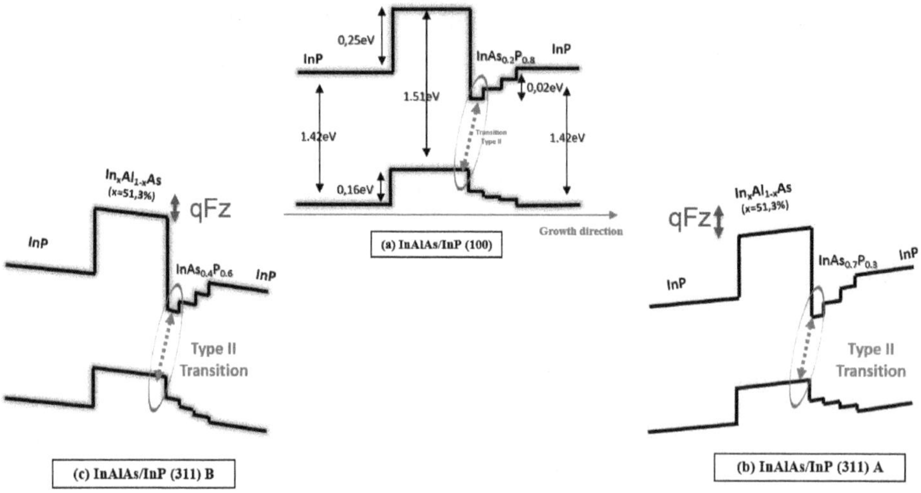

Figure 2.
(a) Schematic band diagrams for sample of InAlAs on InP (100), (b) on InP (311) A, and (c) on InP (311) B.

at the interface, resulting in a mixed type I–II transition [21]. We notice that the difference in energy between the two transitions type II corresponds to the difference in arsenic atoms surface segregation. Thus, depending on the growth axis, this energy shift can be attributed to the piezoelectric field [1, 5]. In addition, the existence of defect states may function as non-radiative centers resulting in a decrease in PL intensity. We should note that as mentioned at the beginning of this chapter, these defects originated from the development process as well as doping. Added to this, the sample cultivated on the (311) B surface has a greater residual impurity concentration than the samples cultivated on the (311) A surface [17]. For the (311) A sample, therefore, the interface quality is considered better than that for the (311) B sample.

3.1.2 Effect of V/III flux ratio

To verify the impact of V/III ratio on the optical properties of InAlAs/InP (311) A/B, PL spectra were recorded at low temperature.

Figure 3 shows PL spectra at 10 K of the studied samples on different oriented InP substrate and at various V/III molar ratio. PL spectra are normalized and deconvoluted into various Gaussian curves for the convenience of comparison and to

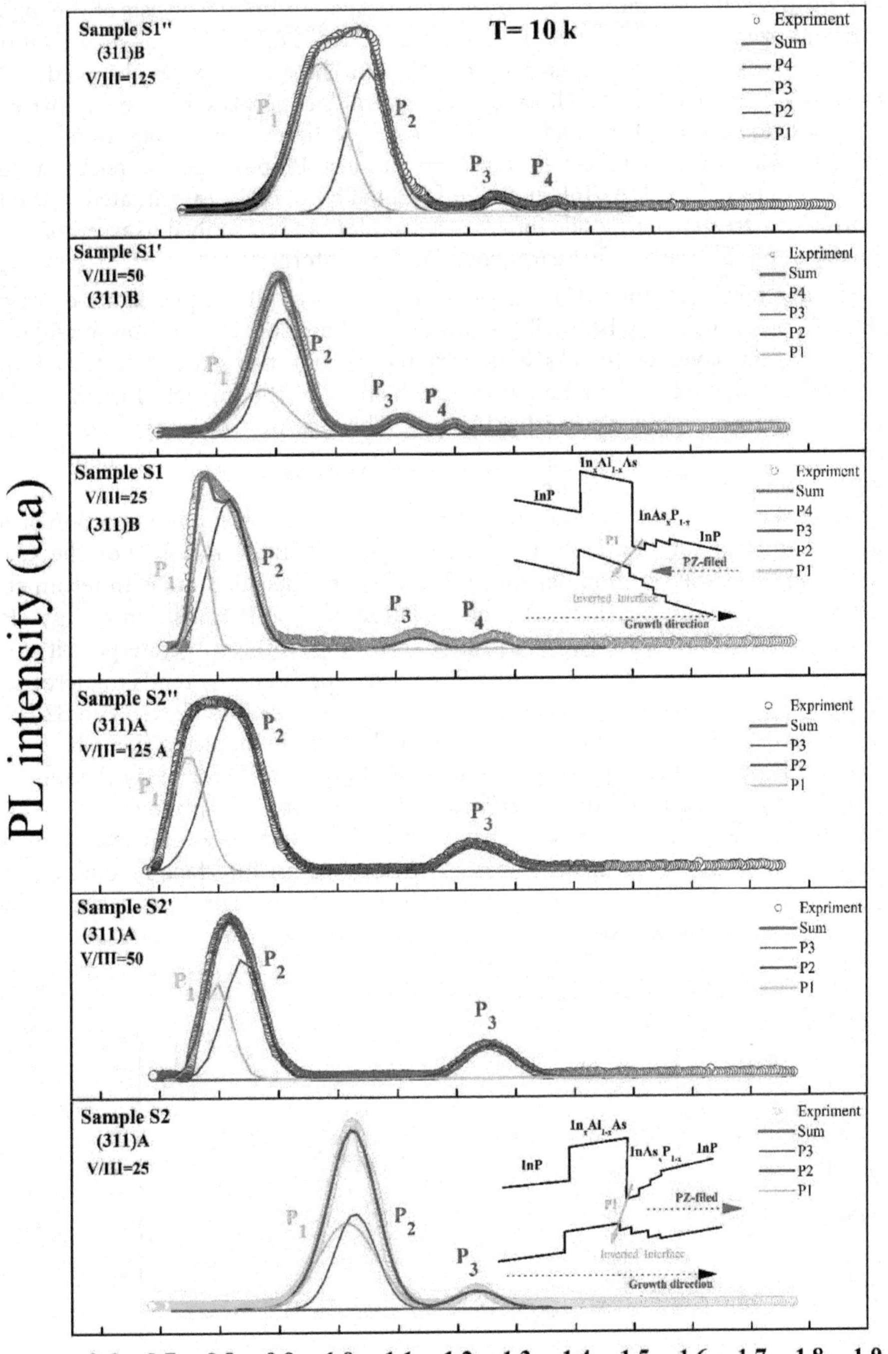

Figure 3.
Low-temperature (10 K) PL for samples S1, S1′, S1″, S2, S2′ and S2″. In order to identify the different emission peaks, the spectra are normalized and deconvoluted by a multiple Gaussian curve fitting. Inset shows the schematic band diagram showing the radiative transitions from the type II transition (P1).

identify the different emission peaks. The spectra show four significant emission peaks marked as P1, P2, P3 and P4 for the (311) B sample. The P1 position is attributed to the type II emission across the interface between the two-dimensional electron states in thin $InAs_xP_{1-x}$ graded layer at the inverted interface and holes located on top of the InAlAs valence band [1]. The intermediate layer of InAsP consists of the higher coefficient of incorporation of As compared to P [1, 5, 18]. Numerous investigations have shown that the type II emissions in InAlAs-InP heterostructures observed in the 0.8–1.25 eV range [1, 20, 22]. This very broad variety conceals some general patterns that appear to be related to conditions of growth. Whereas, the peak P2 emission is associated with a mixed type I–II transition and else P3 peak can be related to the interfacial defects in InAlAs/InP emission [1]. Finally, the P4 peak situated in the PL spectrum at a greater energy side for (311) B samples can be ascribed to acceptor-band recombination [5]. See our first part above for more information.

On the other hand, for (311) A plane, the peak P4 of all samples is not observed in PL spectra, which may be attributed to the existence of composition modulation in the epitaxial layer of InAlAs with varying molar ratio and substrate polarity. Sayari et al. [23] performed a Raman research study of the impact of the V/III flux ratio in heterostructures of InP/InAlAs/InP. The research demonstrates that the quantity of clustering is envisaged to depend on conditions of development such as substrate polarity, V/III ratio, etc.

Figure 4 exhibits the variation in the type II transition energy as a function of the V/III molar ratio for samples cultivated on (311) A and (311) B. For the (311) orientation, the peak position depends upon whether the substrate is In terminated (A-face) or P terminated (B-face). The variation of type II transition energy (P1) was acting differently with the V/III ratio, according to the substrate polarity. For the B-face (samples S1, S1′ and S1), the P1 emission peak energy tends to increase as the V/III ratio decreases [5]. Whereas for the A-face (samples S2, S2′ and S2′), we noted a decrease in the P1 emission peak as the V/III ratio increases (i.e. a blue shift is observed for (311) B samples [5], while red shift is noted for (311) A samples). These findings could possibly be clarified by the meaning of a piezoelectric field in InAlAs/InP heterostructures resulting from the difference between the atomic terminated surface [A or B] in InP substrate differences in interface reconstructions [1]. The PL shift can simply be attributed to the type of atoms present on the surface,

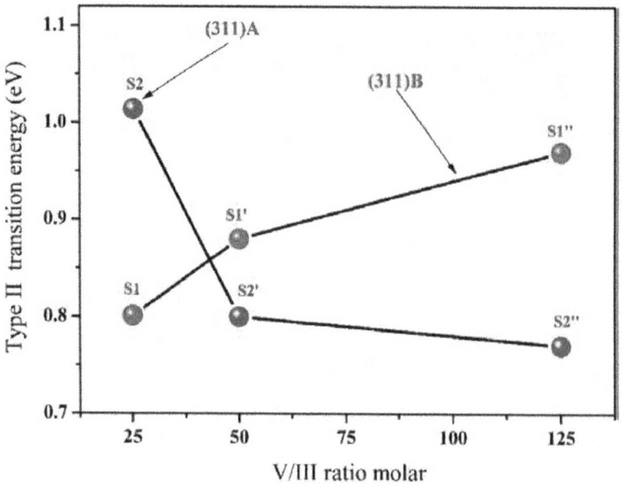

Figure 4.
Variation of the type-II transition energy as a function of V/III ratio for the (311) A and (311) B substrate orientation.

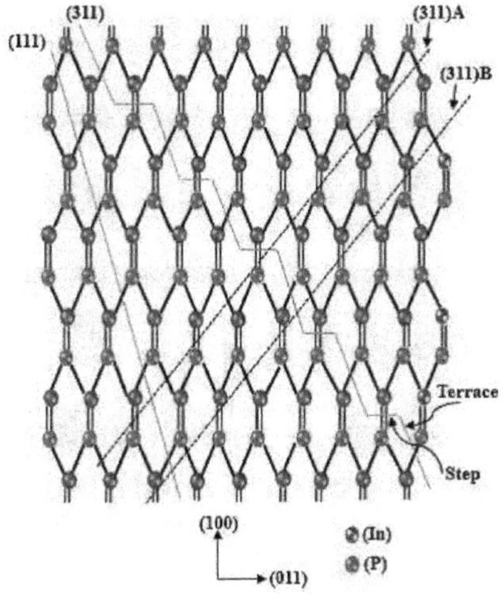

Figure 5.
Illustration of (311) A and (311) B planes of InP.

resulting in different levels of confinement. There are two types of sites on the surface (311): one is the double bond site found in the direction [100] and the other is a single bond site found in the direction [111], their densities being precisely the same [5]. On a (311) A surface, the double dangling bond sites are In sites and the single dangling bond sites are P sites, but on a (311) B surface the double dangling bond sites are P sites and the single dangling bond sites are In sites (see **Figure 5**) [19, 24, 25]. These completely distinct configurations of plane bonding are eventually responsible for the meaning of the PZ field on the different planes. Furthermore, the study of the impacts of InP substrate polarity shows that As incorporation (arsenic diffusion) may be improved on (311) A samples but may be decreased on (311) B samples with an increased V/III ratio. It can be shown from Ref. [20] that the increase in arsenic composition (InAsP) will reduce the transition energy of type II.

3.1.3 Power-dependent photoluminescence (PDPL)

We studied the evolution of the PL peak energy as PDPL, as shown in **Figure 6**, to verify these assignments. The energy blue changes with increased excitation

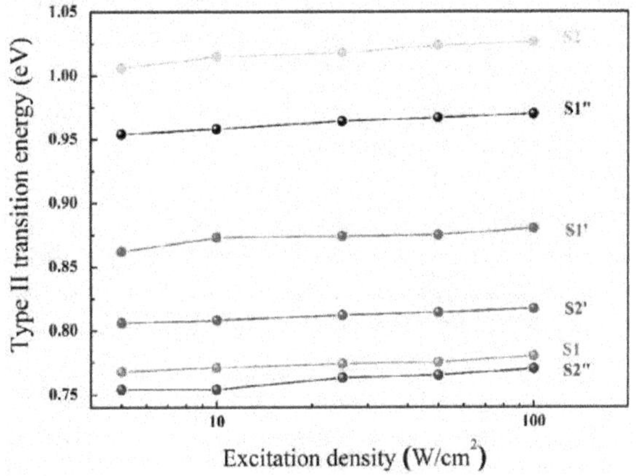

Figure 6.
The PL emission energy for the type II transition of all samples as a function of excitation power densities.

power intensity for type transition. This conduct is associated with recombination of type II via an interface with other material structures such as GaAsSb/InP [26, 27] and GaSb/InGaAs [28].

The following Eq. (1) was used to estimate the nature of the recombination around the InP/InAlAs inverse interface [1]:

$$I_{PL} = (P_{exc})^n \tag{1}$$

where P_{exc} is the power excitation, IPL is the integrated PL intensity and n is an exponent.

We found that the exponent n close to the unity. At higher power of excitation, there is no saturation. This shows that this PL transition is not attributable to impurity or defects, but is intrinsic recombination (band-to-band) [1]. **Figure 6** indicates a logarithmic linear dependence of the PL peak energy of the inverse interface with the density of excitation power. Hallara et al. [6] noted this behavior. Additionally, **Figure 6** display that the blue-shifted with the increase of the excitation power density. The offset is approximately 7 meV. This shows that the change of emission energy with increasing excitation power is also proof to verify our hypothesis (carrier localization) and PZ-filed presence in our structures [18].

A final analysis based on the PL temperature was created to further verify our hypothesis. In both samples, we can gain greater insight into the carrier localization and the mechanism of luminescence.

3.1.4 Temperature-dependent photoluminescence (TDPL)

Scientific study shows that two factors are noted for conventional orientation (100), at low temperature 10 K and with aspect ratio V/III equivalent to 50, two factors are observed: (i) clusters formation and (ii) composition modulation leading to natural superlattice [6]. Some scientists discovered that the latter factor disappears and if the ratio V/III changes [5, 6, 22], the clustering impact remains only.

It related to the phenomenon of exciton location. Carefully research as a function of PL temperature is conducted to demonstrate our hypothesis.

In both samples (S1 and S2) mentioned above, the PL peak energy-temperature reliance obtained from PL spectra is shown in **Figure 7(a)** and **(b)**. The PL spectrum (inset **Figure** 7) is revealed between 10 and 300 K. The so-called S-shaped temperature dependence of emission energy is obviously shown as a successive reform to low-high-low energy. It displays anomalous behavior as a temperature function. This behavior is characteristic of localization effects, and has been already observed in InAlAs alloys on InP [29, 30], GaAsSb/InP layers [31], BGaAs/GaAs layers [32, 33] and In0.21Ga0.79As MQWs on GaAs (311) A [16].

The S-shaped shape in PL peak energy can be separated into three primary intervals of temperature and is interpreted as follows: at the low-temperature range, the excitons should be situated in the levels whose distribution goes into the material's prohibited band. This is called band tails, associated with cluster appearance in the InAlAs [18]. With growing temperature, the excitons get enough thermal energy to reach deeper localized states and recombine primarily from low energy levels, resulting in a dramatic red shift (part I). Indeed, when the temperature increases, the thermal energy becomes adequate for the excitons in the tails to attain the corners of the stripes, where they are delocalized to high levels of energy. A blue shift is noted as a consequence (part II). All carriers are delocalized to the continuum at the high-temperature region, where band-band transitions are favored (part III) [18].

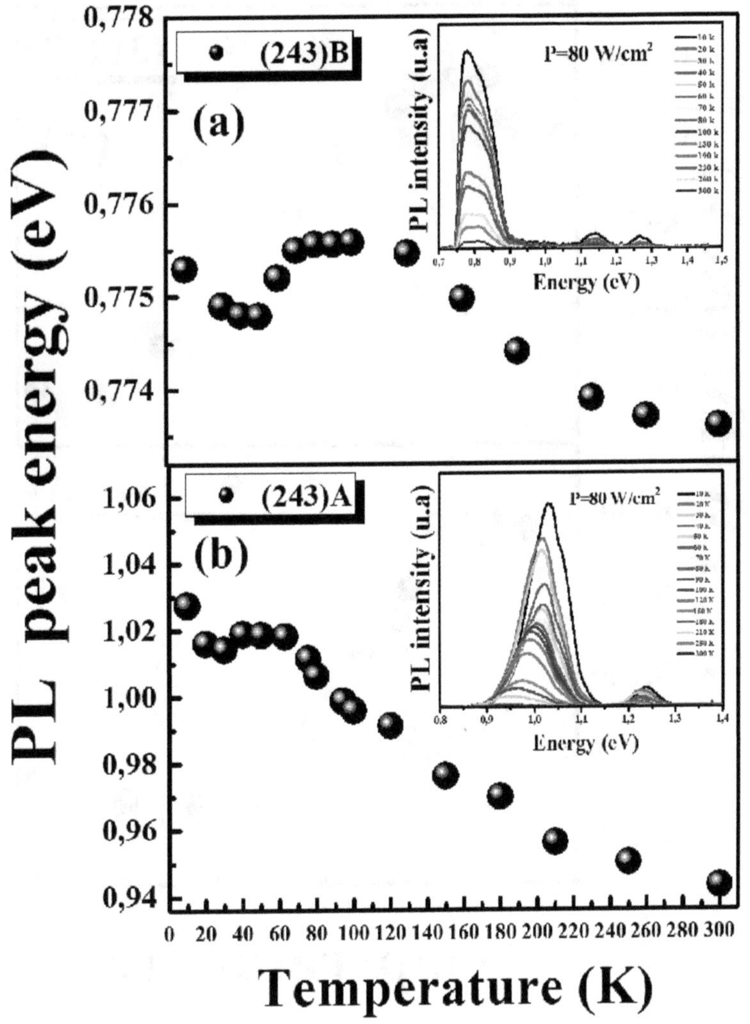

Figure 7.
(a) and (b) experiment data for the temperature dependence of Transition type II emission in Ino.513Alo.487As grown on InP (311) B and on InP (311) A substrates, respectively.

Another main parameter of the presence of the localization phenomenon is the Full width at Half Maximum (FWHM) behavior. Indeed, compared to the classic IIIV semiconductor alloys, it demonstrates atypical behavior. It shows an inverted "N-shape" (decrease-increase-decrease) (see **Figure 8**). The excitons are localized in the potential minima at cryogenic temperature. The carriers gain more thermal energy at intermediate temperatures to overcome the tiny energy barriers and attain greater energy states. As the temperature increases further, the FWHM broadening is described as the interaction of the electron-phonon. Finally, up to room temperature, the line width decreases continuously. It demonstrates the inverse trend of expanding optical phonons [1, 18]. This behavior can be clarified by the thermo-activation of the carriers and their transfer between nearby fluctuation potentials induced by the inhomogeneous distribution. Indeed, the carriers that are being thermally activated into a small potential can be further retrapped by the large one [18].

As a function of temperature, the S-shape in PL peak energy is not clear in polarity B as polarity A. The reason for this distinction can be clarified by

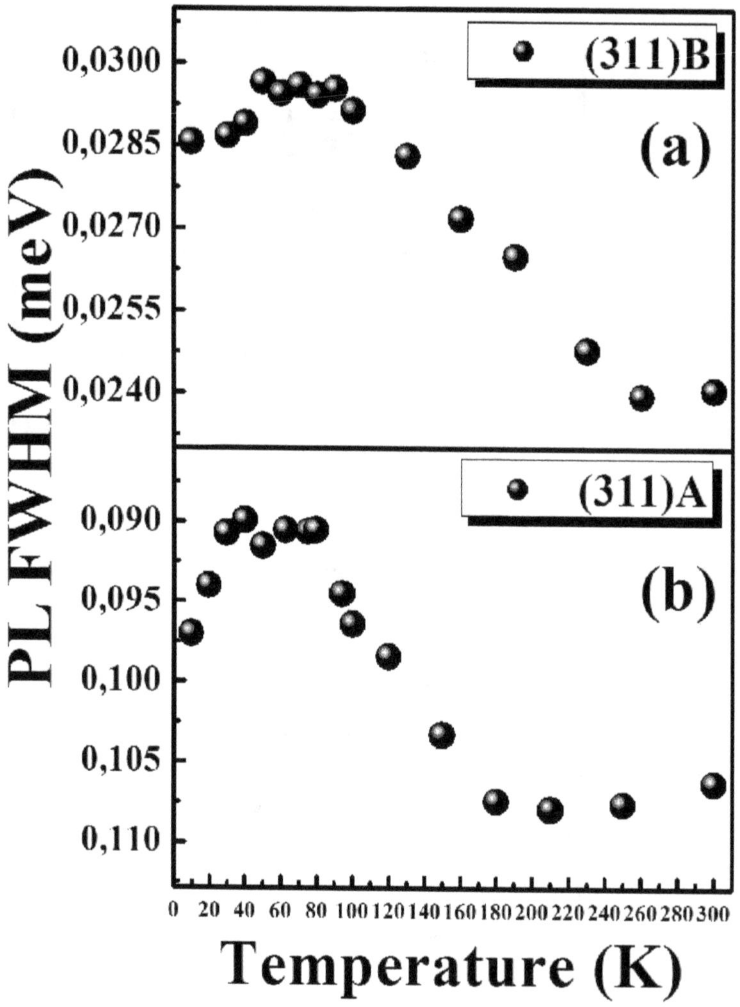

Figure 8.
(a) and (b) Evolution of the FWHM dependence of transition type II emission in $In_{0.513}Al_{0.487}As$ *grown on InP (311) B and on InP (311) A substrates, respectively.*

considering the polarity of the surface [In-rich (A) or P-rich (B)] from which the distinction of the PZ field within both structures [1, 18]. In the growth of the high index plane (311), the PZ field and an internal field are along the same direction (polarity A) [18, 19]. In fact, carriers can be easily delocalized due to the strong phonon coupling carrier, which is an important channel for carrier transfer in-plane hopping effect [24]. In comparison, field direction is opposite in polarity B (see **Figure 3**) [25, 34]. Due to the PZ field effect, the impact of localization is affected by interface undulation in this phase. Furthermore, the amount of alloy fluctuations in the composition of the material and the exchange of P-As.

3.2 Photoreflectance study

PR measurements were performed at room temperature to explore the evolution of the optical properties of the InAlAs/InP heterostructures during the development phase. The received PR spectra are described in **Figure 9**, with a distinct V/III ratio and substrate polarity. The PR spectra show a transition varying between 1.18 and 1.21 eV for samples S1, S1′, S2 and S2′. We suggest that this peak is related to the

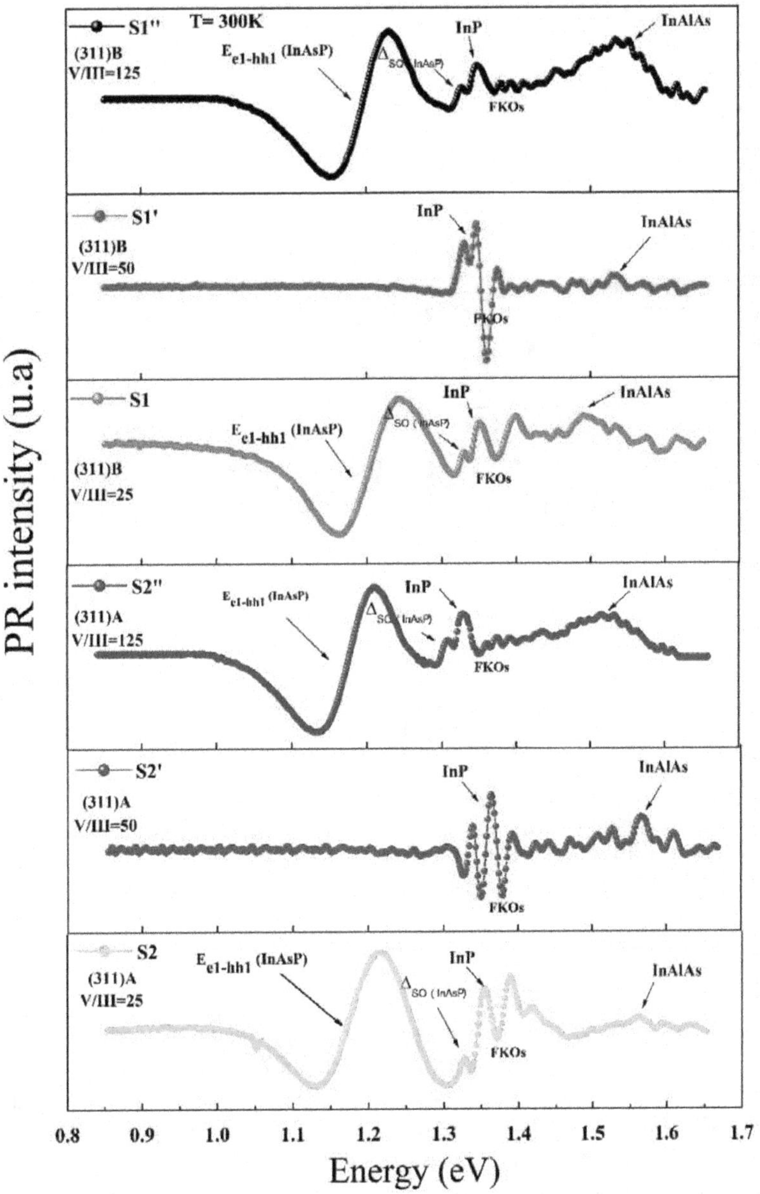

Figure 9.
Room-temperature PR spectra for InAlAs layers grown on (311) B (samples S1, S1′ and S1″) and (311) A (samples S2, S2′ and S2″) InP substrates with different V/III ratio molar.

emission (1e–1h) between electron sub band and light-hole sub-band of quantum well (QW) InAsP [5]. While the spin split-off band of the InAsP layers lies at about 1.32 eV. On the other side, for the samples (S1′, S2′), these two peaks totally disap-pear with a V/III ratio of 50 [5]. This result can be related during the growth to the uniformity of the inverse interface between InAlAs and InP. Furthermore, we noted that PR spectra of all samples show an additional above-band-gap (E_0 InP) charac-teristics that are Franz-Keldysh oscillations (FKO), reflecting the presence of a built-in electric field in the InP substrates [5, 35]. Finally, it is possible to attribute the peak exposed between 1.5 and 1.56 eV to the InAlAs band gap transition [19]. The origin of this very wide range is associated with the existence of the clustering phenomenon in the layer $In_xAl_{1-x}As$ caused by the fluctuation of the indium composition [19].

In order to assess the electrical field quality of each InP substrate, the FKO period is evaluated using an asymptotic expression for the PR spectrum provided in [36, 37]. In this strategy, InP FKO is indicated by:

$$\frac{\Delta R}{R} \propto \frac{1}{E^2(E - E_0)} exp\left[-2(E - E_0)^{1/2}\frac{\Gamma}{(\hbar\Omega)^{3/2}}\right] cos\left[\frac{4}{3}\left(\frac{E - E_0}{\hbar\Omega}\right)^{3/2} + \phi\right] \quad (2)$$

where E_0 is the transition energy, ϕ is an arbitrary phase factor, Γ is the linewidth broadening.

According to the asymptotic Franz-Keldysh model [37], the energy of oscillation extrema is given by:

$$E_m = (\hbar\Omega)H_m + E_0 \quad (3)$$

where $H_m = \left[\left(\frac{3\Pi}{2}\right)(m - 1/2)\right]^{2/3}$ and m $(m \geq 1)$ denotes the mth FKO extremum.

E_0 is the band-gap energy (InP) and $\hbar\Omega$ is the electro-optic energy expressed as takes after:

$$\hbar\Omega = \left(\frac{q^2\hbar^2 Fint^2}{8\mu}\right)^{1/3} \quad (4)$$

Noting that q is the electron charge, μ is the reduced inter-band effective mass in the electric field direction, $\hbar$ is the reduced Planck constant, and F_{int} is the internal electric field.

Thus, by plotting E_m as a function of H_m, from linear fitting, we can decide $\hbar\Omega$ and E_0 from the slope and the intersection with the ordinate at the origin, respectively. Recalling Eq. (4), F_{int} can be calculated by:

$$F_{int} = \sqrt{\frac{8\mu(\hbar\Omega)^3}{q^2\hbar^2}} \quad (5)$$

Plotted in **Figure 10** is the intercept E_0 of the best linear adjustments of the variation of E_n as a function of H_n for all samples [5]. Then, E_0 was the intercept defined from the best linear adjustment of En (Hn) plot, whereas $\hbar\Omega$ is simply the slope of the linear fit. **Table 2** shows the acquired values of transitions E_0 (InP) discovered by this fitting procedure and the FKOs analysis method.

From Eq. (3), a plot of E_m vs H_m is a straight line, and the electric field strength F_{int} can be acquired from the plot slope. **Figure 10** shows the graphs of E_m vs H_m for all samples with their corresponding linear fits. The residual strain in (311) A/B-oriented samples will generate a PZ field contributing to the F_{int} electrical field built-in InP. We assumed F_{int} was the sum of the PZ and built-in electrical fields. So, with regard to built-in electrical fields, the PZ field component predominates. The integrated electric field F_{int} is therefore equal to the PZ areas [5].

The values acquired from the above-explained FKO evaluation of the built-in electrical field strength are provided in **Table 2**. We discovered that when the orientation of the substrate changes, the F_{int} is modified, which suggests PZ impacts in the samples. The estimated PZ field is of about 36.27, 20.53, 8.11, 25.44, and 32.86 KV/cm for samples S1, S1′, S1″, S2, and S2′, respectively. Our findings will show that the PZ field is very susceptible to growth conditions in the InAlAs/InP samples [5].

Figure 11 shows the dependency of the PZ field in the studied samples on the polarity of the InP substrate and the V/III ratio. It is obviously noted that the PZ

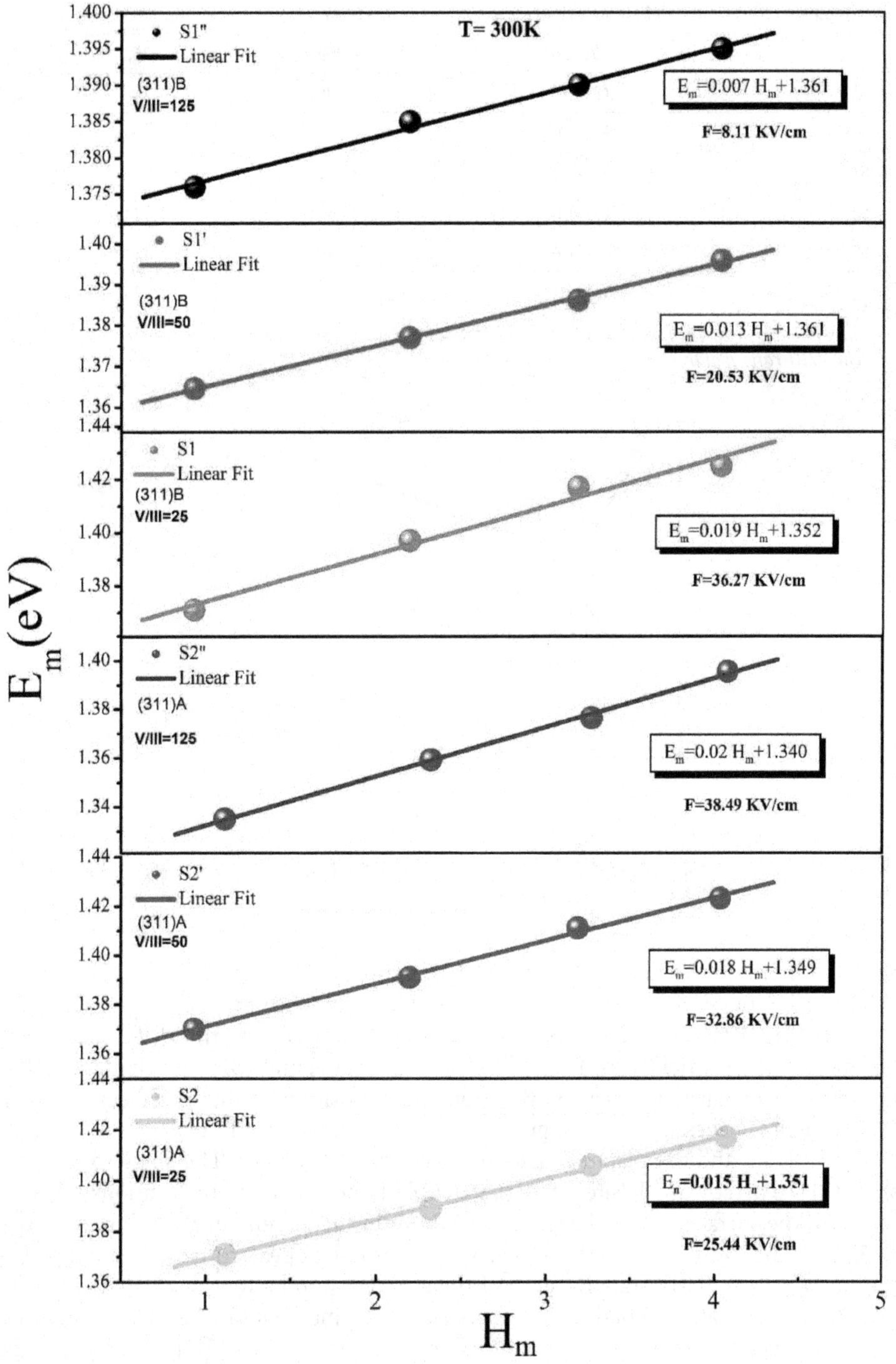

Figure 10.
FKO analysis for samples S1, S1′, S1″, S2, S2′, and S2″. Symbols are the energies of the FKO Extrema En as a function of $H_n = \left[\left(\frac{3\Pi}{2}\right)(n - 1/2)\right]^{2/3}$. *Solid lines are the linear fit using Eq. (2).*

field drops with increasing the V/III ratio molar for the orientation (311) B [5]. The S1 sample indicates a comparatively powerful PZ field. As the V/III ratio increases, the PZ field decreases rapidly to reach 20.53 KV/cm for sample S1′ (V/III = 50) and 8.11 KV/cm for sample S1″ (V/III = 125). In comparison, for (311) A-surface, we

Samples	Substrate orientation	E_0 (InP) (eV)	F_{int} (KV/cm)
S1	(311) B	1.352	36.27
S1′	(311) B	1.396	20.53
S1″	(311) B	1.361	8.11
S2	(311) A	1.351	25.44
S2′	(311) A	1.349	32.86
S2″	(311) A	1.340	38.49

Table 2.
Summary of energy gap (Eg) and the piezoelectric field (Fpz) values obtained from photoreflectance PR with different V/III ratio molar.

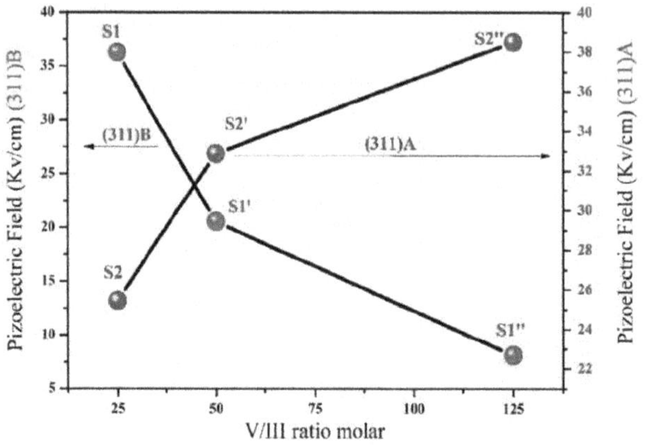

Figure 11.
Piezoelectric field dependence on the V/III ratio molar for (311) A (closed circles blue) and (311) B (closed circles red) substrate orientation.

noted an increase in the PZ field as the V/III ratio increases. It increases from 25.44 to 32.86 KV/cm when the V/III ratio increases almost twice [5].

From the surface kinetics aspect, it is possible to understand this difference in the variation of the PZ field with the V/III ratio for (311) A and (311) B orientation. The sign of the PZ charge is based on the atomic composition of the interfaces [38–40]. Therefore, it is necessary to distinguish between the substrates In planes, alluded to as (N11) A plane, and the planes P, alluded to as (N11) B-planes. The single-bond sites on a (311) surface are P sites and the double-bond sites are In locations. On the contrary, the surface (311) B has an inverse bonding arrangement as shown in **Figure 5** [5, 19]. Therefore, for (311) A-face, less As atoms appear to close the cultivated surface at a small V/III ratio. Because of the stronger bond strength of Al-As than In-As, additional As atoms bond to Al statistically, bringing about more AlAs. Therefore, when V/III decreases, the coefficient of incorporation decreases [5]. By contrast, if the V/III ratio increases, though the tendency of As to combining with Al is still high, because of this oversupply, In atoms have more opportunity to bond with As atoms. In this way, the In incorporation coefficient increases with V/III ratio [5], which creates an increase in the PZ field with V/III ratio for the (311) A-oriented samples (S2 and S2′). Furthermore, we noted the opposite phenomenon of what is occurring for (311) B-face. The PZ field in our samples cultivated on (311) A substrate has the opposite direction than for (311) B [5].

We plotted the type II transition energy with the field for both the A and B polarity substrates in order to estimate the role of the PZ field (see **Figure 11**). The shift in energy owing to the PZ field has a distinct impact on the termination of the substrates A and B [1, 5]. This is because the PZ field may increase or decrease the space separation of the electron/hole that already exists because of the strain [41]. Consequently, the band structures of the InP/InAlAs/InP samples are plotted schematically of the inset of **Figure 2**. Based on the orientation of the substrate, the band diagrams are shifted to the left side while for (311) A and shifted to the right side for (311) B due to the presence of the PZ field [1, 5]. Finally, based on the discourse in previous parts, it can be concluded that the red shift in the type II recombination for the (311) A and the blue shift for the (311) B orientation with an increasing V/III ratio is related to the meaning of the PZ field. Another possible raison for the shift of type-II transition is the difference of exchange As/P at the InAlAs/InP interface resulting from a different polarity of InP [1, 5].

4. Conclusion

In summary, InAlAs/InP type-II heterostructures with a varying V/III ratio grown successfully on (311) A or B Fe-doped InP substrates by MOCVD were investigated and the optical properties of the grown structures were examined. The different optical properties of the samples grown on (311) A or B substrates are caused from the difference of their plane-bonding configurations. In particular, the optical properties of InAlAs-InP interface display a significant reliance on AsH3 overpressure and substrate polarity. PL and PR measurements indicated that substrate orientation and V/III ratio molar have an important impact on the quality of inverted interface. The measurements of excitation power density PL confirm the intrinsic transition of type II. A red shift of the type-II transition was noted at low temperature with an increased V/III ratio of the polarity A samples and a blue shift for polarity B samples. These findings could be clarified by the opposite field of PZ at the heterostructures of InAlAs/InP resulting from distinct polarity of the InP substrate. We acquired the InP field built-in PZ in the heterostructures from an assessment by the FKO. We have made an explanation of the transition shift from these values. Additionally, the temperature variety shows an anomalous S-shaped dependence that is typical of carrier localization in the material. The optical properties are significantly influenced by the PZ field in our samples. Therefore, the impact of the PZ field on the design and manufacture of greater quality instruments should be taken into consideration.

Author details

Badreddine Smiri*, Faouzi Saidi and Hassen Maaref
Laboratoire des Micro-optoélectroniques et Nanostructures, Faculté des Sciences Monastir, Université de Monastir, Monastir, Tunisia

*Address all correspondence to: badreddinesmiri@gmail.com

References

[1] Smiri B, Fraj I, Saidi F, Mghaïth R, Maaref H. Effect of piezoelectric field on type II transition in InAlAs/InP (311) alloys with different substrate polarity. Journal of Alloys and Compounds. 2018; **736**:29-34

[2] Cornet C, Schliwa A, Even J, Doré F, Celebi C, Létoublon A, et al. Electronic and optical properties of InAs/InP quantum dots on InP(100) and InP(311) B substrates: Theory and experiment. Physical Review B. 2006;**74**:035312

[3] Kawamura Y, Kamada A, Yoshimatsu K, Kobayashi H, Iwamura H, Inoue N. Photoluminescence study of interfaces between heavily doped AlInAs:Si layers and InP (Fe) substrates. Institute of Physics Conference Series. 1997;**155**:129

[4] Yerino CD, Liang B, Huffaker DL, Simmonds J, Lee ML. Molecular beam epitaxy of lattice-matched InAlAs and InGaAs layers on InP (111) A, (111) B, and (110). Journal of Vacuum Science and Technology B. 2017;**35**:010801

[5] Smiri B, Fraj I, Bouzidi M, Saidi F, Rebey A, Maaref H. Effect of V/III ratio on the optical properties of (311) A and (311) B oriented InAlAs/InP heterostructures. Results in Physics. 2019;**12**:2175-2182

[6] Hellara J, Borgi K, Maaref H, Souliere V, Monteil Y. Optical properties of InP/InAlAs/InP grown by MOCVD on (100) substrates: Influence of V/III molar ratio. Journal of Materials Science and Engineering C. 2002;**21**:231-236

[7] Wang CA, Goyal A, Huang R, Donnelly J, Calawa D, Turner G, et al. Strain-compensated GaInAs/AlInAs/InP quantum cascade laser materials. Journal of Crystal Growth. 2010;**312**: 1157-1164

[8] Bohrer J, Krost A, Heitz R, Heinrichsdorff F, Eckey L, Bimberg D, et al. Interface inequivalence of the InP/InAlAs/InP staggered double heterostructure grown by metalorganic chemical vapor deposition. Journal of Applied Physics Letters. 1996;**68**:1072

[9] Demir I, Elagoz S. V/III ratio effects on high quality InAlAs for quantum cascade laser structures. Journal of Superlattices and Microstructures. 2017; **104**:140-148

[10] Czuba K, Jurenczyk J, Kaniewski J. A study of InGaAs/InAlAs/InP avalanche photodiode. Solid-State Electronics. 2015;**104**:109-115

[11] Chevallier CB, Baltagi Y, Guillot G, Hong K, Pavlidis D. Application of Photoreflectance spectroscopy to the study of interface roughness in InGaAs/InAlAsInGaAs/InAlAs heterointerfaces. Journal of Applied Physics. 1998;**84**:5291

[12] Ajayan J, Nirmal D. A review of InP/InAlAs/InGaAs based transistors for high frequency applications. Journal of Superlattices and Microstructures. 2015; **86**:1-19

[13] Reynolds DC, Bajaj KK, Litton CW, Yu PW, Maselink WT, Rischer R, et al. Sharp-line photoluminescence spectra from GaAs-GaAlAs multiple-quantum-well structures. Journal of Physics. Review B. 1984;**29**:7038-7041

[14] Kuriharaa K, Takashimab M, Sakatab K, Uedab R, Takaharab M, Ikedab H, et al. Phase separation in InAlAs grown by MOVPE with a low growth temperature. Journal of Crystal Growth. 2004;**271**:341-347

[15] Vignaud D, Wallart X, Mollot F, Sermage B. Photoluminescence study of the interface in type II InAlAs-InP heterostructures. Journal of Applied Physical. 1998;**84**:2138-2145

[16] Fraj I, Hidouri T, Saidi F, Maaref H. Carrier localization localization in

In0.21Ga0.79As/GaAs multiple quantum wells: A modified Passler model for the S-shaped temperature dependence of photoluminescence energy. Superlattices and Microstructures. 2017;**102**:351-358

[17] Teodoro MD, Dias IFL, Laureto E, Duarte JL, González-Borrero PP, Lourenço SA, et al. Substrate orientation effect on potential fluctuations in multiquantum wells of GaAs/AlGaAs. Journal of Applied Physics. 2008;**103**:093-508

[18] Smiri B, Hidouri T, Hassen Maaref SF. Carriers' localization and thermal redistribution in InAlAs/InP grown by MOCVD on (311)A- and (311)B-InP substrates. Applied Physics A: Materials Science & Processing. 2019;**125**:134

[19] Smiri B, Saidi F, Mlayah A, Maaref H. Effect of substrate polarity on the optical and vibrational properties of (311)A and (311) B oriented InAlAs/InP heterostructures. Physica E: Low-dimensional Systems and Nanostructures. 2019;**112**:121-127

[20] Abraham P, Monteil Y, Sacilotti M. Optical studies of InP/InAlAs/InP interface recombinations. Journal of Applied Surface Science. 1993;**65-66**: 777-783

[21] Poças LC, Duarte JL, Dias IFL, Laureto E, Lourenço SA, Toginho Filho DO, et al. Photoluminescence study of interfaces between heavily doped Al0.48In0.52As: Si layers and InP (Fe) substrates. Journal of Applied Physics. 2002;**91**:8999

[22] Ezzedini M, Bouzidi M, Qaid MM, Chine Z, Rebey A, Sfaxi L. Comprehensive study of the structural, optical and electrical properties of InAlAs: Mg films lattice matched to InP grown by MOVPE. Journal of Materials Science: Materials in Electronics. 2017; **18**:18221-18227

[23] Sayari A, Yahyaoui N, Meftah A, Sfaxi A, Oueslati M. Residual strain and alloying effects on the vibrational properties of step-graded InxAl1-xAs layers grown on GaAs. Journal of Luminescence. 2009;**129**:105-109

[24] Li Y, Niewcza M. Strain relaxation in (100) and (311) GaP/GaAs thin films. Journal of Applied Physics. 2007; **101**:064910

[25] Tromson-Carli A, Patriarche G, Druilhe R, Lusson A, Marfaing Y, Triboulet R, et al. Effect of the {h11} orientations and polarities of GaAs substrates CdTe buffer layer structural properties. Journal of Materials Science and Engineering B. 1993;**16**:145-150

[26] Peter M, Herres N, Fuchs F, Winkler K, Bachem K-H, Wagner J. Band gaps and band offsets in strained GaAs(1−y)Sb(y) on InP grown by metalorganic chemical vapor deposition. Applied Physics Letters. 1999;**74**: 410-412

[27] Hu J, Xu G, Stotz J, Watkins S, Curzon A, Thewalt M, et al. Type II photoluminescence and conduction band offsets of GaAsSb/InGaAs and GaAsSb/InP heterostructures grown by metalorganic vapor phase epitaxy. Applied Physics Letters. 1998;**73**:2799

[28] Shuhui Z, Lu W, Zhenwu S, Yanxiang C, Haitao T, Huaiju G, et al. Lattice parameter accommodation between GaAs(111) nanowires and Si (111) substrate after growth via Au-assisted molecular beam epitaxy. Nanoscale Research Letters. 2012; **7**(1):109

[29] Ferguson I, Cheng T, Sotomayor Torres C, Murray R. Photoluminescence of molecular beam epitaxial grown Al0.48In0.52As. Journal of Vacuum Science and Technology B. 1994;**12**:1319

[30] Merkel K, Bright V, Cerny C, Shcuermeyer F, Solomon J, Kaspi R.

Beryllium ion implantation in GaAsSb epilayers on InP. Journal of Applied Physics. 1995;**79**:699

[31] Chouaib H, Bru-Chevallier C, Guillot G, Lahreche H, Bove P. Photoreflectance study of GaAsSb/InP heterostructures. Journal of Applied Physics. 2005;**98**:123524

[32] Saidi F, Hamila R, Maaref H, Rodriguez P, Auvray L, Monteil Y. Structural and optical study of BxInyGa1−x−yAs/GaAs and InyGa1−yAs/GaAs QW's grown by MOCVD. Journal of Alloys and Compounds. 2010;**491**:45-48

[33] Hamila R, Saidi F, Maaref H, Rodriguez P, Auvray L. Photoluminescence properties and high resolution X-ray diffraction investigation of BInGaAs/GaAs grown by the metalorganic vapour phase epitaxy method. Journal of Applied Physics. 2012;**112**:063109

[34] Hou HQ, Tu CW. Field screening in (111)B InAsP/lnP strained quantum wells. Journal of Electronic Materials. 1996;**25**:1019-1022

[35] Nukeaw J, Matsubara N, Fujiwara Y, Takeda Y. Characterization of InP δ-doped with Er by FFT photoreflectance. Applied Surface Science. 1997;**117/118**: 776-780

[36] Bouzidi M, Benzarti Z, Halidou I, Soltani S, Chine Z, Jani BEL. Photoreflectance investigation of band gap renormalization and the Burstein-Moss effect in Si doped GaN grown by MOVPE. Materials Science in Semiconductor Processing. 2016;**42**: 273-276

[37] Bilel C, Fitouri H, Zaied I, Bchetnia A, Rebey A, El Jani B. Photoreflectance characterization of vanadium-doped GaAs layers grown by metalorganic vapor phase epitaxy. Journal of Materials Science in Semiconductor Processing. 2015;**31**:100-105

[38] Sanguinetti S, Gurioli M, Grilli E, Guzzi M, Henini M. Piezoelectric-induced quantum-confined Stark effect in self-assembled InAs quantum dots grown on (N11) GaAs substrates. Journal of Applied Physics Letters. 2000;**77**:1982

[39] Anufriev R, Chauvin N, Khmissi H, Naji K, Patriarche G, Gendry M, et al. Piezoelectric effect in InAs/InP quantum rod nanowires grown on silicon substrate. Journal of Applied Physics Letters. 2014;**104**:183101

[40] Kwak JS, Lee KY, Han JY, Cho J, Chae S, Nam OH, et al. Crystal-polarity dependence of Ti/Al contacts to freestanding n-GaNn-GaN substrate. Journal of Applied Physics Letters. 2001;**79**:3254

[41] Sacilotti M, Chaumont D, Brainer Mota C, Vasconcelos T, Nunes FD, Pompelli MF, et al. Interface recombination & emission applied to explain photosynthetic mechanisms for (e−, h+) charges' separation. World Journal of Nano Science and Engineering. 2012;**2**:58-87

11

Synthesis Techniques and Applications of Perovskite Materials

Dinesh Kumar, Ram Sagar Yadav, Monika, Akhilesh Kumar Singh and Shyam Bahadur Rai

Abstract

Perovskite material is a material with chemical formula ABX_3-type, which exhibits a similar crystal structure of $CaTiO_3$. In this material, A and B are metal cations with ionic valences combined to +6, e.g., Li^+:Nb^{5+}; Ba^{2+}:Ti^{4+}; Sr^{2+}:Mn^{4+}; La^{3+}:Fe^{3+}, and X is an electronegative anion with ionic valence (−2), such as O^{2-}, S^{2-}, etc. The properties of a perovskite material strongly depend on the synthesis route of materials. The perovskite materials may be oxides (ABO_3:$CaMnO_3$), halides (ABX_3: X = Cl, Br, I), nitrides (ABN_3:$CaMoN_3$), sulfides (ABS_3:$LaYS_3$), etc., and they may exist in different forms, such as powders, thin films, etc. There are various routes for the synthesis of several perovskites, such as solid-state synthesis, liquid-state synthesis, gas-state synthesis, etc. In this chapter, we discuss various techniques for the synthesis of oxide perovskites in powder form using solid-, liquid-, and gas-state synthesis methods, and we also present an overview on the other type of perovskite materials. The X-ray diffraction, scanning electron microscopy, and optical techniques are used to study the purity of crystallographic phase, morphology, and photoluminescence properties of the perovskite materials. Some applications of different perovskite materials are also discussed.

Keywords: XRD, perovskite, Rietveld refinement, FullProf Suite, lanthanide phosphor

1. Introduction

The general chemical formula of a perovskite material is ABX_3, which contains a crystal structure similar to $CaTiO_3$. It was initially discovered by German geologist Gustav Rose in 1839 in Ural Mountains, and named after Russian mineralogist Lev Perovski [1, 2]. In ABX_3 perovskite, A and B are termed as metal cations having ionic valences combined to +6, e.g., (Li^+:Nb^{5+}; Ba^{2+}:Ti^{4+}; Sr^{2+}:Mn^{4+}; La^{3+}:Fe^{3+}) and X is an electronegative anion with ionic valence −2 such as O^{2-}, S^{2-} etc. [3–6]. The perovskite materials may be oxides, halides, nitrides, sulfides, etc., and they may exist in different forms, such as powders, thin films, etc. [7–10]. The perovskite material has attracted our attention as it can house up a variety of cations at A- and B-sites individually and/or simultaneously along with anions at X-site [11, 12]. The perovskite materials can be classified in ideal and distorted perovskite materials.

An ideal perovskite material crystallizes into a simple cubic structure with $Pm\bar{3}m$ space group. In the $Pm\bar{3}m$ space group with perovskite structure, A atoms occupy *1(a)* site at (0, 0, 0) and B atoms occupy *1(b)* site at (1/2, 1/2, 1/2) whereas X atoms occupy *3(c)* site at (1/2, 1/2, 0). However, equivalently A, B, and X atoms can also occupy 1(*a*) site at (1/2, 1/2, 1/2), 1(*b*) site at (0, 0, 0) and *3(c)* site at (0, 0, 1/2), respectively, as shown in **Figure 1**. In this figure, A, B, and X are presented in terms of ionic radii [13, 14]. In the unit cell of a perovskite, the cation "B" forms octahedral arrangement with X-anions, i.e., BX_6 and the cation "A" occupies cuboc-tahedral site with X-anions, i.e., AX_{12}.

The family of perovskite material includes numerous types of oxide forms, such as transition metal oxides with the general formula of ABO_3. The oxide perovskite materials are widely synthesized and are studied for wide applications in various technological fields. In light of these properties, we describe oxide perovskites in more detail.

Victor Moritz Goldschmidt presented an empirical relationship among the ionic radii of A, B, and O, known as tolerance factor (t) to estimate the stability of a perovskite structure. This relation is valid for the relevant ionic radii at room temperature [15]. The numerical value of the tolerance factor can be found by Eq. (1):

$$t = \frac{r_A + r_o}{\sqrt{2}\left(r_B + r_o\right)} \quad (1)$$

where, the term r_A is the ionic radius of cation A and that of r_B is ionic radius of B cation whereas r_O is the ionic radius of oxygen anion (O^{2-}). The ionic radius of A cation is always larger than that of the B cation. The tolerance factor provides an idea about the selection of combination of A and B cations in order to prepare an ideal perovskite material. Eq. (1) can also be expressed in other form, which may be valid for any temperature as given by Eq. (2):

$$t = \frac{d_{A\text{-}o}}{\sqrt{2}\, d_{B\text{-}o}} \quad (2)$$

where d_A-O and d_B-O are average bond-lengths between A-O and B-O, respectively [16].

The distorted perovskite materials are those materials, which crystallize into other than the cubic structures. As far as we know that the perovskite material can accommodate different ions at the A- and B-sites. The variation in the A- and/or B-sites cations causes a variation in the tolerance factor. The variation in tolerance factor leads to a change in the perovskite structure from cubic to non-cubic distorted perovskite structure. For a stable perovskite, the value of tolerance factor should lie in the range of 0.88–1.09 [17]. An ideal perovskite crystal exhibits tolerance factor equal to unity (i.e., $t = 1$). For $t < 1$, the perovskite materials show the rhombohedral or monoclinic structure while in the case of $t > 1$; it reveals tetragonal or orthorhombic structure [18]. Due to distortion in the perovskite system, the BO_6 octahedral led tilted from an ideal situation and causes a change/enhancement in unit cell volume. Thus, the tolerance factor is a measure of the extent of distortion in the perovskite structure. **Figure 2** shows unit cells for some distorted perovskite structures.

There are two general requirements for the formation of a perovskite material, which are given as:

1. **Ionic radii:** the average ionic radii of A- and B-sites cations should be greater than 0.90 Å and 0.51 Å, respectively, and the value of tolerance factor should lie in the range of 0.88–1.09 [19, 20].

2. **Electro-neutrality:** the chemical formula of perovskite material should have neutral balanced charge; consequently, the sum of total charges at A- and B-sites cations must be equal to total charges at O-site (oxygen) of anion(s). A suitable charge distribution is to be achieved in the forms of $A^{3+}B^{3+}O_3$ or $A^{4+}B^{2+}O_3$ or $A^{1+}B^{5+}O_3$ [14, 19].

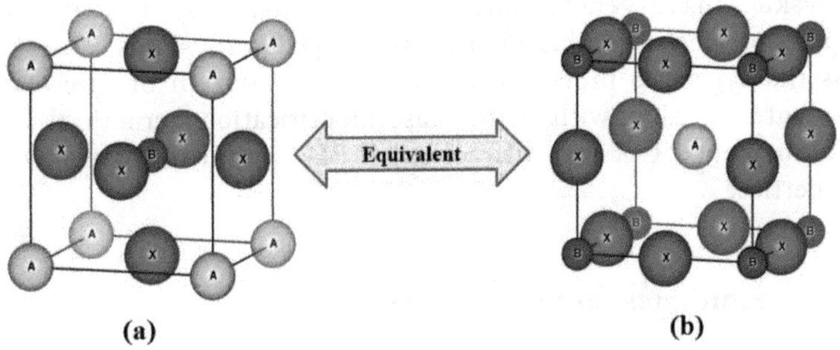

Figure 1.
Molecular structures for ABX_3 perovskite with $Pm\bar{3}m$ space group along with the positions of different atoms in a single unit cell. The figures (a) and (b) are equivalent structures to each other. In the figure (a), A and B take the positions at the corner and body center of the cubic cell, respectively, and X is at the center of face of the unit cell. However, in the figure (b), A and B occupy at body center and the corner of the cubic cell, respectively, and X lies at the center of edge of the unit cell.

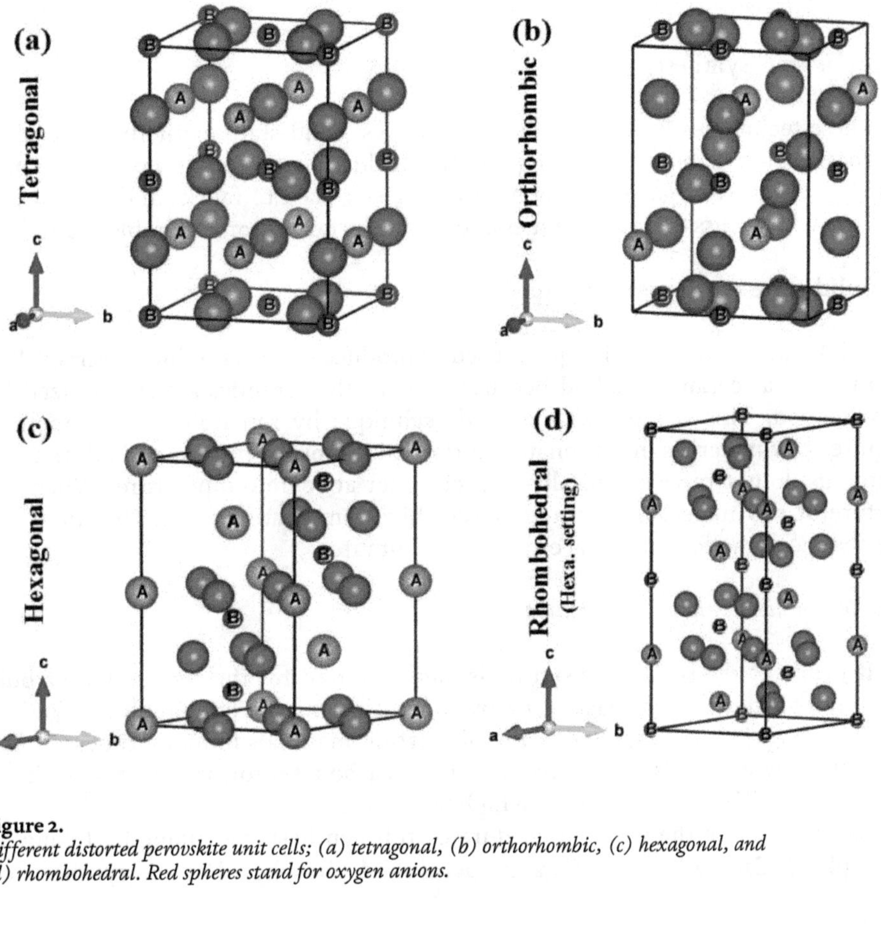

Figure 2.
Different distorted perovskite unit cells; (a) tetragonal, (b) orthorhombic, (c) hexagonal, and (d) rhombohedral. Red spheres stand for oxygen anions.

The pure perovskite materials (ABO_3) do not always provide the desired properties. In order to make them useful, the doping at A- and/or B-sites is required. Doping at A- and/or B-sites improves the properties and also generates very interesting phenomena due to change in crystal structure, bond-lengths, ionic states, etc. The general chemical formula of A- and B-sites after doping in the perovskite matrix may be found in the form of A_1-xA'_xBO_3 ($0 < x < 1$) and AB_1-yB'_yO_3 ($0 < y < 1$), respectively. However, the simultaneous A- and B-sites doped oxide perovskites have general formula A_1-xA'_xB_1-yB'_yO_3. Recently, there are several reports on lanthanide-based rare-earth doped perovskites [21–24]. Initially, we discuss the synthesis process used for the preparation of rare-earth doped perovskite materials alongwith their phase identification, structural analysis by Rietveld refinement of X-ray diffraction (XRD) patterns, morphological and optical properties.

2. Synthesis techniques for perovskite materials

As we know that the physical, chemical, and optical properties of the perovskite materials are strongly synthesis route dependent. One has to choose a suitable synthesis method to obtain the desired properties from the prepared materials. Synthesis techniques also affect crystal structure and morphology of the samples [25]. The synthesis techniques can be divided into three main classes as given below:

1. Solid-state synthesis

2. Liquid-state synthesis

3. Gas-state synthesis

These techniques have their own advantages. Solid-state methods are used to synthesize bulk materials, while liquid-state techniques are used to produce nano-materials. However, gas-state methods are mostly used to fabricate thin films. In this, we will discuss these techniques one by one by taking some suitable examples.

2.1 Solid-state synthesis technique

Solid-state synthesis technique is used to produce polycrystalline materials. It is also known as ceramic method because most of the ceramics are synthesized by this method. This is most widely used technique by researchers. This method requires raw materials in carbonates and/or oxides forms. In this method, the raw materials do not react chemically to each-other at room temperature. When the nixture of raw materials is heated at very high temperatures (i.e., 700–1500°C), the chemical reaction takes place at a significant rate.

2.1.1 Mechanical ball-milling method

It is one of the solid-state synthesis methods used for the production of bulk perovskite materials. Synthesis of perovskites using this method involves grinding, hand mixing, ball milling, and firing of starting materials in many times. In this case, the raw materials used in oxides and/or carbonates forms are grinded, hand mixed, ball milled, and calcined at a high temperature.

Let us describe the whole procedure of this synthesis technique by taking an example of $Ca_{0.97}$-x$TiO_3 : 3Yb^{3+}$, xBi^{3+} perovskite phosphor. For the synthesis of

$CaCO_3$, TiO_2, Yb_2O_3, and Bi_2O_3 were used as the raw materials. First of all, these materials were weighed in the stoichiometric amounts and grinded and mixed by hand in mortor by using pestle for 1–2 h. After hand mixing and grinding of all the starting materials, the mixture was ball milled in a planetary ball milling system to further get homogeneous mixture in the presence of acetone/alcohol as a mixing media for 4–12 h at a nominal rpm (round per minute) of 25–100 in clockwise and anticlockwise directions. After the ball milling, the homogeneous mixture was dried at ordinary temperature and divided into various parts for calcination at different temperatures from 600 to 1500°C for 4–30 h to optimize the pure phase. These phosphors were then structurally and optically characterized [26]. The perovskites prepared using this method have particles size in the submicrometer range [27]. Some other rare earths doped $R_{0.5}Ba_{0.5}CoO_3$ (R = La, Pr, Nd, Sm, Eu, Gd, Tb, Dy) perovskites were also synthesized using this method [28].

2.1.2 High energy ball-milling method

This technique is very similar to that of mechanical ball-milling technique. Only difference is that this method used very high rpm from few hundreds to few thousands for milling with very small sized balls. This technique uses low temperature for the synthesis of the oxide materials. This method produces generally the nanoparticles. This method only takes metal oxides because there is very high possibility of chemical reactions during high energy ball-milling, which may yield different toxic gases.

Now, let us describe the whole process by considering an example of $0.7BiFeO_3$–$0.3PbTiO_3$ (BF-PT) solid solution [29]. The stoichiometric amounts of Bi_2O_3, Fe_2O_3, PbO, and TiO_2 were taken as raw materials to prepare BF-PT and were mixed using agate mortar and pestle. Then the raw materials were ball-milled for 12–30 h with zirconia balls using alcohol as mixing media in the ratio of 1:10 with the sample and balls at 300–1000 rpm. Here, acetone or acid is not used for the purpose of mixing of raw materials as it may damage the milling jars and O-rings. The material dried at 90°C was annealed at various temperatures (i.e., 400–900°C) to optimize the pure phase of the sample. The XRD pattern of the samples was recorded to check the purity of the synthesized perovskite materials.

For additional characterizations, such as dc poling and dielectric measurements, the annealed sample was mixed in 2% aqueous solution of PVA (polyvinyl alcohol) as binder and a pellet was made. The pellet of the perovskites was fired at 500°C for 10–12 h to burn out the PVA binder. Then the pellet was sintered at quite higher temperatures than annealing temperature to obtain the highly dense perovskite material. To further check the purity of the samples after sintering, the XRD pattern was recorded [30]. Sometimes, this method is also used to decrease the particles size from a micrometric scale to nanometric scale. **Figure 3** represents flow chart of the preparation procedures for the synthesis of perovskite materials using solid-state synthesis technique.

2.2 Liquid-state synthesis technique

The liquid-state synthesis technique is a method of synthesis of nanomaterials. It is most widely used by researchers and scientists for the production of nanoparticles of the oxide materials. In this method, the raw materials may be in the form of nitrates, acetates or oxalates, which may react to each other at an ordinary temperature. Auto-combustion, sol-gel, co-precipitation, etc. are different liquid-state synthesis techniques used for the preparation of perovskite nanomaterials, which are to be described in more detail [19].

2.2.1 Auto-combustion method

The auto-combustion synthesis method is a low-cost and very facile technique for the production of perovskite nanomaterials. In this technique, the starting materials are used in oxalates and/or acetates and/or nitrates forms, which are easily soluble in de-ionized water. It involves some organic fuel, such as urea, citric acid, and glycine to assist the combustion.

Let us describe the process of synthesis of the perovskite materials using auto - combustion method (see **Figure 4 (a)**) with the help of an example

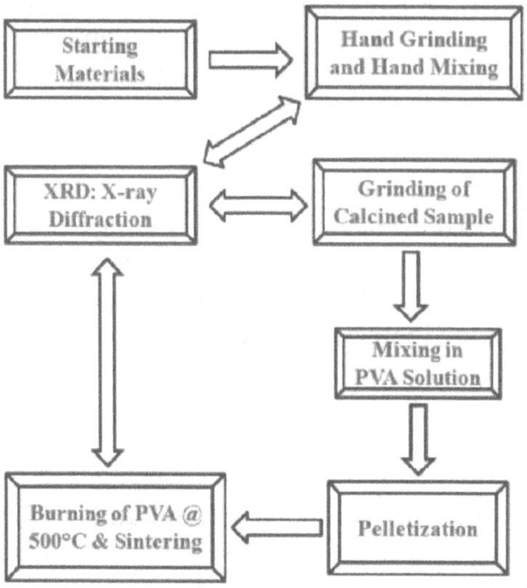

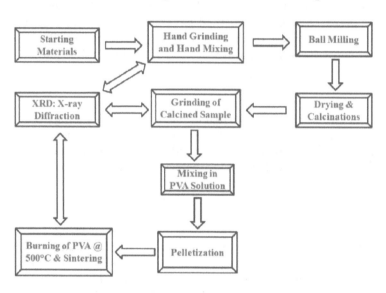

Figure 3.
A representative flow-chart for the synthesis of oxide materials using ball milling technique.

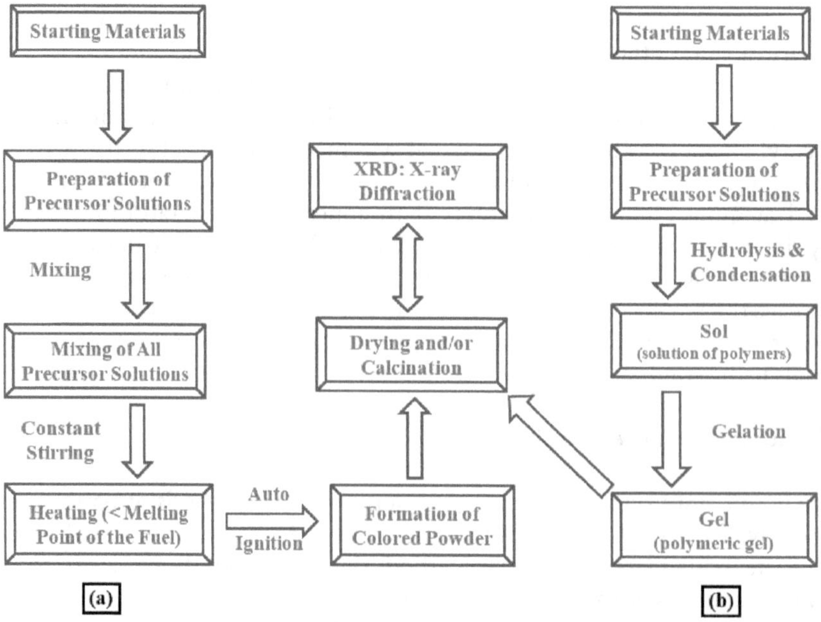

Figure 4.
Schematic flow charts for the synthesis of oxide nanomaterials via (a) auto-combustion and (b) sol-gel methods.

of $La_{0.7}Sr_{0.3}MnO_3$ manganite. Kumar et al. have synthesized $La_{0.7}Sr_{0.3}MnO_3$ perovskite manganite using the starting materials as La_2O_3, $SrCO_3$, and $Mn(CH_3COO)_2.4H_2O$ and glycine was used as fuel. First of all, La_2O_3 and $SrCO_3$ were dissolved in diluted nitric acid to prepare their respective nitrates and then $Mn(CH_3COO)_2.4H_2O$ and glycine were dissolved into distilled water. All the prepared precursor solutions of the materials were dissolved/prepared independently and at the end, they were mixed simultaneously under nonstop stirring in a large beaker and heated on a magnetic stirrer at 175–200°C [31]. After continuous stirring of 6–7 h, the mixed solution became thicker and converted into gel, with further increase in time of stirring, an auto-ignition takes place resulting in flame that evolves huge amount of different gases. During the ignition process, the temperature of the whole mixture may reach upto 800–1000°C for very short time duration. The obtained blackish-brown powder was collected from the bea-ker and divided into various parts for the calcination at different temperatures. The $La_{0.7}Sr_{0.3}MnO_3$ perovskite comes up with different particles size range. They have also prepared bulk samples of $Ba_{1-x}Sr_xMnO_3$ perovskites using the same method [32]. The $CaTiO_3$:Pr^{3+}, Al^{3+} phosphor was also prepared by Yin et al. using combustion method [33].

2.2.2 Sol-gel method

The sol-gel method is used by most of the chemists for producing nanomaterials. This technique comprises both types of processes (physical and chemical) associated with the following, such as hydrolysis, condensation, polymerization, gelation, drying, and densification [34]. In this technique, the starting materials are used in the form of metal alkoxides. The general chemical formula of metal alkoxides is $M(OR)_x$. Metal alkoxides can be assumed either a derivative of the alcohol ROH, where R is an alkyl group or a derivative of metal hydroxide $M(OH)_x$ [35]. The stoichiometric amounts of metal alkoxides are weighed and dissolved in alcohol or in de-ionized water at a temperature of 60–80°C under steady stirring. It is very important to control the pH value of the metal alkoxides solutions to avoid the formation of the precipitation and to form the homogeneous gel that can be achieved by using basic or acidic solutions. This is known as hydrolysis and condensation leads to form the polymeric chains. The progress of the polymeric chains eventually results to a perceptible improvement in the viscosity of the reaction mixture and the formation of a gel. The obtained gel is to be dried in temperature range between 150 and 200°C to remove the unwanted contents from the gel [35, 36]. After the removal of contents, the obtained gel was annealed at various temperatures in 400–800°C range to obtain the pure phased materials.

Let us discuss the practical processes involve in this technique by taking an example of $La_{0.6}Ca_{0.4}MnO_3$ perovskite. Andrade et al. have synthesized nanotubes and nanoparticles of $La_{0.6}Ca_{0.4}MnO_3$ perovskite manganite using sol-gel method following calcination at different temperatures [37]. They have used stoichiometric amounts of $La(NO_3)_3.6H_2O$, $CaCO_3$ and $Mn(CH_3COO)_2.4H_2O$ for the synthesis of $La_{0.6}Ca_{0.4}MnO_3$ perovskite. The $CaCO_3$ was dissolved in nitric acid to convert into calcium nitrate, while $La(NO_3)_3.6H_2O$ and $Mn(CH_3COO)_2.4H_2O$ were dissolved in distilled water. All the solutions were mixed together in a beaker. The appropriate amount of polyethylene glycol (PEG) was incorporated to the precursor solutions playing the role of polymerizing agent. Then the solution was heated at 70°C for 6 h to complete the polymerization process. At last, the whole solution was converted in yellow viscous gel, which was calcined at different temperatures from 700 to 1000°C. **Figure 4(b)** displays a representative flow chart of the processes involved in the synthesis of perovskite materials using sol-gel method.

2.2.3 Co-precipitation method

The co-precipitation method is also one of the methods used of the production of nanomaterials. This method needs raw materials of metal cations from a general medium and precipitates in the form of oxalates, carbonates, citrates or hydroxides [38–41]. The resultant precipitates are several times washed with distilled water and then obtained product was calcined at various temperatures to acquire the pure phase of the desired materials in the polycrystalline form. This method can yield almost homogeneous polycrystalline powders. The solubility of the used compounds should be very close to each other for a proper precipitation [42]. The precursor solutions are mixed at atomic level resulting lower particles size and it requires very low calcination temperature to get a pure material [43]. The controlling pH of the precursor solution, stirring speed, concentration, and temperature of the mixture are important parameters for the co-precipitation method [44].

The $LaMn_{1-x}Fe_xO_3$ (x = 0, 0.1, 0.2) perovskite synthesized by Geetha et al. is an example of the co-precipitation method [45]. The stoichiometric amounts of $La(NO_3)_3.6H_2O$, $Fe(NO_3)_3.9H_2O$ and $MnCl_2.4H_2O$ were dissolved in distilled water. These solutions were mixed at one platform and stirred continuously at 50°C for 30 min. After this, NaOH solution was added slowly until the pH of solution is attained to 13. The mixed solution of the precursors was again stirred constantly till the formation of black precipitate. The precipitate was collected and washed many times to remove the excess of chlorides and kept in an oven to dry at 50°C. Thus, the final product was calcined at 800°C for 6 h.

It was observed that liquid-state technique produces nanoparticles of perovskite materials at very low temperature. However, the sub-micron sized perovskite materials can be obtained by firing them at higher temperature similar to solid-state technique.

2.3 Gas-state synthesis technique

Gas-state synthesis technique is also used to prepare the nanoparticles. It contains various procedures for the synthesis of powder oxide materials, viz. furnaces, flames, plasmas, and lasers. The basics of thermodynamics and kinetics of the reaction are very similar to each other; however, their reactors are different. These methods provide narrow distribution of the nanoparticles. The dispersion must be reduced for narrow distribution of the nanoparticles as it leads to increase the particles size [25]. Gas-state synthesis technique is a bottom-up method for the synthesis of multifunctional nanoparticles. Method of bottom-up nanofabrication is based on the gathering of nanomaterials from smaller components.

A variety of electronic devices and solar cells of perovskite materials are prepared in terms of thin films via these techniques and they are entirely different from the other synthesis techniques. There are various techniques for the preparation of thin films, such as chemical vapor deposition [46], molecular beam epitaxy [47], laser ablation [48], DC sputtering [49], magnetron sputtering [50], thermal evaporation [51], and electron beam evaporation [52]. The gas-state synthesis method needs particular set up for a good quality of samples to offer the preferred properties. The gas-state synthesis techniques are categorized into three types:

a. Fabrication at the crystallization temperature under a suitable atmosphere condition of temperature.

b. Fabrication in an intermediate temperature range (500–800°C) followed by post-annealing treatment at higher temperatures.

c. Fabrication at a very low substrate temperature followed by post-annealing treatment at very high temperature.

The perovskite materials fabricated using gas-state techniques have variety of applications ranging from optical and anticorrosion coatings to photocatalysts and solar cells, from semiconductor devices to capacitor dielectrics, from bio-implantable devices to chemical reactors and catalysts. One can undoubtedly say that the industrial interest for preparing nanomaterial-based technologies through gas-state synthesis is going to increase in the upcoming years.

3. Structural and optical properties

3.1 Identification of phase purity: X-ray diffraction studies

It is very important to check the phase and its purity of synthesized perovskite materials. Without knowing the phase purity, one cannot come to any conclusion about the properties exhibited by the perovskite materials. The XRD technique is a suitable tool to identify the phase of perovskites. From the XRD data, one can find out relative phase fractions of different phases present in the prepared samples. One can also find out lattice constants, unit cell volume, crystallite size, lattice strain, and theoretical density from Rietveld refinement of the XRD pat-tern. The XRD technique is also used to optimize the synthesis conditions for the perovskite materials. By matching the XRD pattern of the synthesized material with the standard XRD pattern of the cubic phase of $CaTiO_3$ (CT) perovskite, one can conclude about the phase purity, i.e., whether, the perovskite is pure or has some amount of impurity phase(s) or crystallizes in the distorted perovskite. **Figure 5** shows the XRD pattern of the cubic CT perovskite with JCPDS File No. 75-2100 using X-ray radiation of 1.5406 Å wavelength. The analysis of XRD pattern gives unit cell lattice parameter a = 3.795 Å and unit cell volume V = 54.656 Å^3. An ideal perovskite material displays reflections from all allowed planes of the primitive unit cell. It shows most intense XRD peak for (1 1 0) plane in the 2θ range of 32–34° and first singlet reflection corresponding to (1 0 0) plane between 22 and 24°.

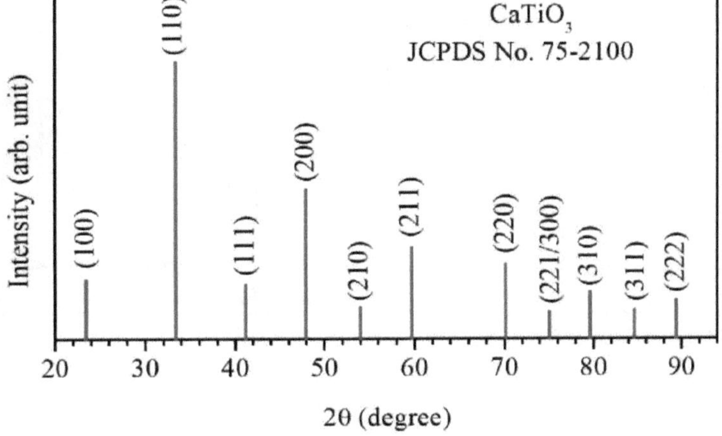

Figure 5.
The standard XRD pattern for the cubic phase of $CaTiO_3$ perovskite.

As we know that the different materials display different XRD patterns as like the finger print of the humans. However, a certain prototype of materials gives the XRD pattern in a well-defined manner. In order to make sure that the synthesized sample is perovskite or not, we have to match the XRD pattern of the synthesized sample with the XRD pattern of CT. If it matches it forms a perovskite phase otherwise not.

Figure 6(a) shows the room-temperature XRD pattern for $BaTiO_3$ (BT) synthesized by mechanical ball-milling method followed by calcination at 950°C for 5 h and it is compared with XRD pattern of CT. The comparison of both the XRD patterns reveals that the XRD pattern of BT matches with that of the CT. This further confirms that BT is a perovskite material. The Bragg's peaks in XRD pattern of BT perovskite is shifted towards lower angle side compared to CT. This indicates that the lattice parameters of BT are larger than that of the CT. Furthermore, the Bragg peak around 22° is asymmetric in lower angle side, which shows doublet nature in the peak, i.e., the lattice plane (1 0 0) of CT split into (0 0 1) and (1 0 0) of BT is shown as inset in **Figure 6(a)**. Similarly, the other Bragg's peaks also show asymmetrical behavior. All these observations reveal that the crystal structure of BT is distorted from an ideal perovskite structure. The Rietveld refinement of XRD pattern of BT reveals that the BT crystallizes into tetragonally distorted structure with *P4mm* space group [53]. It has been also observed that the presence of superlattice reflection(s) in the XRD pattern indicates the formation of distorted perovskite structure from an ideal cubic [54, 55]. The structural analysis of the known or unknown perovskite materials can be identified by Rietveld refinement of the powder XRD pattern with the help of some standard software [56]. Most of researchers use FullProf Suite to identify the structure of various perovskite materials [57]. From Rietveld refinement, one can get lattice parameters, atomic coordinates of the constituent atoms present in the unit cell, average bond-lengths and angles, draw molecular structure to see the surroundings of the A- and B-sites' cations with anions, and many more [29, 33, 58]. One can also find out magnetic structure from the Rietveld refinement of the neutron diffraction data [59].

We performed the Rietveld refinement of XRD pattern of BT using *P4mm* space group of the tetragonal structure. In *P4mm* space group, Ti^{4+}-ions and Ba^{2+}-ions occupy *1(a)* site at (0, 0, 0) and *1(b)* site at (0.5, 0.5, 0.5 + δz), respec-tively, while, O^{2-}(1) and O^{2-}(2) anions substitute *1(a)* site at (0, 0, 0.5 + δz) and *2(c)* site at (0.5, 0, δz), respectively [53, 60]. The structural analysis provides

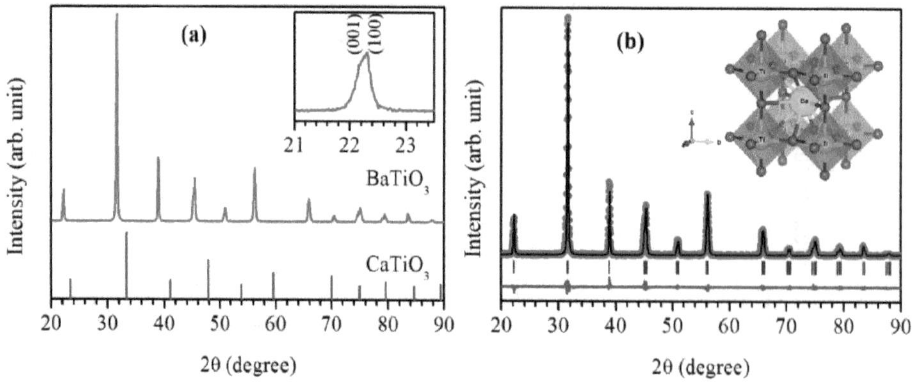

Figure 6.
(a) Room temperature XRD pattern of $BaTiO_3$ compared with the $CaTiO_3$. The inset in (a) shows selected Bragg's peak between 21.0 and 23.5°. (b) Rietveld fit of the XRD pattern of $BaTiO_3$. The inset in (b) shows ball and stick molecular model for the unit cell of $BaTiO_3$ perovskite, where Ba, Ti, and O are present in their atomic sizes.

lattice constants a = b = 4.0005(1) Å and c = 4.0255(2) Å and unit cell volume V = 64.423(4) $Å^3$ with tetragonality (c/a) of 1.006 close to the earlier reported value [60]. **Figure 6(b)** shows Rietveld fit of the XRD pattern for the BT perovskite. In this figure, the scattered dots are used to present experimental XRD pattern and continuous line over experimental pattern is to show the simulated XRD pattern. Lower continuous curve shows difference between experimental and simulated XRD patterns. The vertical bars represent positions of the Bragg's reflection. The inset of **Figure 6(b)** demonstrates ball and stick molecular model for the unit cell of *P4mm* space group for BT in terms of atomic sizes. It is clear that Ba atom forms octahedral with oxygen atoms (BaO_6) and Ti atom forms cubocta-hedral with oxygen atoms (TiO_{12}).

3.2 Morphological studies

The scanning electron microscopy (SEM) and transmission electron microscopy (TEM) can be used to study the microstructure nature of the perovskite materials. We have recorded SEM micrographs of $Nd_{0.4}Sr_{0.6}MnO_3$ manganite samples calcined at 800°C and 1200°C synthesized by auto-combustion method using FEI, Nova Nano SEM for the $Nd_{0.4}Sr_{0.6}MnO_3$ perovskite and they are shown in **Figure 7**. The average values of the particle size were analyzed by ImageJ software and they are found to be 50 and 425 nm for the $Nd_{0.4}Sr_{0.6}MnO_3$ perovskites calcined at 800°C and 1200°C, respectively. This clearly shows that the average particle size improves on increasing the calcination temperature. It confirms that the sample calcined at 800° C produces nanomaterial whereas that of at 1200°C gives bulk material.

3.3 Optical studies

Figure 8 shows photoluminescence (PL) excitation and PL emission spectra of the $Nd_{0.4}Sr_{0.6}MnO_3$ manganite calcined at 800°C and 1200°C. **Figure 8(a)** shows the excitation spectra of the $Nd_{0.4}Sr_{0.6}MnO_3$ manganite in both the cases and contains an intense peak at 355 nm [45]. It was monitored at an emission wavelength of 646 nm. The intensity of bulk material is two times larger than the nanomaterial. When both the materials were excited with 355 nm they give strong red color at 646 nm as shown in **Figure 8(b)**. It also contains weak peaks at 483 and 582 nm. The emission intensity obtained in the case of bulk material is further two times larger than the nanomaterial. This is due to the increase in crystallinity and particles size of the materials. Thus, the calcination affects the morphology and optical properties of perovskites even though the perovskite material was synthesized by auto-combustion method.

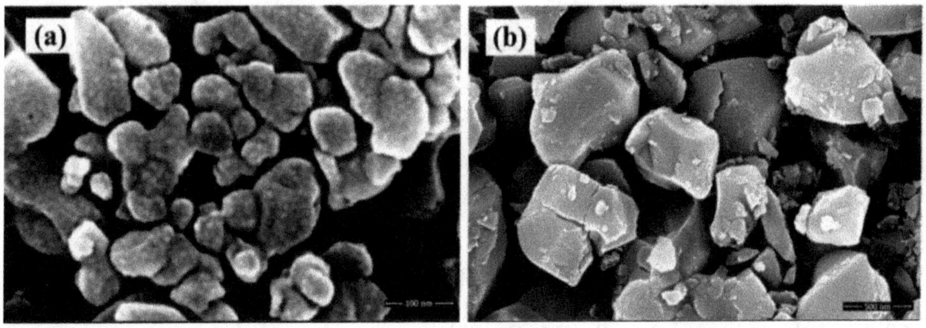

Figure 7.
SEM micrographs of $Nd_{0.4}Sr_{0.6}MnO_3$ manganite calcined at (a) 800°C and (b) 1200°C temperatures.

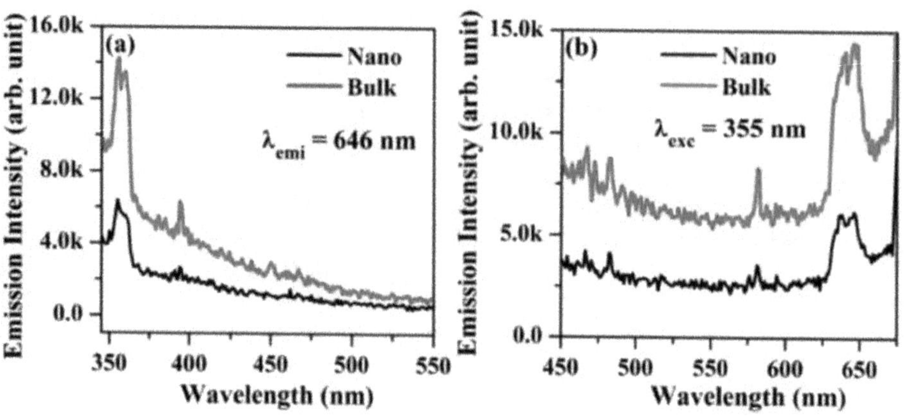

Figure 8.
(a) PL excitation and (b) PL emission spectra of the $Nd_{0.4}Sr_{0.6}MnO_3$ manganite calcined at 800°C and 1200°C temperatures.

4. Applications of perovskite materials

The perovskite materials are extensively studied by researchers due to their attractive properties. The perovskite materials have wide applications in various fields, which are listed below:

1. Photocatalytic activity; e.g., $LaFeO_3$ [61].

2. Photovoltaic solar cells; e.g., $LaVO_3$ [62].

3. Phosphor materials in photoluminescence; e.g., $Ho^{3+}/Yb^{3+}/Mg^{2+}$ doped $CaZrO_3$ [63].

4. Solid oxide fuel cells; e.g., $Gd_{0.7}Ca_{0.3}Co_{1-y}Mn_yO_3$ [64].

5. Sensors and actuators; e.g., $PbZr_xTi_1$-xO_3 [65].

6. Magnetic memory devices; e.g., $Pt/La_2Co_{0.8}Mn_{1.2}O_6/Nb{:}SrTiO_3$ [66].

7. Magnetic field sensors; e.g., $La_{0.67}Sr_{0.33}MnO_3$ and $La_{0.67}Ba_{0.33}MnO_3$ [67].

8. Electric field effect devices; e.g., hetrostructure of $Pb(Zr_{0.2}Ti_{0.8})O_3/La_{0.8}Ca_{0.2}MnO_3$ [68].

9. Ferroelectric and piezoelectric devices; e.g., $BaTiO_3$, $PbTiO_3$ [69].

10. Semiconducting electronic devices; e.g., $La_{0.7}Ca_{0.3}MnO_3/SrTiO_3/La_{0.7}Ce_{0.3}MnO_3$ [70].

11. High dielectric constant; e.g., Bi_1-xSr_xMnO_3 ($x = 0.4, 0.5$) [71].

12. High temperature superconductor; e.g., $BaPb_1$-xBi_xO_3 [72].

13. Hypothermia; e.g., $La_{0.7}Sr_{0.3}MnO_3$ [73].

14. Supercapacitor; e.g., $KNi_{0.8}Co_{0.2}F_3$ [74].

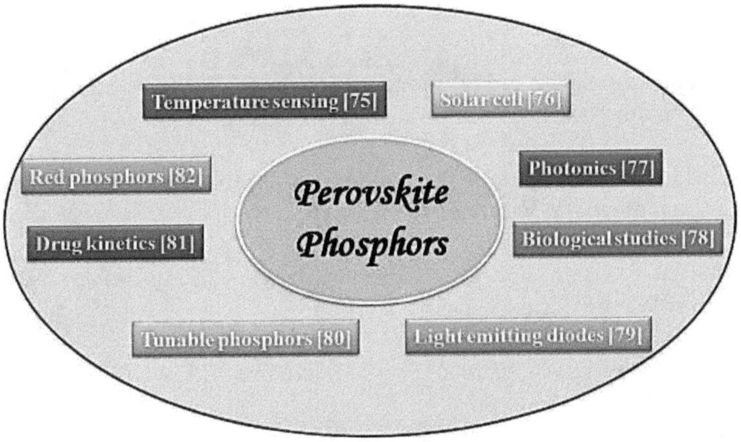

Figure 9.
Block diagram of the lanthanide based perovskite phosphor materials including their applications in various fields [75–82].

The lanthanide based perovskite phosphors are one of them and have unique photolumincence properties. The lanthanide ions are known for their narrow and sharp band emissions. They give emissions in the entire range of visible spectrum along with ultraviolet (UV) and near infrared (NIR) regions. These emissions are observed due to presence of meta-stable energy levels in the lanthanide ions. These levels have long lifetime and are responsible for very strong emissions. The lanthanide ions possess upconversion, downconversion, and quantum cutting phenomena. In upconversion, the two or more than two low energy photons are converted into high energy photons. When a high energy photon is converted into low energy photons it is termed as downconversion process. However, in quantun cutting, a high energy photon is converted into two low energy NIR photons. These processes lead numerous technological applications in different fields, such as red phosphors, light emitting diodes (LEDs), photonics, solar cells, tunable phosphors, temperature sensing, biological studies, drug kinetics, etc. [63, 75–82]. **Figure 9** shows a block diagram for the lanthanide based perovskite phosphors, which shows their applications in various fields.

5. Conclusions

This chapter summarizes the basics of the perovskite structure, its stability and distortion. We have discussed the novelty of the perovskite materials, which accommodate different cations at A- and/or B-sites individually and/or simulta-neously. We have also discussed various routes such as solid, liquid and gas-state synthesis for the preparation of perovskite materials mostly in the oxide powder forms. We have also briefly described the phase identification of the perovskites and their structural analysis using Rietveld refinement of the XRD data by tak-ing an example of tetragonal $BaTiO_3$ perovskite. The morphological and optical studies were also incorporated. We have also briefly listed various applications of perovskite materials including lanthanide based perovskite phosphors in various fields.

Conflict of interest

The authors declare no conflict of interest in the chapter.

Author details

Dinesh Kumar[1], Ram Sagar Yadav[2*], Monika[3], Akhilesh Kumar Singh[1] and Shyam Bahadur Rai[3]

1 School of Materials Science and Technology, Indian Institute of Technology, Banaras Hindu University, Varanasi, Uttar Pradesh, India

2 Department of Zoology, Institute of Science, Banaras Hindu University, Varanasi, Uttar Pradesh, India

3 Department of Physics, Institute of Science, Banaras Hindu University, Varanasi, Uttar Pradesh, India

*Address all correspondence to: ramsagaryadav@gmail.com

References

[1] Roth RS. Classification of perovskite and other ABO_3-type compounds. Journal of Research of the National Bureau of Standards. 1957;**58**:75-88

[2] Yashima M, Ali R. Structural phase transition and octahedral tilting in the calcium titanate perovskite $CaTiO_3$. Solid State Ionics. 2009;**180**:120-126

[3] Weis RS, Gaylord TK. Lithium niobate: Summary of physical properties and crystal structure. Applied Physics A: Materials Science & Processing. 1985;**37**:191-203

[4] Kwei GH, Lawson AC, Billinge SJL, Cheong SW. Structures of the ferroelectric phases of barium titanate. The Journal of Physical Chemistry. 1993;**97**:2368-2377

[5] Sakai H, Ishiwata S, Okuyama D, Nakao A, Nakao H, Murakami Y, et al. Electron doping in the cubic perovskite $SrMnO_3$: Isotropic metal versus chainlike ordering of Jahn-Teller polarons. Physical Review B. 2010;**82**:180409R–180412R

[6] Wang Y, Zhu J, Zhang L, Yang X, Lu L, Wang X. Preparation and characterization of perovskite $LaFeO_3$ nanocrystals. Materials Letters. 2006;**60**:1767-1770

[7] Kubicek M, Bork AH, Rupp JLM. Perovskite oxides—A review on a versatile material class for solar-to-fuel conversion processes. Journal of Materials Chemistry A. 2017;**5**:11983-12000

[8] Protesescu L, Yakunin S, Bodnarchuk MI, Krieg F, Caputo R, Hendon CH, et al. Nanocrystals of cesium lead halide perovskites ($CsPbX_3$, X = Cl, Br, and I): Novel optoelectronic materials showing bright emission with wide color gamut. Nano Letters. 2015;**15**:3692-3696

[9] Perez RS, Cerqueira TFT, Korbel S, Botti S, Marques MAL. Prediction of stable nitride perovskites. Chemistry of Materials. 2015;**27**:5957-5963

[10] Kuhar K, Crovetto A, Pandey M, Thygesen KS, Seger B, Vesborg PCK, et al. Sulfide perovskites for solar energy conversion applications: Computational screening and synthesis of the selected compound $LaYS_3$. Energy & Environmental Science. 2017;**10**:2579-2593

[11] Kansara SB, Dhruv D, Kataria B, Thaker CM, Rayaprol S, Prajapat CL, et al. Structural, transport and magnetic properties of monovalent doped $La_{1-x}Na_xMnO_3$ manganites. Ceramics International. 2015;**41**:7162-7173

[12] Pandey R, Pillutla RK, Shankar U, Singh AK. Absence of tetragonal distortion in (1-x)$SrTiO_{3-x}Bi(Zn_{1/2}Ti_{1/2})O_3$ solid solution. Journal of Applied Physics. 2013;**113**:184109-184114

[13] Ali Z, Ahmad I, Amin B, Maqbool M, Murtaza G, Khan I, et al. Theoretical studies of structural and magnetic properties of cubic perovskites $PrCoO_3$ and $NdCoO_3$. Physica B. 2011;**406**:3800-3804

[14] Johnsson M, Lemmens P. Crystallography and chemistry of perovskites. In: Kronmuller H, Parkin S, editors. Handbook of Magnetism and Advanced Magnetic Materials. Vol. 4. USA: John Wiley & Sons; 2005. pp. 1-11

[15] Goldschmidt VM. Die Gesetze der Krystallochemie. Naturwissenschaften. 1926;**14**:477-485

[16] Kumar D, Verma NK, Singh CB, Singh AK. Evolution of structural characteristics of $Nd_{0.7}Ba_{0.3}MnO_3$ perovskite manganite as a function of crystallite size. AIP Conference Proceedings. 2018 2009:020013-4

[17] Duan R. High curie temperature bismuth- and indium-substituted lead titanate [thesis]. Georgia Institute of Technology: School of Materials Science and Engineering; 2004

[18] Kumar D, Singh AK. Investigation of structural and magnetic properties of $Nd_{0.7}Ba_{0.3}Mn_{1-x}Ti_xO_3$ (x = 0.05, 0.15 and 0.25) manganites synthesized through a single-step process. Journal of Magnetism and Magnetic Materials. 2019;**469**:264-273

[19] Atta NF, Galal A, El-Ads EH. Synthesis, characterization and applications. In: Pan L, Zhu G, editors. Perovskite Nanomaterials. London: IntechOpen; 2016. pp. 107-151. Ch. 4

[20] Porta P, De RS, Faticanti M, Minelli G, Pettiti I, Lisi L, et al. Perovskite-type oxides: I. Structural, magnetic, and morphological properties of $LaMn_{1-x}Cu_xO_3$ and $LaCo_{1-x}Cu_xO_3$ solid solutions with large surface area. Journal of Solid State Chemistry. 1999;**146**:291-304

[21] Chen M, Zhang H, Liu T, Jiang H, Chang A. Preparation, structure and electrical properties of $La_{1-x}Ba_xCrO_3$NTC ceramics. Journal of Materials Science: Materials in Electronics. 2017;**28**:18873-18188

[22] Yang J, Ma YQ , Zhang RL, Zhao BC, Ang R, Song WH, et al. Structural, transport, and magnetic properties in the Ti-doped manganites $LaMn_{1-x}Ti_xO_3$ $(0 \leq x \leq 0.2)$. Solid State Communications. 2005;**136**:268-272

[23] Kameli P, Salamati H, Heidarian A, Bahrami H. Ferromagnetic insulating and reentrant spin glass behavior in Mg doped $La_{0.75}Sr_{0.25}MnO_3$ manganites. Journal of Non-Crystalline Solids. 2009;**355**:917-921

[24] Choudhary N, Verma MK, Sharma ND, Sharma S, Singh D. Correlation between magnetic and transport properties of rare earth doped perovskite manganites $La_{0.6}R_{0.1}Ca_{0.3}MnO_3$ (R = La, Nd, Sm, Gd, and Dy) synthesized by Pechini process. Materials Chemistry and Physics. 2020;**242**:122482 pp. 11

[25] Ring TA. Fundamentals of Ceramic Powder Processing and Synthesis: Ceramic Powder Synthesis. United States: Academic Press; 1996

[26] Lin L-T, Chen J-Q, Deng C, Tang L, Chen D-J, Meng J-X, et al. Broadband near-infrared quantum-cutting by cooperative energy transfer in Yb^{3+}–Bi^{3+} co-doped $CaTiO_3$ for solar cells. Journal of Alloys and Compounds. 2015;**640**:280-284

[27] Pandey R, Tiwari A, Upadhyay A, Singh AK. Phase coexistence and the structure of the morphotropic phase boundary region in (1-x)$Bi(Mg_{1/2}Zr_{1/2})O_3$-x$PbTiO_3$ piezoceramics. Acta Materialia. 2014;**76**:198-206

[28] Troyanchuk IO, Kasper NV, Khalyavin DD, Szymczak H, Szymczak R, Baran M. Magnetic and electrical transport properties of orthocobaltites $R_{0.5}Ba_{0.5}CoO_3$ (R = La, Pr, Nd, Sm, Eu, Gd, Tb, Dy). Physical Review B. 1998;**58**:2418-2421

[29] Zhang L, Xu Z, Cao L, Yao X. Synthesis of BF-PT perovskite powders by high-energy ball milling. Materials Letters. 2007;**61**:1130-1133

[30] Upadhyay A, Pandey R, Singh AK. Origin of ferroelectric P-E loop in cubic compositions and structure of poled (1-x)$Bi(Mg_{1/2}Zr_{1/2})O_3$-x$PbTiO_3$ piezoceramics. Journal of the American Ceramic Society. 2017;**100**:1743-1750

[31] Kumar D, Verma NK, Singh CB, Singh AK. Crystallite size strain analysis of nanocrystalline $La_{0.7}Sr_{0.3}MnO_3$ perovskite by Williamson-Hall plot method. AIP Conference Proceedings. 2018 1942:050024-4

[32] Kumar D, Singh CB, Verma NK, Singh AK. Synthesis and structural investigations on multiferroic $Ba_{1-x}Sr_xMnO_3$ perovskite manganites. Ferroelectrics. 2017;**518**:191-195

[33] Yin S, Chen D, Tang W. Combustion synthesis and luminescent properties of $CaTiO_3$:Pr, Al persistent phosphors. Journal of Alloys and Compounds. 2007;**441**:327-331

[34] Brinker CJ, Scherer GW. Sol-Gel Science: The Physics and the Chemistry of Sol-Gel Processing. London: Academic Press Inc.; 1990

[35] Rahaman MN. Ceramic Processing and Sintering. 2nd ed. New York: Marcel Dekker Inc.; 2003

[36] Rajaeiyan A, Mohagheghi MMB. Comparison of sol-gel and co-precipitation methods on the structural properties and phase transformation of γ and α-Al_2O_3 nanoparticles. Advanced Manufacturing. 2013;**1**:176-182

[37] Andrade VM, Vivas RJC, Pedro SS, Tedesco JCG, Rossi AL, Coelho AA, et al. Magnetic and magnetocaloric properties of $La_{0.6}Ca_{0.4}MnO_3$ tunable by particle size and dimensionality. Acta Materialia. 2016;**102**:49-55

[38] Pei RR, Chen X, Suo Y, Xiao T, Ge QQ , Yao HC, et al. Synthesis of $La_{0.85}Sr_{0.15}Ga_{0.8}Mg_{0.2}O_3$-δ powder by carbonate co-precipitation combining with azeotropic-distillation process. Solid State Ionics. 2012;**219**:34-40

[39] Uskokovic V, Drofenik M. Four novel co-precipitation procedures for the synthesis of lanthanum-strontium manganites. Materials and Design. 2007;**28**:667-672

[40] Cho TH, Shiosaki Y, Noguchi H. Preparation and characterization of layered $LiMn_{1/3}Ni_{1/3}Co_{1/3}O_2$ as a cathode material by an oxalate co-precipitation method. Journal of Power Sources. 2006;**159**:1322-1327

[41] Wei Y, Han B, Hu X, Lin Y, Wang X, Deng X. Synthesis of Fe_3O_4 nanoparticles and their magnetic properties. Procedia Engineering. 2012;**27**:632-637

[42] West AR. Solid State Chemistry and its Applications. USA: John Wiley & Sons; 2005

[43] Gaikwad AB, Navale SC, Samuel V, Murugan AV, Ravi V. A co-precipitation technique to prepare $BiNbO_4$, $MgTiO_3$ and $Mg_4Ta_2O_9$ powders. Materials Research Bulletin. 2006;**41**:347-353

[44] Zawrah MF, Hamaad H, Meky S. Synthesis and characterization of nano $MgAl_2O_4$ spinel by the co-precipitated method. Ceramics International. 2007;**33**:969-978

[45] Geetha N, Senthil KV, Prakash D. Synthesis and characterization of $LaMn_{1-x}Fe_xO_3$ (x = 0, 0.1, 0.2) by coprecipitation route. Journal of Physical Chemistry & Biophysics. 2018;**8**:273-278

[46] Kwak BS, Zhang K, Boyd EP, Erbil A, Wilkens BJ. Metalorganic chemical vapor deposition of $BaTiO_3$ thin films. Journal of Applied Physics. 1991;**69**:767-772

[47] Yu Z, Ramdani J, Curless JA, Finder JM, Overgaard CD, Droopad R, et al. Epitaxial perovskite thin films grown on silicon by molecular beam epitaxy. Journal of Vacuum Science and Technology B. 2000;**18**:1653-1657

[48] Imai T, Okuyama M, Hamakawa Y. $PbTiO_3$ thin films deposited by laser ablation. Japanese Journal of Applied Physics. 1991;**30**:2163-2166

[49] Bangchao Y, Wang JY, Jia YM, Yongjie H. Preparation of $PbTiO_3$ thin film by dc single-target magnetron

sputtering. Proceedings of the SPIE. 1991;**1519**:725-728

[50] Lu CJ, Shen HM, Wang YN. Preparation and crystallization of $Pb(Zr_{0.95}Ti_{0.05})O_3$ thin films deposited by radio-frequency magnetron sputtering with a stoichiometric ceramic target. Applied Physics A: Materials Science and Processing. 1998;**67**:253-258

[51] Li Y, Xu X, Wang C, Wang C, Xie F, Yang J, et al. Investigation on thermal evaporated $CH_3NH_3PbI_3$ thin films. AIP Advances. 2015;**5**:097111-097116

[52] Pae SR, Byun S, Kim J, Kim M, Gereige I, Shin B. Improving uniformity and reproducibility of hybrid perovskite solar cells via a low-temperature vacuum deposition process for NiO_x hole transport layers. ACS Applied Materials & Interfaces. 2018;**10**:534-540

[53] Xiao CJ, Jin CQ, Wang XH. Crystal structure of dense nanocrystalline $BaTiO_3$ ceramics. Materials Chemistry and Physics. 2008;**111**:209-212

[54] Geller S. Crystal structure of gadolinium orthoferrite $GdFeO_3$. The Journal of Chemical Physics. 1956;**24**:1236-1239

[55] Jakymiw C, Vocadlo L, Dobson DP, Bailey E, Thomson AR, Brodholt JP, et al. The phase diagrams of $KCaF_3$ and $NaMgF_3$ by *ab initio* simulations. Physics and Chemistry of Minerals. 2018;**45**:311-322

[56] Young RA. The Rietveld Method. International Union of Crystallography: Oxford University Press; 1993

[57] Carvajal JR. 'FULLPROF' Program: Rietveld Pattern Matching Analysis of Powder Patterns-ILL. Grenoble; 1990

[58] Yadav RS, Kumar D, Singh AK, Rai E, Rai SB. Effect of Bi^{3+} ion on upconversion-based induced optical heating and temperature sensing characteristics in the Er^{3+}/Yb^{3+} co-doped La_2O_3 nano-phosphor. RSC Advances. 2018;**8**:34699-34711

[59] Carvajal JR. Recent advances in magnetic structure determination by neutron powder diffraction. Physica B. 1993;**192**:55-69

[60] Villars P. Pauling file in: Inorganic Solid Phases. Springer Materials (online database). Springer Materials. $BaTiO_3$ tetragonal ($BaTiO_3$ rt) Crystal Structure. Heidelberg: Springer. Available from: https://materials.springer.com/isp/crystallographic/docs/sd_1626689

[61] Afifah N, Saleh R. Enhancement of photocatalytic activities of perovskite $LaFeO_3$ composite by incorporating nanographene platelets. IOP Conference Series: Materials Science and Engineering. 2017;**188**:012054-012058

[62] Wang L, Li Y, Bera A, Ma C, Jin F, Yuan K, et al. Device performance of the Mott insulator $LaVO_3$ as a photovoltaic material. Physical Review Applied. 2015;**3**:064015-064029

[63] Maurya A, Yadav RS, Yadav RV, Rai SB, Bahadur A. Enhanced green upconversion photoluminescence from Ho^{3+}/Yb^{3+} co-doped $CaZrO_3$ phosphor via Mg^{2+} doping. RSC Advances. 2016;**6**:113469-113477

[64] Skinner SJ. Recent advances in perovskite-type materials for solid oxide fuel cell cathodes. International Journal of Inorganic Materials. 2001;**3**:113-121

[65] Uchino K. Glory of piezoelectric perovskites. Science and Technology of Advanced Materials. 2015;**16**:046001-0460016

[66] Mir LL, Frontera C, Aramberri H, Bouzehouane K, Fernandez JC, Bozzo B, et al. Anisotropic sensor and memory device with a ferromagnetic tunnel

barrier as the only magnetic element. Scientific Reports. 2018;**8**:861-870

[67] Xu Y, Memmert U, Hartmann U. Magnetic field sensors from polycrystalline manganites. Sensors and Actuators A. 2001;**91**:26-29

[68] Zhao T, Ogale SB, Shinde SR, Ramesh R, Droopad R, Yu J, et al. Colossal magnetoresistive manganite-based ferroelectric field-effect transistor on Si. Applied Physics Letters. 2004;**84**:750-752

[69] Nuraje N, Su K. Perovskite ferroelectric nanomaterials. Nanoscale. 2013;**5**:8752-8780

[70] Mitra C, Raychaudhuri P, Kobernik G, Dorr K, Muller KH, Schultz L, et al. p-n diode with hole- and electron-doped lanthanum manganites. Applied Physics Letters. 2001;**79**:2408-2410

[71] Munoz JLG, Frontera C, Murias BR, Mira J. Dielectric properties of $Bi_{1-x}Sr_xMnO_3$ (x = 0.40, 0.50) manganites: Influence of room temperature charge order. Journal of Applied Physics. 2009;**105**:084116-084120

[72] Sleight AW, Gillson JL, Bierstedt PE. High-temperature superconductivity in the $BaPb_{1-x}Bi_xO_3$ system. Solid State Communications. 1993;**88**:841-842

[73] Manh DH, Phong PT, Nam PH, Tung DK, Phuc NX, Lee IJ. Structural and magnetic study of $La_{0.7}Sr_{0.3}MnO_3$ nanoparticles and AC magnetic heating characteristics for hyperthermia applications. Physica B. 2014;**444**:94-102

[74] Ding R, Li X, Shi W, Xu Q, Han X, Zhou Y, et al. Perovskite $KNi_{0.8}Co_{0.2}F_3$ nanocrystals for supercapacitors. Materials Chemistry A. 2017;**5**:17822-17827

[75] Mahata MK, Koppe T, Mondal T, Brusewitz C, Kumar K, Rai VK, et al. Incorporation of Zn^{2+} ions into $BaTiO_3$:Er^{3+}/Yb^{3+} nanophosphor: An effective way to enhance upconversion, defect luminescence and temperature sensing. Physical Chemistry Chemical Physics. 2015;**17**:20741-20753

[76] Ho WJ, Li GY, Liu JJ, Lin ZX, You BJ, Ho CH. Photovoltaic performance of textured silicon solar cells with $MAPbBr_3$ perovskite nanophosphors to induce luminescent down-shifting. Applied Surface Science. 2018;**436**:927-933

[77] Singh DK, Manam J. Structural and photoluminescence studies of red emitting $CaTiO_3$:Eu^{3+} perovskite nanophosphors for lighting applications. Journal of Materials Science: Materials in Electronics. 2016;**27**:10371-10381

[78] Kumar KN, Vijayalakshmi L, Choi J. Investigation of upconversion photoluminescence of Yb^{3+}/Er^{3+}:$NaLaMgWO_6$ noncytotoxic double-perovskite nanophosphors. Inorganic Chemistry. 2019;**58**:2001-2011

[79] Tian Y, Zhou C, Worku M, Wang X, Ling Y, Gao H, et al. Highly efficient spectrally stable red perovskite light-emitting diodes. Advanced Materials. 2018;**30**:1707093-1707097

[80] Jain N, Singh RK, Sinha S, Singh RA, Singh J. Color tunable emission through energy transfer from Yb^{3+} co-doped $SrSnO_3$:Ho^{3+} perovskite nano-phosphor. Applied Nanoscience. 2018;**8**:1267-1278

[81] Li X, Zhang Q, Ahmad Z, Huang J, Ren Z, Weng W, et al. Near-infrared luminescent $CaTiO_3$:Nd^{3+} nanofibers with tunable and trackable drug release kinetics. Journal of Materials Chemistry B. 2015;**3**:7449-7456

[82] Itoh S, Toki H, Tamura K, Kataoka F. A new red-emitting phosphor, $SrTiO_3$:Pr^{3+} for low-voltage electron excitation. Japanese Journal of Applied Physics. 1999;**38**:6387-6391

12

Energy Harvesting Prediction from Piezoelectric Materials with a Dynamic System Model

José Carlos de Carvalho Pereira

Abstract

Piezoelectric vibration energy harvesting has been investigated for different applications due to the amount of wasted vibration from dynamic systems. In the case of piezoelectric materials, this energy lost to the environment can be recovered through the vibration of energy harvesting devices, which convert mechanical vibration into useful electrical energy. In this context, this chapter aims to present the mechanical/electrical coupling on a simple dynamic system model in which a linear piezoelectric material model is incorporated. For this purpose, a mechanical/electrical element of a piezoelectric disk is developed and integrated into a lumped mass, viscous damping, and spring assembling, similar to a quarter car suspension system. Equations of motion for this dynamic system in the time domain can be solved using the finite element method. The recovered electric power and energy density for PZT (Lead Zirconate Titanate) from the wasted vibration can be predicted considering that the road roughness is introduced as an input mode.

Keywords: harvesting energy, wasted vibration, dynamic system, linear piezoelectric model, PZT

1. Introduction

Energy use is widely discussed nowadays, as energy conversion and management. In this way, new sources of energy are required to be investigated. Thus, one of the energy sources that can be used is from vibration systems, which can be subject to different excitations. This wasted energy to the environment can be recovered through vibration energy harvesting devices, which convert mechanical vibration into useful electrical energy in a way that low power devices may utilize.

Piezoelectric materials are known to have the electro/mechanical coupling effect. This property has a large range of applications in engineering. Currently, they are extensively used as sensors and actuators in vibration control systems. As a sensor, it can monitor the vibrations when bonded to a flexible structure. As an actuator, it can control the vibration level by introducing a restored force or by adding damping to the system.

In the context of recovered energy from mechanical vibrations based on the conversion of piezoelectric harvesting devices and its application on powering electronic devices, this subject has received the attention from various researchers [1–3]. An example of this type of recovered energy is the suspension system vibration for use of the vehicle itself, such as an energetic source for an active and semi-active suspension.

On a typical road, vehicles suffer accelerations due to its roughness, which excite undesired vibration. Some recently conducted reviews mentioned the potential of recovering a few hundred watts for a passenger car driven in experimental tests as well as some mathematical models [4, 5]. One of the ways to convert the mechanical energy from the vehicle suspension to electric energy is through piezoelectric materials [6]. Therefore, the objective of this chapter is to present the coupling between a piezoelectric element and a dynamic system in the context of predicting the recovered electric power and energy density for piezoelectric materials, especially the PZT (Lead Zirconate Titanate).

2. Piezoelectric material modelling

Mathematical models for predicting the harvesting energy in piezoelectric materials submitted to axial loads consider its geometric properties, diameter D_p and thickness h_p, and its mechanical and electrical properties, Young's modulus is c^E_{zz}, the piezoelectric constant is e_{zz} and the dielectric constant is ϵ^S_{zz}, as shown in **Figure 1**. Points 1 and 2 represent the two faces of the piezoelectric disk, and w and V are the mechanical displacement and electric potential, respectively, at these two points.

The electromechanical coupling effect of the piezoelectric material can be described using a set of basic equations as given in the IEEE Standard on Piezoelectricity [7]:

$$\begin{aligned} \sigma_z &= c^E_{zz}\frac{\partial w}{\partial z} - e_{zz}\frac{\partial V}{\partial z} \\ D_z &= e_{zz}\frac{\partial w}{\partial z} + \epsilon^S_{zz}\frac{\partial V}{\partial z} \end{aligned} \tag{1}$$

Where σ_z is the normal stress, D_z is the electric flux density, both in direction z. As seen in the above equations, the electro/mechanical coupling occurs due to the piezoelectric constant e_{zz}.

The mechanical strain energy U_m and the electric energy U_e of the piezoelectric material are written as [7]:

$$\begin{aligned} U_m &= \frac{1}{2}\int_V \sigma_z \frac{\partial w}{\partial z} dV = \frac{1}{2}\int_V \left(c^E_{zz}\frac{\partial w}{\partial z} - e_{zz}\frac{\partial V}{\partial z}\right)\frac{\partial w}{\partial z} dV \\ U_e &= \frac{1}{2}\int_V D_z \frac{\partial V}{\partial z} dV = \frac{1}{2}\int_V \left(e_{zz}\frac{\partial w}{\partial z} + \epsilon^S_z\frac{\partial V}{\partial z}\right)\frac{\partial V}{\partial z} dV \end{aligned} \tag{2}$$

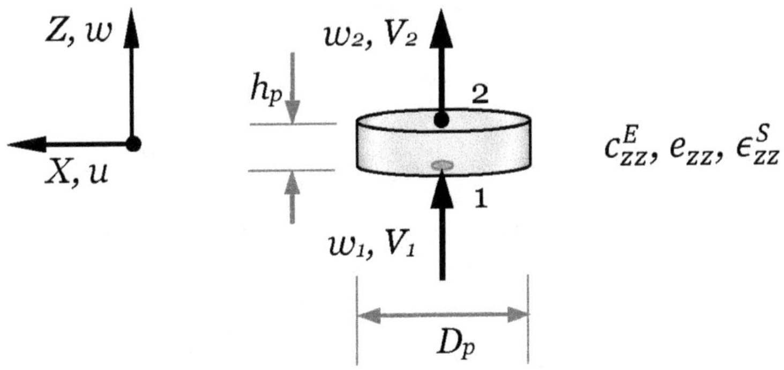

Figure 1.
The piezoelectric material model.

The electrical power can be calculated as the partial derivative of the electric energy, presented by Eq. (2), in respect to time:

$$P_e = \frac{\partial U_e}{\partial t} = \frac{1}{2}\int_V \frac{\partial}{\partial t}\left(D_z \frac{\partial V}{\partial z}\right) dV = \frac{1}{2}\int_V \frac{\partial}{\partial t}\left(e_{zz}\frac{\partial w}{\partial z} + \epsilon_z^S \frac{\partial V}{\partial z}\right)\frac{\partial V}{\partial z} dV \tag{3}$$

3. Dynamic system modelling

This chapter aims to predict the wasted energy from vibration systems that can be further transformed into electrical energy. A typical vibration system can be described as mass, spring and damper elements, and that can represent a suspension system assembly.

The suspension system is an assembly of suspension arms or linkages, springs and shock absorbers that connect the wheels to the vehicle's chassis in order to isolate passengers from vibrations due to bumps and roughness of the road. Furthermore, it must maintain the contact of the wheels with the road to ensure drivability. Thus, the suspension system is the mechanical system where the stability and handling of the vehicle, besides energy harvesting, must be equilibrated.

Mathematical models were initially developed for vertical vehicle performance, and the one-dimensional quarter car model is the simplest from the frequently used suspension system [8]. It is composed of the sprung mass m_s, which represents ¼ of the vehicle's body and the unsprung mass m_u, which represents the wheel assembling mass. Both are considered rigid bodies. Its displacements are w_s and w_u, respectively and both are vertically aligned. There are some studies that include a third degree in the system to describe road roughness excitation w_r. In this case, only the bounce input mode, or the vertical displacement can be implemented. Other elements of the suspension system are included, such as tire stiffness k_t, suspension stiffness k_s, and viscous damping c_s. All vertical displacements are a function of the independent variable t that represents the time. **Figure 2** illustrates this 1D quarter car model, which could represent both the front and rear of the vehicle.

The expressions of kinetical energy from the masses, strain energy and dissipation function from the shock absorber, and the virtual work from the road roughness excitation are written as:

$$\begin{aligned}
T &= \frac{1}{2}m_s\dot{w}_s^2 + \frac{1}{2}m_u\dot{w}_u^2 \\
U &= \frac{1}{2}k_s(w_s - w_u)^2 \\
R &= \frac{1}{2}c_s\left(\dot{w}_s^2 - \dot{w}_u^2\right) \\
\delta W &= F_r(t)\delta w_u = k_t w_r(t)\delta w_u
\end{aligned} \tag{4}$$

4. The suspension system and piezoelectric disk coupling

The conversion of the mechanical energy from the vehicle suspension to electric energy through piezoelectric materials can be predicted by piezoelectric disk coupling illustrated in **Figure 1** and the suspension system illustrated in **Figure 2**. Since the conversion of mechanical energy to electrical energy in this system is produced by compressive efforts, the piezoelectric disk is located between the shock absorber

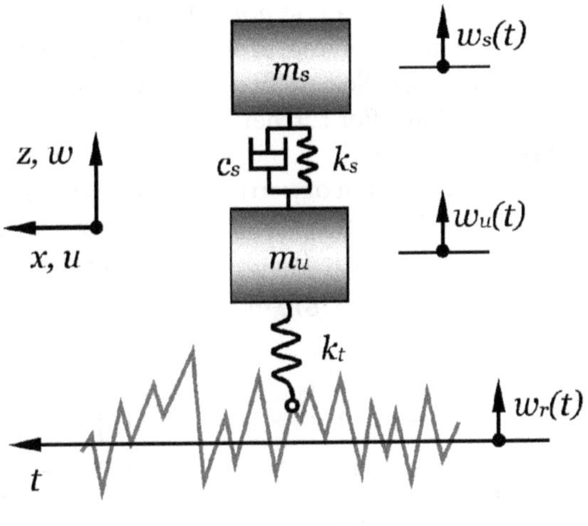

Figure 2.
1D quarter car model.

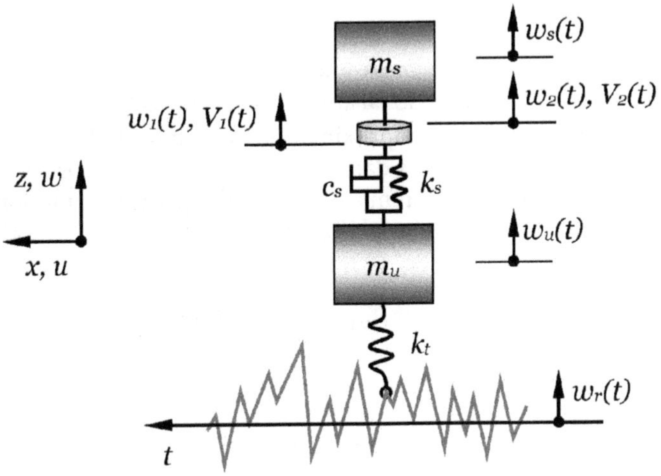

Figure 3.
The suspension system and piezoelectric coupling model.

system composed by stiffness k_s and viscous damping c_s on the bottom side, along with the sprung mass m_s on the upper side, as illustrated in **Figure 3**. As stated previously, vertical displacements w and now the electric potential V, are all a function of the independent variable t.

5. The vertical displacement and electric potential approach

Understanding physical problems can be accomplished when the numerical simulation of equations within the variables that describe them are represented. In the case of the electric energy prediction from the vehicle suspension system, the variables are vertical displacement w and electric potential V of different points, or nodes, and the domain within the model is considered valid.

Some numerical methods are often used to understand physical problems, and among them, the most widely used in engineering is the finite element method. Within this method, the variables in the equations that describe the physical problem must be approximated to the polynomial functions of these variables. For the application of this method in solving problems, all elements that compose it must be represented by matrices, called elementary matrices, which are the result of the adopted polynomial functions. For further details on this method, the following references [9, 10] are recommended.

If the linear approximation function of vertical displacement w and electric potential V over thickness h_p in direction z of the piezoelectric disk are considered, stiffness elementary matrices of the piezoelectric finite element can be obtained by applying the Lagrange equations [11] over the energy expressions presented by Eq. (2).

$$\begin{aligned}
[K_m] &= \frac{c_{zz}^E S_p}{h_p}\begin{bmatrix} 1 & -1 \\ -1 & 1 \end{bmatrix} = k_m \begin{bmatrix} 1 & -1 \\ -1 & 1 \end{bmatrix} \\
[K_{m_el}] &= -\frac{e_{zz} S_p}{2h_p}\begin{bmatrix} 1 & -1 \\ -1 & 1 \end{bmatrix} = -k_{m_el} \begin{bmatrix} 1 & -1 \\ -1 & 1 \end{bmatrix} \\
[K_{el}] &= \frac{\epsilon_z^S S_p}{h_p}\begin{bmatrix} 1 & -1 \\ -1 & 1 \end{bmatrix} = k_{el} \begin{bmatrix} 1 & -1 \\ -1 & 1 \end{bmatrix}
\end{aligned} \tag{5}$$

Where $K_m[]$, $K_{m_el}[]$and $K_{el}[]$are mechanical elementary stiffness, electromechanical coupling elementary stiffness and electric elementary stiffness matrices, respectively, and S_p is the cross-sectional area of the piezoelectric disk.

Moreover, if the same linear approximation function of the vertical displacement w over stiffness k_s and viscous damping c_s of the shock absorber in direction z are considered, the suspension system's differential equation of motion, illustrated in **Figure 2**, can be obtained by applying the Lagrange equations [11] over the energy expressions presented by Eq. (4).

$$[M]\{\ddot{w}\} + [C]\{\dot{w}\} + [K]\{w\} = \{F(t)\} \tag{6}$$

Where the elementary matrices and the vectors are:

$$\begin{aligned}
[M] &= \begin{bmatrix} m_s & 0 \\ 0 & m_u \end{bmatrix} \\
[K] &= \begin{bmatrix} k_s & -k_s \\ -k_s & k_s \end{bmatrix} \\
[C] &= \begin{bmatrix} c_s & 0 \\ 0 & c_s \end{bmatrix} \\
\{w\} &= \begin{Bmatrix} w_s(t) \\ w_u(t) \end{Bmatrix} \\
\{\dot{w}\} &= \begin{Bmatrix} \dot{w}_s(t) \\ \dot{w}_u(t) \end{Bmatrix} \\
\{\ddot{w}\} &= \begin{Bmatrix} \ddot{w}_s(t) \\ \ddot{w}_u(t) \end{Bmatrix} \\
\{F(t)\} &= \begin{Bmatrix} 0 \\ k_t w_r(t) \end{Bmatrix}
\end{aligned} \tag{7}$$

Using the technique of assembling the elementary matrices of the finite element method, the suspension system's differential equation of motion and piezoelectric disk coupled model, as illustrated in **Figure 3**, is as shown in Eq. (8):

$$[M_m]\{\ddot{w}\} + [C_m]\{\dot{w}\} + [K_m]\{w\} + [K_{m\_el}]\{V\} = \{F(t)\} \tag{8}$$

$$[K_{m\_el}]^t\{w\} + [K_{el}]\{V\} = \{0\} \tag{9}$$

Eq. (9) can be manipulated and substituted for Eq. (8). Thus, the final equation of motion as a function of only mechanical variables is:

$$[M_m]\{\ddot{w}\} + [C_m]\{\dot{w}\} + \left[[K_m] - [K_{m\_el}][K_{el}]^{-1}[K_{m\_el}]^t\right]\{w\} = \{F(t)\} \tag{10}$$

Eq. (10) can be solved in the time domain with an integration method, such as the Newmark Method [12]. The mechanical force is due to road roughness as well as the tire characteristics of the wheel as shown in Eq. (7). These data are used to apply the condition in each time step in solving Eq. (10), in which all variables $w_s = w_2$, w_1, and w_u and their time derivatives are obtained.

The response in the time domain in respect to vertical displacements is obtained by using Eq. (10) and subsequently, the electric potential is obtained as:

$$\{V(t)\} = -[K_{el}]^{-1}[K_{mel}]^t\{w(t)\} \tag{11}$$

The electrical energy can be calculated as presented by Eq. (2). The development of this equation follows:

$$U_e = \frac{1}{2}\int_0^{h_p}\int_{S_p}\left[e_{zz}\left(\frac{\partial w}{\partial z}\right)\frac{\partial V}{\partial z} + \epsilon_z^S\left(\frac{\partial V}{\partial z}\right)^2\right]dS\,dz \tag{12}$$

Thus, the expression of the electrical energy due to the piezoelectric disk on this suspension system is as:

$$U_e = \frac{1}{2}\frac{e_{zz}S_p}{h_p}\,(w_2 - w_1)\,(V_2 - V_1) + \frac{\epsilon_z^S S_p}{h_p}\,(V_2 - V_1)^2 \tag{13}$$

In addition, the electrical power can be calculated as presented by Eq. (3) and its development is as follows:

$$P_e = \frac{1}{2}\int_0^{h_p}\int_{S_p}\left[e_{zz}\left(\frac{\partial}{\partial t}\left(\frac{\partial w}{\partial z}\right)\frac{\partial V}{\partial z} + \frac{\partial w}{\partial z}\frac{\partial}{\partial t}\left(\frac{\partial V}{\partial z}\right)\right) + \epsilon_z^S\frac{\partial}{\partial t}\left(\frac{\partial V}{\partial z}\right)^2\right]dS\,dz \tag{14}$$

Thus, the expression of the electrical power due to the piezoelectric disk on this suspension system is as:

$$P_e = \frac{1}{2}\frac{e_{zz}S_p}{h_p}\left[(\dot{w}_2 - \dot{w}_1)\,(V_2 - V_1) + (w_2 - w_1)\,(\dot{V}_2 - \dot{V}_1)\right] + \frac{\epsilon_z^S S_p}{h_p}\,(V_2 - V_1)\,(\dot{V}_2 - \dot{V}_1) \tag{15}$$

Where $\dot{w}_1$, $\dot{w}_2$, $\dot{V}_1$ and $\dot{V}_2$ are time derivatives of the displacements and electric potential of nodes 1 and 2 of the piezoelectric disk.

6. Application example

For a simpler demonstration of the potential for harvesting energy in a suspension system by means of piezoelectric material, a MATLAB® code to obtain the results was developed and presented below.

The dimensions of the piezoelectric disk are diameter $D_p = 0.065\,m$ and thickness $h_p = 0.025\,m$. The vehicle data and the properties of piezoelectric material PZT- 5H are shown in **Tables 1** and **2**, as indicated in references [13, 14], respectively.

Property description	Value
Body and wheel mass	
¼ Vehicle sprung mass (m_u)	362.5 kg
Unsprung mass (m_s)	39 kg
Tire stiffness (k_t)	200 kN/m
Shock absorber suspension	
Suspension stiffness (k_s)	30 kN/m
Suspension damping (c_s)	4 kN s/m

Table 1.
Vehicle data.

	PZT-5H
Piezoelectric constant – e_{zz} [C/m^2]	23.30
Dielectric constant – ϵ_{zz}^S [F/m]	1.30 x 10^{-8}
Density – ρ [kg/m^3]	7500
Young's modulus – c_{zz}^E [GPa]	23.0

Table 2.
Properties of piezoelectric material.

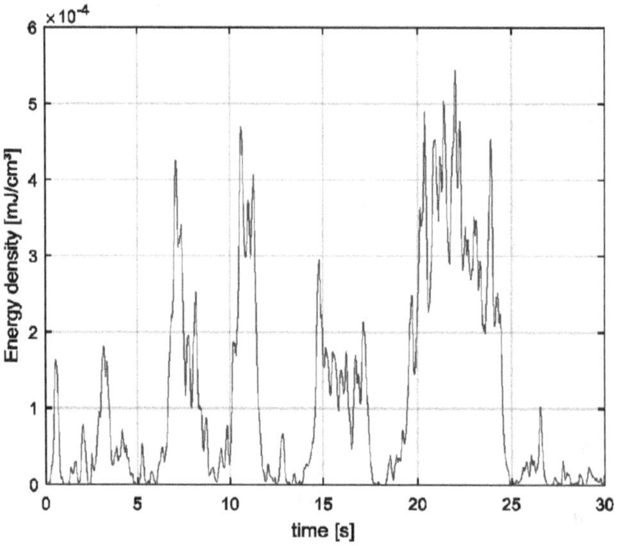

Figure 4.
Electric energy density response for PZT-5H.

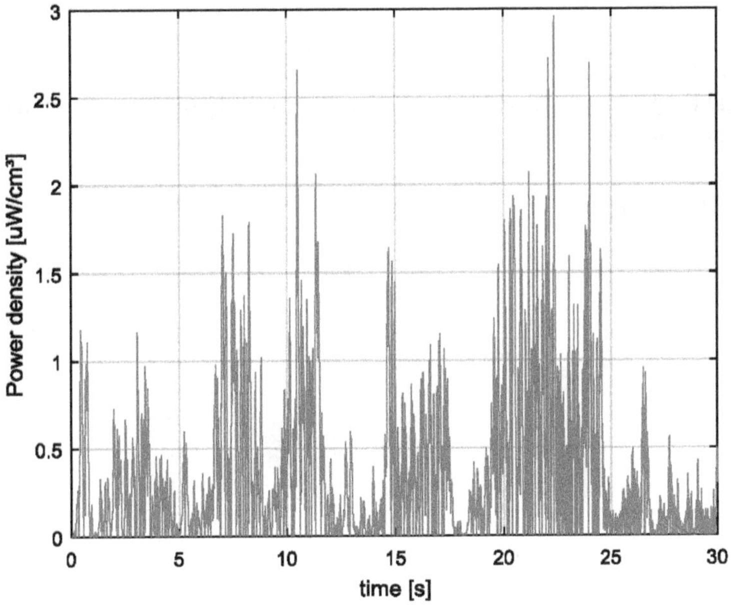

Figure 5.
Electric power density response for PZT-5H.

The road excitation is as depicted in [6]. The excitation is the bounce input mode vertical position of the road $w_r(t)$ with class D road (poor) and a driving speed for a vehicle of 20 m/s.

In **Figure 4**, the instantaneous electric energy density for piezoelectric material PZT-5H calculated by Eq. (13), with a simulation time of 30 s and a fixed step time of 0.00001 s.

Figure 5 exhibits the overall instantaneous electric power density response for piezoelectric material PZT-5H calculated by Eq. (15).

7. Conclusions

This chapter proposes a coupled suspension system and piezoelectric model to predict the potential of harvested electric power in vehicle suspension systems. The performance of piezoelectric material PZT-5H was investigated, in respect to harvesting energy based on energy density and electric power density.

The approach presented in this chapter is a way to simulate the electric power generated in vehicle suspension systems established by piezoelectric harvesting. Nonetheless, these results would need to be compared with experimental results to demonstrate the validity of the proposed model.

Author details

José Carlos de Carvalho Pereira
Mechanical Engineering Department, Federal University of Santa Catarina, Florianópolis, Brazil

*Address all correspondence to: carlosp@emc.ufsc.br

References

[1] Shenck NS, Paradiso JA. Energy scavenging with shoe-mounted piezoelectrics. IEEE Micro. 2001;**21**: 30-41

[2] Ottman GK, Hofmann HF, Lesieutre GA. Optimized piezoelectric energy harvesting circuit using step-down converter in discontinuous conduction mode. IEEE Transactions Power Electronics. 2003;**18**:696-703

[3] Roundy S, Wright PK. A piezoelectric vibration based generator for wireless electronics. Smart Materials and Structures. 2004;**13**:1131-1142

[4] Zhang Y, Guo K, Wang D, et al. Energy conversion mechanism and regenerative potential of vehicle suspensions. Energy. 2017;**119**:961-970

[5] Addelkareem MAA, Xu L, Guo X, et al. Energy harvesting sensitivity analysis and assessment of the potential power and full car dynamics for different road modes. Mechanical Systems and Signal. 2018;**110**:307-332

[6] Morangueira YLA, Pereira JCC. Energy Harvesting Assessment with a Coupled Full Car and Piezoelectric Model. Energy. 2020;**210**:1-13

[7] IEEE Standard on Piezoelectricity. ANSI Standard 176; 1987.

[8] Zuo L, Zhang P. Energy harvesting, ride comfort, and road handling of regenerative vehicle suspensions. Journal of Vibration and Acoustics. 2013; 135: 011002–1–011002-8.

[9] Yang TY. Finite Element Structural Analysis. ed. In: Prentice-Hall. 1986

[10] Bathe KJ. Finite Element Procedure. ed. In: Prentice Hall. 1996

[11] Meirovitch L. Methods of analytical Dynamics. ed. McGraw-Hill, Inc.; 1988.

[12] Pereira JCC. Fundamentos da Análise de Sistemas Mecânicos. ed. In: UFSC. 2017

[13] Addelkareem MAA, Xu L, Guo X, et al. Energy Harvesting Sensitivity Analysis and Assessment of the Potential Power and Full Car Dynamics For Different Road Modes. Mechanical Systems and Signal. 2018;**110**:307-332

[14] Kocbach J. Finite Element Modeling of Ultrasonic Piezoelectric Transducers. University of Bergen. Tech. Rep; 2000

Permissions

All chapters in this book were first published by InTech Open; hereby published with permission under the Creative Commons Attribution License or equivalent. Every chapter published in this book has been scrutinized by our experts. Their significance has been extensively debated. The topics covered herein carry significant findings which will fuel the growth of the discipline. They may even be implemented as practical applications or may be referred to as a beginning point for another development.

The contributors of this book come from diverse backgrounds, making this book a truly international effort. This book will bring forth new frontiers with its revolutionizing research information and detailed analysis of the nascent developments around the world.

We would like to thank all the contributing authors for lending their expertise to make the book truly unique. They have played a crucial role in the development of this book. Without their invaluable contributions this book wouldn't have been possible. They have made vital efforts to compile up to date information on the varied aspects of this subject to make this book a valuable addition to the collection of many professionals and students.

This book was conceptualized with the vision of imparting up-to-date information and advanced data in this field. To ensure the same, a matchless editorial board was set up. Every individual on the board went through rigorous rounds of assessment to prove their worth. After which they invested a large part of their time researching and compiling the most relevant data for our readers.

The editorial board has been involved in producing this book since its inception. They have spent rigorous hours researching and exploring the diverse topics which have resulted in the successful publishing of this book. They have passed on their knowledge of decades through this book. To expedite this challenging task, the publisher supported the team at every step. A small team of assistant editors was also appointed to further simplify the editing procedure and attain best results for the readers.

Apart from the editorial board, the designing team has also invested a significant amount of their time in understanding the subject and creating the most relevant covers. They scrutinized every image to scout for the most suitable representation of the subject and create an appropriate cover for the book.

The publishing team has been an ardent support to the editorial, designing and production team. Their endless efforts to recruit the best for this project, has resulted in the accomplishment of this book. They are a veteran in the field of academics and their pool of knowledge is as vast as their experience in printing. Their expertise and guidance has proved useful at every step. Their uncompromising quality standards have made this book an exceptional effort. Their encouragement from time to time has been an inspiration for everyone.

The publisher and the editorial board hope that this book will prove to be a valuable piece of knowledge for researchers, students, practitioners and scholars across the globe.

List of Contributors

Manojit De
Department of Pure and Applied Physics, Guru Ghasidas Vishwavidyalaya, Bilaspur, India
Department of Applied Physics, Chouksey Engineering College, Bilaspur, India

Sergey Shornikov
Vernadsky Institute of Geochemistry and Analytical Chemistry of Russian Academy of Sciences, Moscow, Russia

Someshwar Pola and Ramesh Gade
Department of Chemistry, University College of Science, Osmania University, Hyderabad, Telangana, India

Dang Anh Tuan
Ha Nam Provincial Department of Science and Technology, Vietnam

Vo Thanh Tung, Le Tran Uyen Tu and Truong Van Chuong
University of Sciences, Hue University, Vietnam

Burak Gultekin and Shirin Siyahjani
Ege University, Solar Energy Institute, Izmir, Turkey

Ali Kemal Havare
Faculty of Engineering, Electric and Electronics Engineering, Photoelectronic Lab. (PEL), Toros University, Mersin, Turkey

Halil Ibrahim Ciftci
Faculty of Life Sciences, Medicinal and Biological Chemistry Science Farm Joint Research Laboratory, Kumamoto University, Kumamoto, Japan

Mustafa Can
Faculty of Engineering and Architecture, Izmir Katip Celebi University, Izmir, Turkey

Abdelouahid Lyhyaoui
Laboratory of Innovative Technologies, Abdelmalek Essaadi University, Tangier, Morocco

José Pelegri-Sebastia
Research Institute for Integrated Management of Coastal Areas, Universitat Politècnica de Valencia, Valencia, Spain

Asma Bakkali
Laboratory of Innovative Technologies, Abdelmalek Essaadi University, Tangier, Morocco
Research Institute for Integrated Management of Coastal Areas, Universitat Politècnica de Valencia, Valencia, Spain

Youssef Laghmich
Polydisplinary Faculty, Hassan 1st University, Khouribga, Morocco

Madeeha Aslam, Tahira Mahmood and Abdul Naeem
National Centre of Excellence in Physical Chemistry, University of Peshawar, Peshawar, Pakistan

Chung Nguyen Thai and Thuy Le Xuan
Le Quy Don Technical University, Hanoi, Vietnam

Thinh Tran Ich
Hanoi University of Science and Technology, Hanoi, Vietnam

José Antonio Alonso
Instituto de Ciencia de Materiales de Madrid, CSIC, Madrid, Spain

Carlos Alberto López
Instituto de Ciencia de Materiales de Madrid, CSIC, Madrid, Spain
Instituto de Investigaciones en Tecnología Química (INTEQUI), UNSL, CONICET, Facultad de Química, Bioquímica y Farmacia, UNSL, San Luis, Argentina

María Consuelo Alvarez-Galván
Instituto de Catálisis y Petroleoquímica, CSIC, Madrid, Spain

María Teresa Fernández-Díaz
Institut Laue Langevin, Grenoble, France

Carmen Abia
Instituto de Ciencia de Materiales de Madrid, CSIC, Madrid, Spain
Institut Laue Langevin, Grenoble, France

Dinesh Kumar and Akhilesh Kumar Singh
School of Materials Science and Technology, Indian Institute of Technology, Banaras Hindu University, Varanasi, Uttar Pradesh, India

Badreddine Smiri, Faouzi Saidi and Hassen Maaref
Laboratoire des Micro-optoélectroniques et Nanostructures, Faculté des Sciences Monastir, Université de Monastir, Monastir, Tunisia

Ram Sagar Yadav
Department of Zoology, Institute of Science, Banaras Hindu University, Varanasi, Uttar Pradesh, India

Monika and Shyam Bahadur Rai
School of Materials Science and Technology, Indian Institute of Technology, Banaras Hindu University, Varanasi, Uttar Pradesh, India
Department of Physics, Institute of Science, Banaras Hindu University, Varanasi, Uttar Pradesh, India

José Carlos de Carvalho Pereira
Mechanical Engineering Department, Federal University of Santa Catarina, Florianópolis, Brazil

Index